Ahead of the (

Hidden breakthroughs in the bic

Ahead of the Curve
Hidden breakthroughs in the biosciences

Michael Levin
Tufts University

Dany Spencer Adams
Tufts University

IOP Publishing, Bristol, UK

© IOP Publishing Ltd 2016

All rights reserved. No part of this publication may be reproduced, stored in a retrieval system or transmitted in any form or by any means, electronic, mechanical, photocopying, recording or otherwise, without the prior permission of the publisher, or as expressly permitted by law or under terms agreed with the appropriate rights organization. Multiple copying is permitted in accordance with the terms of licences issued by the Copyright Licensing Agency, the Copyright Clearance Centre and other reproduction rights organisations.

Permission to make use of IOP Publishing content other than as set out above may be sought at permissions@iop.org.

Michael Levin and Dany Spencer Adams have asserted their right to be identified as the authors of this work in accordance with sections 77 and 78 of the Copyright, Designs and Patents Act 1988.

ISBN 978-0-7503-1326-1 (ebook)
ISBN 978-0-7503-1327-8 (print)
ISBN 978-0-7503-1328-5 (mobi)

DOI 10.1088/978-0-7503-1326-1

Version: 20161201

IOP Expanding Physics
ISSN 2053-2563 (online)
ISSN 2054-7315 (print)

British Library Cataloguing-in-Publication Data: A catalogue record for this book is available from the British Library.

Published by IOP Publishing, wholly owned by The Institute of Physics, London

IOP Publishing, Temple Circus, Temple Way, Bristol, BS1 6HG, UK

US Office: IOP Publishing, Inc., 190 North Independence Mall West, Suite 601, Philadelphia, PA 19106, USA

This book is dedicated to all of the unsung pioneers—past, present, and future—who make fundamental discoveries 'Ahead of the Curve', and advance science by bearing the risks that such creativity entails.

Contents

Preface ix

Acknowledgements x

Author biographies xi

Introduction xii

Part I Hindsight

1 Medicine   1-1

    Case Study 1: Wash Your Hands   1-1

    Case Study 2: Chromosomes and Cancer   1-4

Part II Foresight

2 Physics in cell and developmental biology   2-1

    Case Study 3: Development Employs Forces   2-1

    Case Study 4: Depolarization Induces Neuron Division   2-16

    Case Study 5: Cells Talk to Each Other via Electromagnetism   2-29

    Case Study 6: Ion Currents Regulate Cell Polarity   2-39

3 Inheritance   3-1

    Case Study 7: Induced Eye-Defects Can Be Passed On to Future Generations   3-1

    Case Study 8: Heritability Via the Cytoskeleton   3-20

    Case Study 9: Don't Blame Mom   3-47

    Case Study 10: Memory Survives Decapitation   3-59

4 Physiology and cancer   4-1

    Case Study 11: Random Expression Profiles Predict Breast Cancer   4-1

    Case Study 12: The Bright Side of Infection   4-22

    Case Study 13: Innervation Suppresses Tumorigenesis   4-34

5 Mathematics and modeling   5-1

    Case Study 14: Top-Down Causation: Not All The Work Is Done By Molecules   5-1

Case Study 15: Standard Deviation not S.E.M.   5-29
Case Study 16: Field Models of Pattern Formation   5-38
Case Study 17: Noise and Prediction in Biology   5-57

Preface

The popular conception of science is of a continuous, steady, upward climb of progress. The reality is not as simple. Highly significant discoveries may often stay unrecognized for decades, if they conflict with the current paradigm or extend it in ways hard to imagine at the time. Recently, we were bemoaning how important scientific signals get lost in the noise generated by the sheer volume of data and the dominance of the latest trend, whatever it may be. Once we started listing our own favorite unknown classics, and soliciting other titles from colleagues, we realized that there are substantially more than enough such forgotten, never-noticed, or ignored papers to fill a book. Re-viewing of research that identifies exciting and influential new aspects of science (here we focus on biology), well in advance of mainstream thought, provides fascinating and instructive case studies for improving our ability to recognize influential findings that significantly revise our understanding of the world. They serve not only as studies of the history of science, but as windows on the dynamic process by which knowledge progresses; moreover, by examining case histories of good science derailed by something other than evidence, we may learn how to better interpret new ideas and data that contradict our expectations.

We decided to start with evidence that important things really do get missed, by including important things that were, in fact, missed. Part I therefore comprises papers describing theories and findings that are now widely accepted, but that received little or no attention when they were first published. Follow up confirmation took years to decades, and in the case of physician hygiene, over a century to happen. In some cases, the work itself was rediscovered, in others, someone else found the same thing. Part II is a compilation of papers that we believe describe important ideas and results that are, as yet, not widely known or used. We do not claim certainty that all of these papers will someday earn renown or prove useful; our argument is only that the ideas presented are supported by data and deserve closer inspection. Our claim is that all of these papers raise important questions or describe novel and essential ideas. Some go against conventional wisdom, some point out the importance of testing assumptions, still others contain known facts that are underused.

Michael Levin is eternally grateful to his parents, Benjamin and Luba, and his wife Kristin, for all of their support of his efforts to pull 'The Curve' in novel and unusual directions. This book is for Sam and Arthur, with the hopes that your work may someday be featured in a future volume of this series.

Dany Spencer Adams expresses immeasurable gratitude to the always supportive Adams–Olden–Silvan family, to wonderful friends, and to her husband Joe. This work is for Zachary, Ariana, Jeremy, Mia, Simone, Ariele, and Ben, with hope that their curiosity will lead them beyond what anybody else tells them is enough to know.

Acknowledgements

We are very grateful to Scott Gilbert, Ray Keller, Larry Stern, Richard Nuccitelli, Lev Beloussov, Wendy Brandts, Sara Walker, Jack Tuszynski, Susan Ernst, and Edward J Steele for suggestions of papers to include and for contributing their perspectives on the importance of the works. We are also grateful to numerous colleagues who suggested other papers and topics, which hopefully will be covered in subsequent volumes of this series. William F Baga did critical logistical work. We are also grateful for the guidance and expertise provided by IOP Publishing, especially Daniel Heatley, and Jessica Fricchione, as well as Chris Benson and Jacky Mucklow.

Author biographies

Michael Levin

Michael Levin was born in Moscow, Russia in 1969, and emigrated to the North Shore of Boston with his family in 1978. He worked as a software engineer interested in artificial intelligence prior to moving from computer science to biology. He received a PhD from Harvard Medical School in genetics, and did post-doctoral training working on the molecular mechanisms of embryogenesis. His first independent laboratory was at Forsyth Institute in 2000, establishing a novel research program in the biophysics of biological pattern control. The group moved to Tufts University in 2008, where he collaborates with computer scientists, bioengineers, and workers in cognitive neuroscience. His lab (www.drmichaellevin.org) now works on a number of frontier topics, including the communication and computation among non-neural cells that underlie control of biological growth and form, somatic memory and learning outside the brain, and artificial intelligence approaches to helping understand complex biological phenomena. He is currently Vannevar Bush Professor in the department of Biology, directing the Allen Discovery Center at Tufts University (allencenter.tufts.edu).

Dany Spencer Adams

Dany Spencer Adams is a Research Professor in the Department of Biology at Tufts University, a Principle Investigator in the Tufts Center for Regenerative and Developmental Biology, faculty in the EBICS program at MIT, and the author of *Lab Math: A Handbook of Measurements, Calculations, and Other Quantitative Skills for Use at the Bench*. Her movies of bioelectric signaling in developing Xenopus embryos have been seen on Discovery's Curiosity. She blogs about numbers at LabMath.org. She has always been interested in looking at questions that are off the beaten path, specifically in the area of biomechanics and biophysics during morphogenesis. As an undergraduate at UC Berkeley, she did research with Drs Ray Keller and M A R Koehl. She got her PhD from The University of Washington where she studied with Drs Thomas Daniel and Garret Odell. She started her independent career as an assistant professor in the Biology department of Smith College. After several years, she decided to focus on research full-time, starting what would turn out to be a long-term collaboration with Dr Michael Levin at The Forsyth Institute. There she began her studies of craniofacial development, specifically the roles of ion-flux dependent phenomena during differentiation and morphogenesis of cranial neural crest and placode-derived structures. Her current work also touches on ion flux during regeneration and during transformation, with a technical emphasis on adapting ion and membrane voltage imaging techniques for use in vivo in embryos. Because of the caliber and nature of their collaborations, Dr Adams joined Dr Levin for the move to Tufts University in 2008.

Introduction

We created this book for two purposes. First, to highlight some specific topics in biology that we consider especially fascinating; this includes the roles of biophysical forces, approaches to mathematical understanding of living systems, non-genic inheritance, and the relationship between memory and the body. Our second purpose is to provide papers that have instructive lessons for us in the present. Most of our entries are accompanied by a Perspective, written by a current expert in each sub-field, who provides a personal commentary on the significance of the work and why it was not recognized as the advance that it was. Taken as a collection, these stories have lessons for us about why key findings get missed and most importantly, how to spot important breakthroughs in the future to reduce the time before their positive impact is felt. Many of the key studies (even ones that were immediately recognized as major advances) are published in journals that are not considered the "top tier". Especially today, journals are highly stratified, and editors exert a significant filtering function over submissions to try to publish only "high impact" papers; guessing this in advance is a most difficult task and our intuitions can be improved by considering where the process failed to identify gems in the past.

We have endeavored to include papers from a variety of Biological disciplines, although our own interests and knowledge have biased the collection towards Cell and Developmental Biology. With most of the papers, we have included timelines or graphics illustrating the state of the field, i.e. the curve. These trend curves were made using PubMed's Results by Year Tool (Canese K. PubMed Discovery Tools. NLM Tech Bull. 2012 May-Jun;(386):e7) or MLTrends (http://mltrends.ogic.ca; Palidwor et al. (2010) J. Biomed. Discov. Collab. 5, 1-6;). As with any search, the choice of keywords determines what you find. In cases where keywords were not obvious, we had to use our own judgment; the terms used are supplied. We have also indicated, with an arrowhead, the date of the case study, to illustrate how far 'ahead of the curve' each paper is. It is important to remember, however, that definitions evolve, new jargon appears, there is no way to search directly for the influence of an idea, and counting instances does not tell you whether the terms were referred to positively or negatively. In other words, the graphs illustrate something both specific and vague, and should not be over-interpreted.

It is our hope that this compilation will inspire and educate by virtue of the value of the work collected, and that our contributions, of choosing and annotating papers, will facilitate those processes. We apologize to the many other researchers whose significant discoveries could not be covered here. While we can take neither credit nor blame for the science herein, we take full responsibility for our judgment and any errors in interpreting, contextualizing, or encapsulating. We have tried to uncover true breakthroughs, in the sense that an article contains the very first recognition of a topic, but this has not always been possible. Reports of earlier works have been followed up to the best of our abilities, but we have been constrained by factors including issues of translation, copyright, and availability of certain works.

Thus, unfortunately, we may have missed the truly deserving first author of an important idea; the irony of that has not escaped us. If that is found to be the case, we would like to know, and we agree that the fault lies with the other editor.

Michael Levin and Dany Spencer Adams
Medford, MA
November, 2016

Part I

Hindsight

IOP Publishing

Ahead of the Curve
Hidden breakthroughs in the biosciences
Michael Levin and Dany Spencer Adams

Chapter 1

Medicine

Case Study 1: Wash Your Hands

Gordon (1795) A Treatise on the epidemic puerperal fever of Aberdeen. This time line shows the publication dates of papers encouraging better hygiene for surgeons. Lister finally got through.

Perspective on *A Treatise on the Epidemic Puerperal Fever of Aberdeen*

Dany Spencer Adams
Research Professor
Tufts University

In 1795, 72 years before Lister made it standard practice, Dr Gordon of the Aberdeen Dispensary in Scotland published his treatise, full of observations and interpretations, and some recommendations:

Excerpts:

> '.. [T]his disease seized such women only, as were visited, or delivered, by a practitioner, or taken care of by a nurse, who had previously attended patients affected with the disease.'
>
> '[T]he nurses and physicians who have attended patients affected with the puerperal fever, ought carefully to wash themselves and to get their apparel properly fumigated before it be put on again.'

He also mentions the reception he received for his ideas:

> 'The benevolent reader must observe, with displeasure, the ungenerous treatment which I met with from that very sex whose sufferings I was at so much pains to relieve; for, while I was using my best endeavours to mitigate the calamities of many miserable sufferers, several others were very busy traducing my character, who, prompted by prejudice, very uncandidly proclaimed the deaths, and concealed the cures, on purpose to raise an odium against my practice.'

He could have looked for support in the 16th Century work of Girolamo Fracastoro, an Italian scholar who is credited with being the first 'modern' European to articulate a germ theory. While Fracastoro was clearly ahead of Gordon, and his work on syphilis was an epic poem that would be wonderful to reprint, his work was widely accepted and influential when it was published. It fell out of favor some time later, supplanted by Galen's Miasma theory, but, unfortunately *De Contagione et Contagiosis Morbis* does not meet the criterion of being 'hidden.'

Gordon's hypothesis that the disease was infectious and the practitioners themselves were carrying it, likely contributed to the ideas of the slightly more famous Ignaz Semmelweis, an extremely colorful character. In 1847, Semmelweis insisted that his medical students wash their hands, and exhorted other doctors also to do the same. Neither Gordon nor Semmelweis was heeded, despite, in Semmelweis's wards at least, a dramatic reduction in the incidence of puerperal fever (now known to be caused by Group A hemolytic streptococcus). Indeed both were actively persecuted. It was not until 1867, six years after the death of Semmelweis, that Joseph Lister ('the father of antiseptic surgery') successfully urged doctors to wash their hands.

The full text of Gordon's work can be found online.

Sources:

Dunn P M 1998 Perinatal lessons from the past: Dr. Alexander Gordon (1752–99) and contagious puerperal fever *Arch. Dis. Child Fetal Neonatal Ed.* **78** F232–F33

Alexander G 1795 Treatise on the Epidemic Puerpural Fever (London: C.G and J. Robinson, Paternoster Row)

Markel H 2003 The doctor who made his students wash up. *The New York Times* (*Science Times*) October 7, 2003. p D6 http://www.nytimes.com/2003/10/07/health/the-doctor-who-made-his-students-wash-up.html

Pitt D and Aubin J-M 2012 Joseph Lister: father of modern surgery *Can J Surg.* **55** E8–E9 PMCID: PMC3468637

Case Study 2: Chromosomes and Cancer

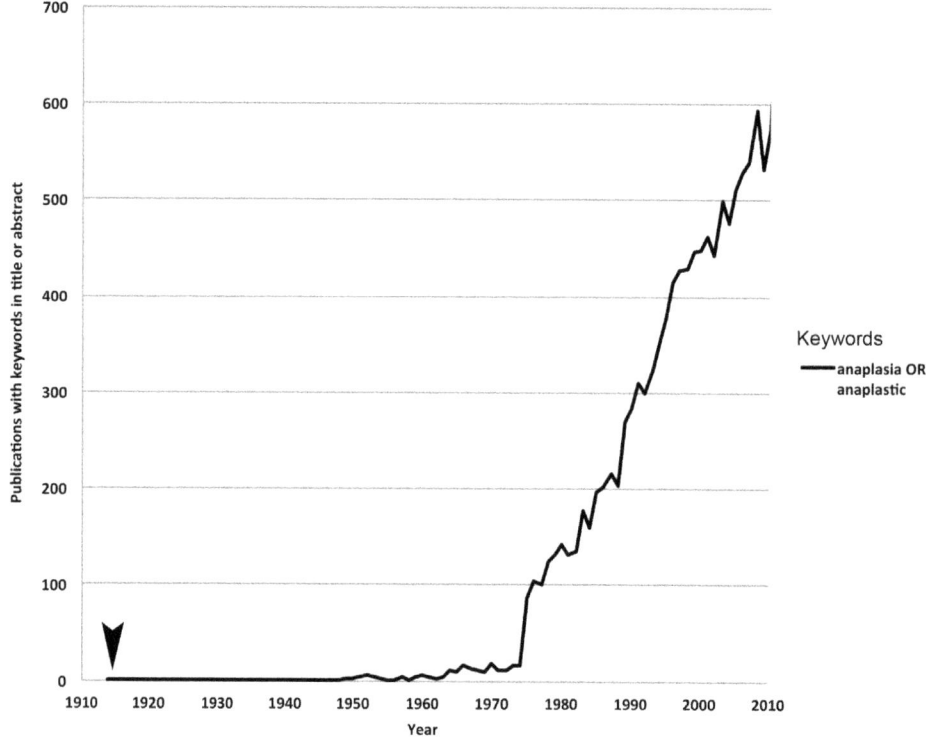

Boveri (1914) Concerning the origin of malignant tumours. This is just one of the many curves that Boveri anticipated by decades.

Perspective on *Concerning the Origin of Malignant Tumors* (*Zur Frage der Enstehung maligner Tumoren*)

Susan G. Ernst
Professor
Tufts University

Theodor Boveri was one of the first people to observe meiosis, to realize that each parent contributes an equal, haploid amount of DNA to the zygote, and to recognize a relationship between Mendel's re-discovered work and cytology. Written in 1914, only one year before his death, his monograph on malignant tumors describes his hypothesis about the role of chromosome abnormalities in tumorigenesis and oncogenesis. This monograph also contains one of the earliest suggestions that genes are arranged linearly on the chromosome. Boveri mainly argues that chromosomal abnormalities contribute to malignancy, that the ability of cells to multiply is inherent, and that proliferation has to be both inhibited and induced. Some of his other remarkable leaps are included almost as afterthoughts, interesting related ideas. Many of those ideas have now been tested and found to have been remarkably close to, or exactly what we now believe.

In this wonderfully readable translation from the original German, you are chaperoned through the train of thought of a remarkable scientist as he ponders the relationships and implications of observations and simple experiments, and places it all in context of what was known at the time. Using only hand-held tools and microscopes, he made observations and interpretations that led him to predict the existence of tumor suppressors, 70 years before Murphree and Benedict described the retinoblastoma gene as a 'suppressor.' Only one year before that, Yunis had described chromosomal abnormalities associated with neoplasia, that he had imaged using high-resolution banding techniques. In the first paragraph of that paper Yunis asserts 'That certain chromosomal defects are consistently associated with some types of human cancer was established in the 1970s' for which he cites a 1980 book. Neither author cites Boveri.

In what is more of a narrative than what we are used to now, Boveri both explains *and* challenges the support for his hypotheses, revealing his own doubts as well as the objections of others. This case study of how to think about data and how to assemble a logic-based argument, should be an inspiration to any scientist.

Cited

Murphree L A and Benedict W F 1984 Retinoblastoma: Clues to Human Oncongenesis *Science* **223** 1028–33

Yunis J J 1983 The Chromosomal Basis of Human Neoplasia *Science* **221** 227–36

Selected excerpts from Theodor Boveri (1914) *Zur Frage der Enstehung maligner Tumoren*. Translation by Henry Harris (2008) The Company of Biologists and Cold Spring Harbor Laboratory Press.

"In the year 1902, I tacked onto the results of my experiments on the development of doubly fertilised sea urchin eggs the speculation that malignant tumours might be the consequence of a certain abnormal chromosome constitution, which in some circumstances can be generated by multipolar mitoses (Boveri, 1902). I had intended, even at that time, to put that assumption on a firmer footing in a separate article. But the scepticism with which my ideas were met when I discussed them with investigators who act as judges in this area induced me to abandon the project.

...

The idea that there might be a connection between abnormal mitoses and malignant tumours has certainly cropped up often enough, but it has always been rejected, indeed so completely that in the recent literature it is curtly dismissed, if mentioned at all.

...

The following observations on the essential nature of tumours seem to me to be best supported by the evidence. The cells of even the most malignant tumours can be formed from normal tissue cells. The determinants of this abnormal behaviour are to be found in the tumour cells themselves, not in their surroundings(5). Although benign and malignant tumours have many properties in common, I have to agree with those authors who draw a sharp line between the two. In my view, there must be a fundamental difference between a tumour that grows in the same way as the tissue from which it is derived and one that does not. It seems to me that the transformation of a benign tumour into a malignant one, so often described, is a phenomenon of the same kind as the appearance of a malignant tumour at some site in normal tissue.

...

The essential elements of my point of view may therefore be summarised as follows. A malignant tumour cell is a cell with a specific defect; it has lost properties that a normal tissue cell retains. In this respect, I am in complete agreement with the concept to which Hansemann has given the name 'anaplasia'. A cell in this drastically altered state reacts differently to its environment, and it is possible that this alone might account for its tendency to multiply without restraint. Such unrestrained proliferation is no doubt a very primitive property of cells.

...

If now we regard the malignant cell as one that has lost certain properties and hence its normal reactivity to the rest of the body, then this change may well be enough to induce an altruistic cell to revert to its egoistical mode and thus release its multiplication from restraint. (Relapse of 'organotypic growth' to 'cytotypic growth', to use the picturesque terminology of R. Hertwig.) But it is also possible

that, in the tissue cells of metazoa, special inhibitory mechanisms have developed that have to be eradicated before unrestrained multiplication can take place.
...

The numerous experiments that have been done to investigate this problem have, in general, reached the conclusion that in every bit of cytoplasm all the properties of the cytoplasm as a whole are latent, or, at least, that under the influence of the nucleus they can be regained. Any fragments taken from a protozoon, so long as they contain a nucleus, will regenerate complete animals, and nucleated fragments taken from eggs will produce normal embryos.
...

However, there are exceptions to this rule. In many kinds of egg, the process of differentiation is such that fragments taken from them generate fragments of embryos. Nonetheless, it is notable that, even in such cases, the cells that grow out of the egg fragment are not abnormal or sick. What happens is that the fragment can generate only certain sorts of cells. And even with eggs that behave in this way, the oocyte from which the egg is derived turns out to be totipotent. Nucleated oocyte fragments give rise to normal dwarf embryos.

There is no reason to believe that tissue cells behave differently. Here too, a fragment containing a nucleus will retain the ability to regenerate the whole cell so that, in all probability, it is impossible—provided the cell survives the operation—to produce a permanent deficit in the cell by removing a bit of its cytoplasm.

In the case of the nucleus, similar experiments yield an entirely different result.
...

Let us label the chromosomes in the one nucleus a, b, c and d. The other nucleus will also contain the same set, a, b, c and d. At fertilisation, the two haploid nuclei are amalgamated into one diploid nucleus, which now contains 2a, 2b, etc. Each chromosome in this duplicate set is then split into two, and a tightly controlled karyokinetic separation(13) of the daughter chromosomes ensures that a complete duplicate set is inherited by each of the two cells produced by the cleavage of the fertilised ovum. In the resultant resting nuclei, the individual chromosomes seem to break down. But we have every reason to believe that, within the stroma of the resting nucleus, every chromosome that contributes to the makeup of that nucleus continues to exist as a discrete region that reappears as the same 'chromosome' when the cell prepares to divide again. (Theory of the individuality of chromosomes.) In this way, the two sets of chromosomes amalgamated at fertilisation are inherited by all the cells of the individual. It is only in the germ cells that the so-called reduction division converts the duplicate set once more into a single set.
...

These observations [of the result of tetrapolar mitoses] are meant to show that, in multipolar mitoses, we have a means of achieving something that can otherwise hardly be done without collateral damage to the cell, namely the production of nuclei in which some parts are missing. And, in this way, we can answer the question that we posed in connection with the cytoplasm. What are the consequences of such a deficit in the case of the nucleus? To put the question more precisely: can a nucleus with a deficit of this kind regenerate what is missing and, if it cannot, can it remain viable without it?

The answer to the first question is that, as far as we know, even bits of chromosomes cannot be replaced. An abnormal chromosome number is inherited by all daughter cells provided that all subsequent mitoses are bipolar. The answer to the second question is that the overwhelming majority of cells with nuclei produced by multipolar mitosis are sick and perish.

...

to explain the extremely variable outcome of double fertilisation, there is only one assumption left, namely that it is the wrong combination of chromosomes that makes dispermy so ruinous for the embryo. Put simply, the individual chromosomes must possess different properties such that only certain combinations permit the cell to function normally or, at least, keep it alive.

...

besides the chromosomes that are indispensable for the life of the cell, there appear to be some whose absence does not limit the viability of the cell, but only destroys or impairs other normal properties, for example the tendency to form connected epithelia or the capacity to stimulate the cells of another tissue.

...

The mitotic figures illustrated by Winiwarter (Winiwarter, 1912) seem to show the best state of preservation. According to him, the diploid chromosome number in man is about 48 and the haploid number is therefore 24 (or 23)20. These 24 chromosomes must embody all the capacities inherent in chromatin. Now, if the hereditary characters about which Mendel has taught us are located in the chromosomes, then each of the small number of 24 chromosomes must contain a large number of such units, most probably aligned along the chromosome in a specific order(21). And if genetic experiments have demonstrated that in some cases certain properties are always linked together when they are inherited, these properties must, in our view, be represented in the same chromosome (22). The manner in which red–green colour blindness is inherited is explained most simply if we suppose that the essential determinant of this defect is linked to the chromosome that Winiwarter regards as the sex chromosome.

So there is increasing evidence that specific segments of chromosomes are involved in every cellular event, whether we are dealing with properties that only become apparent in the cooperative behaviour of a larger complex of cells (e.g. in the shape of the liver) or properties determined by the presence and composition of cytoplasmic constituents (e.g. a pigment or a secretion). From this point it is but a short step to the concept that the intrinsic activity of the cell depends on the cooperation of specific chromosomal elements; this is precisely the conclusion that is best supported by the experiments on doubly fertilised sea urchin eggs.

...

If, in that case, the normal relationship of the cell to its surroundings is permanently disturbed, this, as I mentioned in my introduction, might in itself be enough to precipitate unrestricted multiplication of the cell and its progeny.

Another possibility is that there is a specific inhibitory mechanism in every normal cell that only permits cell division to take place when this mechanism is overcome by some special stimulus(23). It would accord with our basic concept if one assumed that there were specific chromosomes that inhibited cell division. If their inhibitory effect were transitorily overcome, then cell division would resume. A tumour cell that proliferated without restraint would be generated if these 'inhibitory chromosomes' were eliminated(24). In this case, the tumour cell would also lose all the attributes that were located exclusively in the same chromosome as the inhibitory factors.

…

In terms of the hypothesis that there are chromosomes that inhibit cell division, one might see the distinction between the two types of tumour in the idea that in benign tumours these inhibitory chromosomes alone are eliminated, whereas in malignant tumours the combination of chromosomes in the rest of the chromatin also deviates from normality. An objection to this idea is that, in view of the small number of chromosomes, it is hardly plausible to make the extravagant assumption that there are chromosomes that act only as inhibitors of cell division. On the contrary, one might expect that the chromosomes in question also harbour other determinants whose abolition alters the character of the cell. A change of this kind does not seem to take place in benign tumours

The other alternative is that unrestrained proliferation is due either to an excess or to a stable reinforcement of specific stimulatory chromosomes. Because this model does not entail the loss of any chromosome, it might explain the emergence of cells that differ from normal only in their potential to multiply and thus generate tumours that are benign. Malignant tumours would be defined by the fact that, in addition to having an excess of stimulatory chromosomes, they have lost certain other chromosomes.

…

Although I must once again stress the purely hypothetical nature of the arguments I have advanced, mention must nonetheless be made of the fact that there are certainly links between the chromatin and the onset of cell division. In sea urchins, I was able to show in 1902 that too small an amount of chromatin relative to the amount of cytoplasm produces an intracellular stimulus to cell division that persists until further cell divisions re-establish a specified quantitative relationship between nucleus and cytoplasm (R. Hertwig's nucleocytoplasmic ratio).

I recently came upon another remarkable fact about sea urchin embryos. If, in experiments with interspecific crosses like those of Baltzer previously mentioned, one uses eggs that have been enucleated, one can get embryos with no more than four or five chromosomes (instead of 36–40). These embryos do not develop beyond the early blastula stage; but, before they perish, one often finds that the last mitoses are not bipolar but monopolar. This produces a so-called monoaster that cannot give rise to a cell division. So this too is evidence, albeit very indirect, that the composition of the chromatin is a decisive factor in determining whether the cell divides or not.

…

it has to be admitted that agents of a physical or chemical nature that act over longer periods might produce irreparable defects in the cytoplasm. And, indeed, certain specific facts that the study of tumours has revealed, such as the existence of cancers produced by X-rays or cancerous lesions simply called 'chemical' carcinomas, make it seem probable that the hypothetical defects in the tumour cells in these cases do not belong to the category with which I have been solely concerned.

…

If the assumption that every tumour is normally derived from a single cell is correct, this throws an interesting new light on the fact, stressed by all authors, that malignant tumours arise especially frequently at sites where there is chronic irritation. I shall come back to this important observation in another connection; at this point, I should like to consider it from the following point of view. If the irritation was the immediate cause of the tumour, it would then be incomprehensible why tissues irritated uniformly over considerable distances do not undergo the same kind of transformation over the whole region or at least in many places within it. Chronic irritation obviously acts indirectly, which brings us very close to concluding that it gives rise to an inherently variable event that creates the conditions that generate a malignant cell in only a very small number of cases (33). But the requirements of an event such as the one I am now proposing would be satisfied perfectly by abnormal mitoses.

…

3. On the strength of a whole series of investigations, it can be regarded as established that the histological modification of the tumour cell is accompanied by aberrant metabolism. Compared with what is made by the cells of the tissue of origin, it appears that normal products in changed proportions, or completely different products, are synthesised.

Arising out of the particular situation that I have just described, there is yet another possibility that could be considered. If we ask what in interspecific sea urchin crosses determines the fact that some chromosomes divide in the normal way whereas others behave atypically, then the answer is the foreign cytoplasm. In our state of total ignorance concerning the composition of the cytoplasm, we do not know what that means. Nonetheless, we can say, in general terms, that a specific change in the surrounding cytoplasm has a detrimental effect on certain chromosomes but not on others*. But that makes it conceivable that a particular abnormal condition of the cytoplasm might be responsible for the production of a tumour. Furthermore, it does not appear impossible that a change of this sort in the cytoplasm might be determined by an external influence. If so, one cannot reject the idea that diffuse carcinomas, for example those of the stomach lining, might be elicited by external influences that produce a cytoplasmic change in the epithelial cells.

...

if it should be the case that the putative elimination of certain chromosomes is determined by a particular change in the cytoplasm of the cells, then perhaps the onset of the disease does not have to await the occurrence of mitoses at all.

...

To what I have just said I should like to add an observation concerning the heritability of tumours. It is clear from the outset that, in the light of our hypothesis, heritable transmission can only exist in the sense that a particular predisposition is transmitted. It would indeed suffice if, in many individuals, the surrounding tissue found it much more difficult to suppress the emerging tumours than was the case in other individuals and if that property were heritable. But even the specific assumptions that our hypothesis makes do not exclude the possibility that it is a predisposition that is heritable. Thus, for example, diminished cellular resistance to influences that hold back cell division, and therefore lead to multipolar mitoses, might be heritable; and so might a tendency of the centrosomes to undergo simultaneous multiple divisions. Finally, the impairment of certain chromosomes, which I proposed in the previous section as an explanation for certain phenomena, might also be heritable.

...

It seems to be generally accepted that the production of metastases is intimately connected with the effect that tumour cells have on their immediate environment. If the change in the environment results from an alteration in cellular metabolism, as our hypothesis postulates, then the hypothesis has already done enough to help us understand metastasis.

Nonetheless, I should like to draw your attention at this point to an interesting parallel between malignant tumours and the embryos produced by simultaneous multiple division in a doubly fertilised sea urchin egg. It is said to be especially characteristic of many tumours that their cells have a marked tendency to relax or altogether unloosen the normally tight structure of the tissues, an effect that probably makes a substantial contribution to the production of metastases.

...

Apart from changes in malignant tumours that apparently occur suddenly without transitional stages and which, according to our hypothesis, are caused by an additional change in an already abnormal chromosome combination, there seem to be others in which the character of a tumour changes gradually. These gradual changes may be present in the original tumours or first become apparent in the metastases, or, most tellingly, in the secondary tumours formed on transplantation. The reaction of the healthy neighbouring tissue to the presence of the tumour cells might be implicated in this phenomenon(42). But it is also conceivable that the creation of certain abnormal chromosome combinations so perturbs the equilibrium within the nucleus that particular chromosomes go on changing under the influence of changes in other chromosomes. One group of chromosomes might eventually preponderate and perhaps even suppress the activity of others. It is therefore understandable that a malignant tumour that is at first closely similar to its tissue of origin progressively becomes less so and eventually becomes completely unrecognisable.

...

earlier, I described the relationship between nucleus and cytoplasm as a form of symbiosis in that the individual chromosomes colonise the cytoplasm as independent entities. The chromosome cycle makes this comparison acceptable. These little bodies behave like protists, and it is conceivable that the transient inactive phases that have been observed in motile protists, whatever their cause might be, could also affect the individual chromosomes embedded in their cytoplasm. This inactivity might perhaps involve a single chromosome and impair its function. The result would be a change in the character of the cell during the period of inactivity. One might object that such periods of inactivity in individual chromosomes and the consequences that flow from them should also occur in normal tissues, for which we have no evidence; but tumour cells have certain peculiarities that favour the occurrence of such phenomena. First, in a nucleus that is abnormally constituted— and here we touch on a factor that we have already considered—the conditions under which the individual chromosomes operate differ from those present in normal cells. And second, one must remember that a normal cell has two copies of every chromosome, so that the suppression of one of them might pass unnoticed; by contrast, the cells of every malignant tumour have chromosomes present in only one copy and the suppression of this copy would then result in a temporary loss of certain cellular functions.

...

The other observation that I have discussed above—that malignant tumours arise especially often at sites of chronic irritation—is readily understood in terms of our hypothesis after what has just been said about transient traumas. In the first place, chronic injury and the disturbances that it causes induce concomitant regeneration and a great deal of cell multiplication, which obviously presupposes cell division and the occurrence of abnormal divisions(45). In the second place, the noxious influence itself will at the same time greatly facilitate the production of abnormal mitoses.

...

If one compares these agents with chronic irritants that, as is well known, often give rise to cancer, then the correspondence is striking. Carcinoma of the gall bladder in tightly corseted women, especially when combined with gallstones, might illustrate the effect of squeezing. Laceration might well have the same effect, for we learn that in India it is only at the root of the right-hand horn by which the cattle are tethered that skin carcinomas arise. Raised temperature seems to produce carcinoma of the œsophagus in Chinese men who eat their rice as hot as possible, whereas the women let it cool first (46). Cancers resulting from radiation need no more than a mention. The connection between cancer and certain chemical irritants is even clearer than it is between cancer and the physical agents I have mentioned. I need only refer to the cancers of paraffin workers.

...

Parasites of all sorts, from bacteria, moulds and protozoa to worms and arthropods, have been touted as causes of malignant tumours. Most of these claims have been rejected. It does, however, appear to be the case that Distomum haematobium (48) can cause carcinomas and sarcomas, and some nematodes also appear to do so. As far as the latter are concerned, the systematic experiments of Fibiger (49) have recently shown that a species of Spiroptera that parasitises the fundus of the stomach can cause, among other pathological changes, metastatic carcinomas. Fibiger himself is convinced that the parasites exert their effect by means of a poison that they produce, and he classifies this mode of action as a form of chronic irritation. I regard it as highly probable that this interpretation holds for all cases where parasites are thought to be the cause of tumours. A carcinoma produced by parasites is in the same category as a paraffin-worker's cancer or a pipe-smoker's cancer.

...

The incidence of malignant tumours in different tissues is highly variable, and it seems to me in the light of our foregoing observations that this variation in the susceptibility of different tissues to tumour formation runs parallel, on the whole, to the frequency with which cell division takes place within them. In tissues where cells almost never divide, malignant tumours are very rare; they become more frequent if the tissue is forced to increase the rate of cell division to replace dead cells. If chronic irritation stimulates cell division in such tissues even further, then the probability that malignant tumours will occur rises still more. But the best soil for the growth of malignant tumours seems to be when the chronic irritation that leads to cell proliferation at the same time harbours a factor that has the power to produce abnormalities in the process of mitosis.

...

if my assumption is correct, the legacy that a tumour acquires is more or less comparable with the blind withdrawal of a particular number from a sack containing a thousand. One might try to do this hundreds of times and never select the right number, yet someone else might try only once and get it first time. In fact, the element of lottery pertains to the origin of malignant tumours in a number of ways.

...

The very possibility that multipolar mitoses might arise, insofar as they are produced by the inhibition of cell division, is determined in large measure by chance, as it depends on whether the insult is severe enough during the brief period in which the cell is being partitioned. If a tetraploid cell is formed as a result of this insult, it will nonetheless be a matter of chance whether this particular cell receives a stimulus that induces it to divide. And if it gets that far, then the real game of dice begins in the multipolar mitosis, for then the outcome depends on whether a combination of chromosomes happens to come together in such a way in one of the daughter cells that it generates the properties characteristic of malignancy. The more abnormal cell divisions there are in a tissue, the greater the probability will of course be that the required combination will turn up; but it is conceivable that, despite the continuous iteration of abnormal mitoses, it might never do so. By contrast, a single multipolar mitosis taking place in healthy tissue, perhaps the result of simultaneous multiple division of the centrosome, could give rise to the cell that originates the malignant tumour.

...

A final argument that I believe I can regard as a support for my assumption is the outcome of observations on chromosome number and nuclear size in carcinomas. Long ago, Hansemann drew attention to the fact that mitoses with raised or reduced chromosome numbers are to be found in malignant tumours. Borst, too, states that one often observes an easily detectable increase in chromosome number in proliferating tissues, and less often a reduced chromosome number.

...

the nuclei of many carcinomas are larger than those of the tissues from which they were derived, and usually larger than the nuclei of benign epitheliomas. In many carcinomas, however, the nuclei are of normal size and, rarely, they are even smaller than the nuclei of the tissue of origin.

For the cancer researcher seeking a comprehensive morphological feature to characterise tumour cells, this finding, like so many previous ones, must come once again as a disappointment. But, if we look at tumours from the point of view of our hypothesis and accept the proposition that the proportionality between nuclear size and chromosome number holds true for grown men(53), then these results are exactly what you would expect.

...

I would like to think that the explanations given in the previous chapter will have already made it clear to the reader that, according to my hypothesis, atypical mitoses are by no means to be regarded as an essential property of malignant tumours. Indeed, the opposite is the case, a matter that I propose to discuss presently. Moreover, the occurrence of abnormal mitoses of this sort in otherwise healthy or diseased tissues constitutes one of the most important pieces of evidence in support of my hypothesis. For my hypothesis claims no more than that a multipolar or asymmetrical mitosis in any hitherto normal cell might lead to the formation of a malignant tumour, but not at all that it must do so. If then malignant tumours arise preferentially at sites of inflammation, if benign proliferations suddenly become malignant, or if, finally, a carcinoma or a sarcoma arises in normal tissue in which there has been no previous disease, then it is precisely the presence of abnormal mitoses in completely or at least relatively normal tissues that, according to my hypothesis, must be regarded as the origin of the malignant tumours.

...

the cells derived from a tumour cell by multipolar mitosis are in the vast majority of cases no longer tumour cells; paradoxical as it may sound, atypical mitosis that in normal tissue can give rise to this calamitous abnormality can only be regarded as a therapeutic factor in a tumour that has already been established. In fact, one can imagine that if in a malignant tumour multipolar mitoses were the rule, the tumour would gradually undergo involution of its own accord, and if the products of its decomposition were removed quickly enough, a cure might be achieved.

X-rays and radium radiation are said to have both the power to produce cancer and also the power to heal it. This curiously ambivalent effect might be explicable in part by the ability of these rays to produce abnormal mitoses. Under their influence, normal cells give rise to malignant cells, but malignant cells give rise to cells that are no longer viable.

...

The main thesis is admittedly hypothetical, namely whether an abnormal chromosome constitution can be produced such that the cells that harbour it are driven to unrestrained proliferation. This assumption must be made ad hoc, but there is much to be said for it. Above all, I regard it as beyond doubt that the tendency to multiply indefinitely is a primaeval property of cells and that the inhibition of multiplication in metazoan cells occurs secondarily under the influence of the environment(59). Cells normally submit to this inhibition and reassert their original proliferative drive only when there are certain changes in their environment. If this is so, one must presuppose that there exists a specific cellular apparatus that is sensitive to conditions in the environment. Given such an apparatus, one must assume that it is susceptible to aberrations that incur the loss of sensitivity to the conditions of the environment. Then, the inherent proliferative drive of the cell is released and proceeds without taking any notice of the requirements of the rest of the body."

Part II

Foresight

IOP Publishing

Ahead of the Curve
Hidden breakthroughs in the biosciences
Michael Levin and Dany Spencer Adams

Chapter 2

Physics in cell and developmental biology

Case study 3: Development Employs Forces

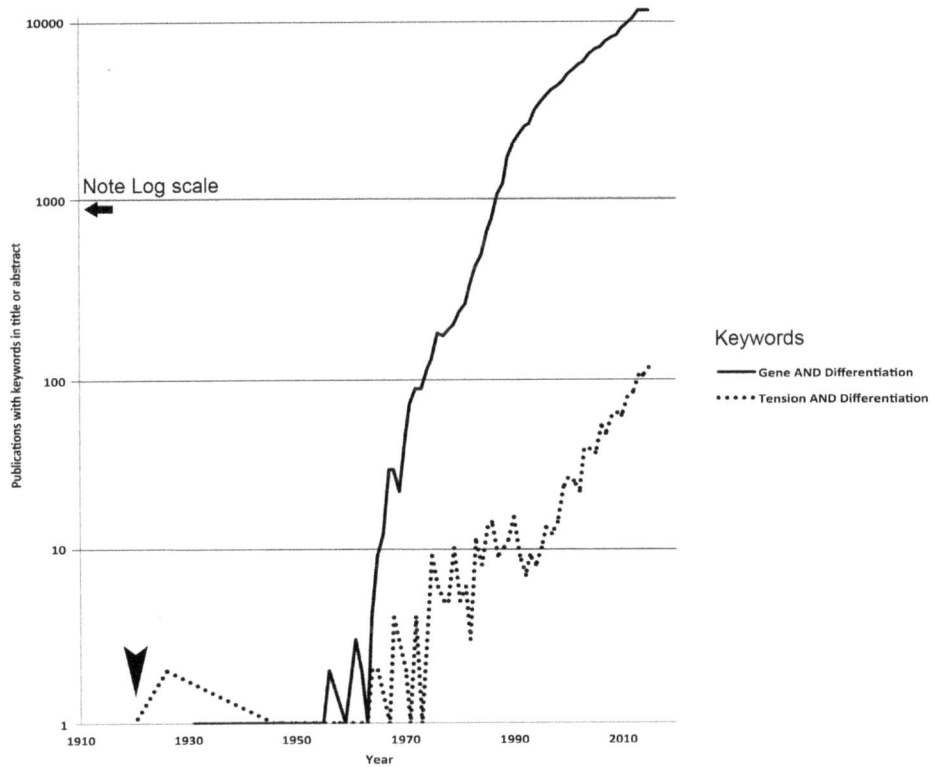

Carey (1920) Studies in the dynamics of histogenesis: tension of differential growth as a stimulus to myogenesis.

Perspective on *Studies in the Dynamics of Histogenesis. I. Tension of Differential Growth as a Stimulus to Myogenesis*

Raymond E Keller
Professor of Biology
University of Virginia

Even today, many biologists do not want to think about morphogenesis and the importance of forces. A more popular idea is 'self assembly' of embryos, a concept that is rather strange: what other choice do embryos have? The idea sprang from the 'self-assembly' of the tobacco moaic virus, spread to self assembly of molecular aggregates of all kinds, and got a big boost from Steinberg's differential adhesion hypothesis, the DAH: genes encode proteins, including cell adhesion molecules, these impart different cell adhesive strengths, and borrowing from mass action chemistry, motility stirs the pot, bringing cells of differing adhesive strengths in contact, and they sort out based on the thermodynamics of adhesive strengths. The DAH kept genetics and molecular biology at center stage—in the minds of many, everything subsequent to the encoding of adhesion was thought to be irrelevant—motility, polarity, cell sociobiology etc were not needed, only generic motility was needed. Overt generation of forces by patterned cell behavior was ignored. Sometime in the 1980s, a then prominent geneticist, told me 'Ray, physics has absolutely nothing to do with morphogenesis'. It was a sign of the times. Cell adhesion is important but it is a regulated property, but one of many among the integrated force-generating behaviors of cells that have rightly taken center stage. Today, the Carey paper looks pretty modern.

STUDIES IN THE DYNAMICS OF HISTOGENESIS

I. Tension of Differential Growth as a Stimulus to Myogenesis.
by Eben J Carey.

(*From the Department of Anatomy, College of Medicine, Creighton University, Omaha.*)
(Received for publication, January 17, 1920.)

The prevalent opinion among embryologists in regard to the origin of muscular tissue is that of self-differentiation. This is due largely to the work of Wilson (1904) on *Dentalium*, and Conklin (1896–97, 1905) on *Cynthia (Styela) partita and Crepidula* in regard to the organ-forming elements of the cytoplasm, and to the experimental work of Harrison and Lewis (1904, 1905). Harrison ablated the spinal cord of the tadpole prior to the growth of the peripheral nerves into the limb buds. This operation eliminated any peculiar formation stimulus emanating from the nervous system. Still the differentiation of the contractile substance took place in the normal manner, as did the grouping of fibers into the individual muscles. Lewis (1910) draws the following conclusion based on Harrison's experiments, in regard to the genesis of cross-striated muscle:

"Thus it is seen that all the constructive processes involved in the production of the specific structure and arrangement of the muscle-fibres take place independently of stimuli from the nervous system and of the functional activity of the muscles themselves. Cross-striated muscle tissue and the individual muscles are thus self-differentiating."

The fact that there is considerable muscular differentiation before nerves establish a connection with their corresponding muscles has been shown by Bardeen (1900, 1906–07), Harrison, and Carey (1918) in the pig embryo. There is also considerable smooth muscle differentiation in the descending colon of the pig before either the myenteric or Auerbach's plexus is detected.

Lewis (1910) endeavored to solve the problem at how early a period in the development of the ovum this power of self-differentiation of muscle tissue begins. He found by transplanting tissue from the lips of the blastopore in the early gastrula stage of the frog that this tissue later on showed muscular differentiation. The conclusion is drawn "that muscle tissue is already predetermined in the early gastrula."

The idea conveyed by the last statement is that muscular tissue is formed, sui generis, by some inherent predetermination and not by the agency of its surroundings nor due to its position in the whole. Lewis' view-point is in accord with Conklin's (1905) as seen in the following statement of the latter observer: "The potencies or prospective values of any blastomere are not primarily a function of its position, but rather of its material substances."

There are three theories regarding cellular differentiation; first, the "mosaic theory" of Roux (1881), later modified by Wilson (1904), Conklin (1905), Zeleny

(1904), and Boveri; second, the "organization theory" of Whitman and more recently elaborated by Child (1915) in his studies on metabolic gradients and individuality; third, "the homogeneity theory" of Driesch (1894, 1899). Driesch considers the peculiar organizing quality of protoplasm as due to the expression of a mysterious force wholly different from any in the inorganic world.

His, Roux (1881, 1892, 1893), Wilson (1892, 1893, 1897, 1904), and Conklin (1905) lay emphasis upon the cell as the key to all ultimate biological problems. Whitman, on the other hand, points out the inadequacy of the cell theory to development. "That organization precedes cell formation and regulates it, rather than the reverse, is a conclusion that forces itself upon us from many sides," is a summary of his studies. Morgan (1895, 1898) had deduced the idea from his studies on regeneration that the multicellular individual is a whole in the same sense that the unicellular form is a whole. Child (1899, 1915) also lays emphasis on the fact that it is the "organism—the individual, which is the unit and not the cell." Differentiation of a single cell, consequently, according to Child and Whitman, is a function of its position in the whole. This view is also upheld by Driesch (1894). Wilson and Conklin, on the other hand, conclude that potencies are functions of the material substance of the cell.

The influence of the organism as a whole in subjugating its dependent parts is convincingly shown by Loeb (1916) in his regeneration experiments on *Bryophyllum calycinum* and in his experiments on *Amblystoma* larva (1897). This influence is exerted through the blood stream by means of "hormones." The sound mechanistic attitude of Loeb toward development may be seen in the following statement: "As soon as we can show that a life phenomenon obeys a simple physical law there is no longer any need for assuming the action of non-physical agencies" (1916).

The object of this paper is not, however, to discuss historically the views of these various authors, but to emphasize certain facts in the development of muscle tissue hitherto overlooked. It is a well known fact that the embryo presents differential rates of growth. It is desired, therefore, to emphasize the fact that in embryological development there are zones of unequal or differential growth, and that the effects of these zones of growth are factors in histogenesis. The active and less active zones are defined with reference to the rate of cell division per mm. of cross-section. This principle was deduced from a series of studies on osteogenesis and myogenesis begun in 1914. Previous reports of a part of this work have been presented to the Association of American Anatomists (Carey, 1917, 1918, 1919).

It will be illuminating to search for the cellular forces outside of the immediate differentiating zone under observation. This search necessitates lower magnifications in order to enlarge our field of view. Heretofore, cytological differentiation has been studied *per se*, with magnifications of 1,000 to 2,000 diameters which considerably reduce our range of view. The higher magnifications are profitable in revealing cytological detail but the interpretation of the process is lost unless, in conjunction with the higher, intermediate magnifications are used. The employment of all magnifications of the microscope in connection with naked eye studies will reveal the interaction of related developing parts.

Early Development of the Descending Colon of the Pig

The attention of the writer was directed to the fact, after plotting hundreds of intestinal epithelial mitotic figures, that these figures were usually confined to some definite region of the circumference of a single section (Figs. 1 to 9). This region was

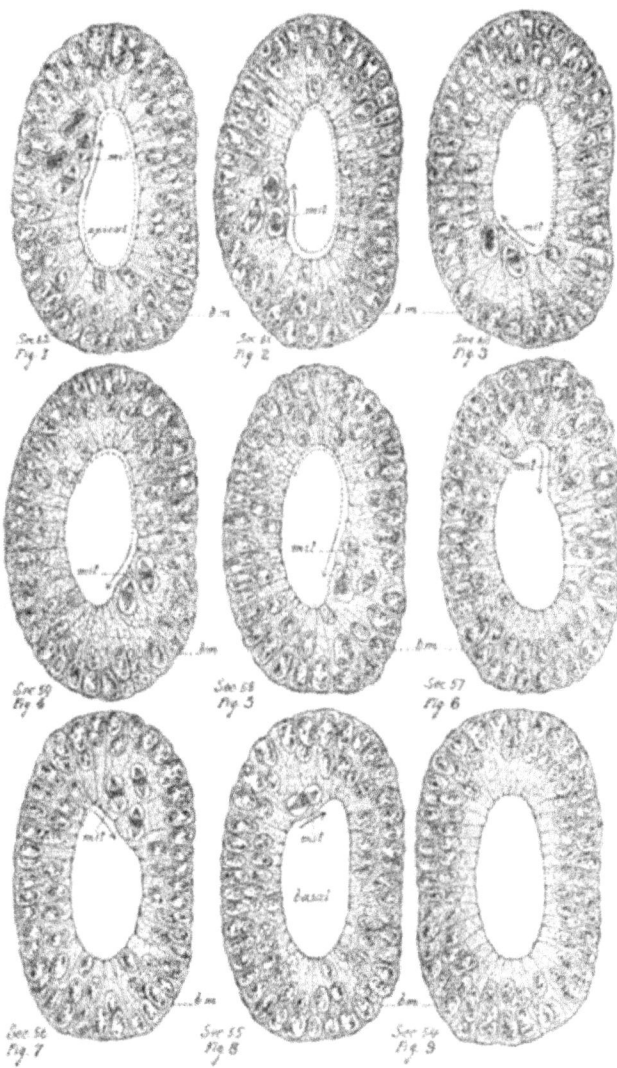

Figs. 1 to 9. Sections of the epithelial tube representing one complete turn in a dexiotropic rotation in a spiral manner of the mitotic division. The primary type is the left-handed helix. Spiral path is directed upward toward the ileocecal valve. Section 62 represents the head or apical end of the mitotic path. Sections 55 and 54 represent the tail or basal end of the mitotic spiral path. *mit*, mitosis; *bm*, basement membrane of the epithelial tube. Drawings are made with the aid of a Spencer Camera lucida. Sections 54 to 62 are from Pig Embryo 19, 24 mm. in length (Creighton Embryological Collection).

found to change at different levels of the serial sections. By graphic reconstruction this plot was found to form the path of a definite spiral. The predominant type was the left-handed helix. In one of the embryos of the twenty that were plotted this spiral was arranged as a right-handed helix (Figs. 1 to 9). The spiral itself presented a head or apical region in which mitotic figures were found to be numerous, and a tail or basal end in which there were fewer and fewer figures. The apical end of the spiral path is always directed towards the ileocecal valve and the basal end towards the rectum. Growth is, therefore, from below upwards in a spiral course. One spiral growth is quickly followed by a second which rifles a path slightly lateral to its predecessor. This in turn is followed by a third in a path still more lateral, and so on around the circumference. This intermittent rhythm of explosive spiral growth may be compared to that of the successive fire balls emitted by a roman candle in fireworks. The paths formed by this explosive spiral growth may be compared to those within the barrel of a Winchester rifle.

The most rapidly growing part of the intestine, therefore, is the epithelial tube. In embryos 10 to 25 mm. in length the descending colon grows relatively more rapidly in diameter than in length (Tables I and II). The increase in diameter is due primarily to the rapid growth of the entodermal epithelial tube and only partially to its surrounding mesenchymal cloak. The latter is relatively passive in growth with respect to the former (Fig. 10). It is during this early increase in diameter that the inner smooth muscle coat is in process of formation. The mesenchymal cells are drawn out gradually in a definite series of concentric rings. These rings appear not unlike those of the planet Saturn and the annular nebula in Lyra.

A definite centripetal force is active in the rapid spiral growth of the intestinal epithelial tube. The surrounding mesenchymal cells are thrown into an obvious series of concentric rings, according to their various densities. Those possessing the greatest density will join the outer ring in the tangential path of the force, whereas the inner ring will be composed of bodies forming a gradient of decreasing densities. The cells forming the outer ring will be most elongated.

As this concentric initial smooth muscle layer becomes differentiated it tends to restrict the diametrical growth of the epithelial tube. The epithelial mitotic figures under this restriction shift their planes of division from a right angle to a parallel position with the smooth muscle cells. This shifting results in an elongation of the intestine.

In embryos 25 to 40 mm. (Tables I and II) in length, the elongation of the descending colon is more rapid in growth than that of the diameter. It is during this period that the outer longitudinal muscular coat is in the process of formation. The rapid growth of the epithelial tube in length tends to elongate the peripheral undifferentiated mesenchymal cells which were not directly involved in the formation of the inner smooth muscular coat (Fig. 11).

The differentiation of the outer longitudinal muscle coat therefore coincides, in time, with the rapid growth in length of the intestinal epithelium. The inner smooth muscle coat, on the other hand, is formed during the period of the rapid growth of

TABLE I. Measurements of Differential Growth of Descending Colon.

Length of embryo. mm.	Thickness of mesenchymal wall.* mm.	Diameter of epithelial tube. mm.	Diameter of descending colon. mm.	Epithelial tubular indices.† per cent
10	0.085	0.048	0.218	6
12	0.099	0.069	0.267	8
13	0.115	0.075	0.305	8.5
14	0.128	0.081	0.337	10
16	0.126	0.089	0.341	14
19	0.124	0.095	0.343	18
20	0.123	0.099	0.345	20
22	0.122	0.119	0.363	25
23	0.121	0.138	0.383	30
24	0.120	0.152	0.392	38
25	0.115	0.164	0.394	40
27	0.109	0.188	0.406	44
30	0.104	0.208	0.416	46
32	0.099	0.220	0.418	47
35	0.098	0.246	0.442	48
37	0.096	0.260	0.452	50
39	0.093	0.279	0.465	52
40	0.092	0.289	0.473	54
42	0.090	0.312	0.482	57
45	0.083	0.321	0.486	61

*The mesenchymal wall begins to diminish in thickness after it reaches a width of 0.128 mm. in the 14 mm. stage of the pig embryo. This diminution is due to the tension caused by the more rapid epithelial tubular growth in diameter. Measurements made with B. and L. filar micrometer, calibrated.

†The ratio of the square of the mean diameter of the epithelial tube to that of the surrounding mesenchyme is referred to as the epithelial tubular index. It has been calculated from the following formula.

$$\left\{ \frac{\frac{(x+y)^2}{(2)} \times 100}{\frac{(X+Y)^2}{(2)} - \frac{(x+y)^2}{(2)}} \right\} = Z$$

x and y are the long and short diameters of the epithelial tube respectively. X and Y are the long and short diameters of the surrounding mesenchymal tube. Z is the ratio of the epithelial tube to the mesenchymal tube.

the intestinal epithelial tube in diameter. Once the formation of the inner circular muscular rings is fairly established a resistance to growth in width is encountered by the cells surrounding the rapidly dilating lumen. These cells then grow primarily along the path of least resistance in a longitudinal manner. At this stage the longitudinal muscle cell, spherical in shape (Fig. 12), is elongated to a spindle-shaped structure (Fig. 13).

TABLE II. Ratio of Diameter to Length of Entire Colon.

Length of embryo. mm.	Diameter of descending colon. mm.	Length of entire colon. mm.	Ratio of diameter to length of entire colon. mm.
10	0.218	1.95	1:9.9
12	0.267	2.00	1:7.5
13	0.305	2.02	1:6.6
14	0.337	2.05	1:6.0
16	0.341	2.10	1:6.1
19	0.343	2.20	1:6.4
20	0.345	2.30	1:6.7
22	0.363	2.75	1:7.5
23	0.383	3.00	1:7.8
24	0.392	3.50	1:8.9
25	0.394	3.95	1:10.0
27	0.406	6.00	1:14.0
30	0.416	8.00	1:19.0
32	0.418	11.00	1:26.0
35	0.442	16.95	1:36.0
37	0.452	18.00	1:39.0
39	0.465	21.00	1:45.0
40	0.473	23.00	1:48.0
42	0.482	27.00	1:56.0
45	0.486	29.00	1:59.0
Adult.	50.000	7,000.00	1:140.0

An interesting correlation in the development of the esophagus in man may be cited. This correlation was detected in the work of Jackson and in that of Keibel and Elze. The former investigator studied the developmental topography of the esophagus, the two latter the histogenesis of the esophagus. Jackson states that the descent of the stomach is accompanied by a great elongation of the esophagus. In a 9.4 mm. specimen, the esophagus measures 1.8 mm. At this proportion it should measure 4.3 mm. in a 22.8 mm. embryo but its actual length is found to be 8 mm. A year previous to this, Keibel and Elze reported that the esophagus in 12.5 mm. embryos shows a circular but no longitudinal muscle layer. In 17 mm. embryos, they find a circular layer with the longitudinal layer faintly indicated. The histogenesis of the outer longitudinal layer of the esophagus as studied by Keibel and Elze coincides in time with the rapid elongation of the esophagus, due to the descent of the stomach, as recorded by Jackson.

Interpretation of Results

The result of the action of a force on an elastic body is the production of a strain. If mechanical forces are at work on organic matter, they tend to produce similar results to these acting upon inert matter. Too frequently the term self-differentiation is

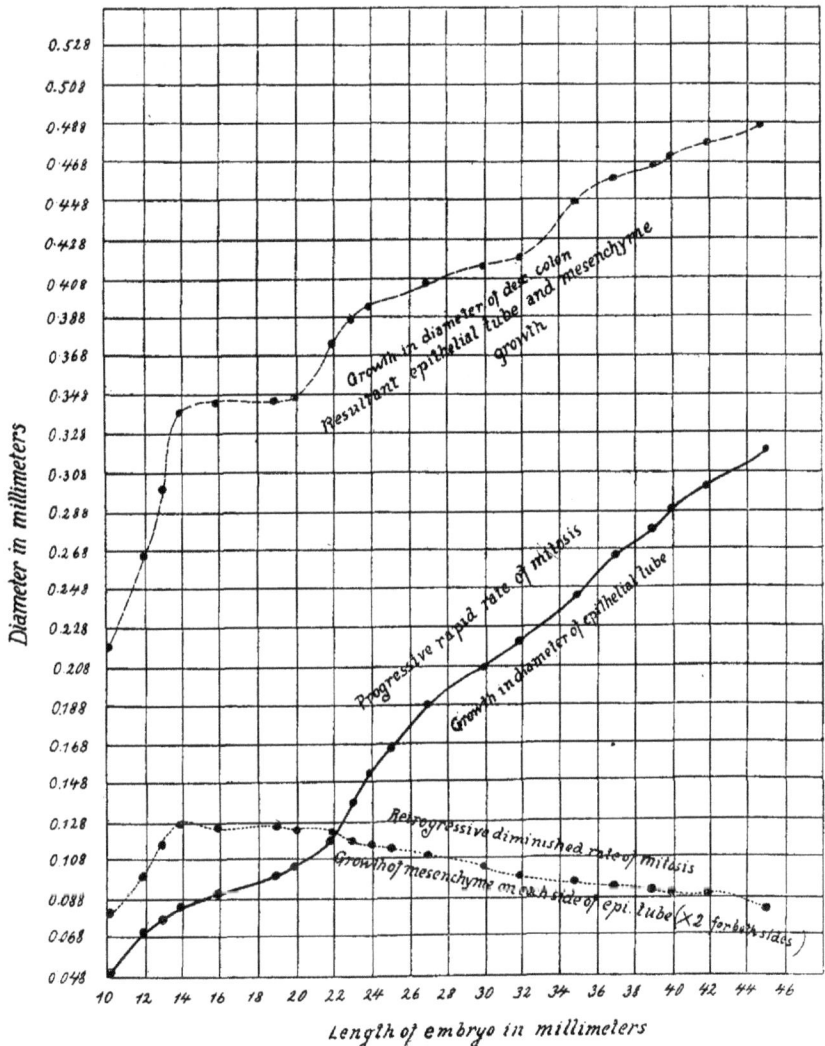

Fig. 10. Curves of differential growth of descending colon. The active growth of epithelial tube is contrasted with the passivity of the mesenchymal wall. An absolute decrease in thickness of the mesenchymal wall is seen after the stage of the 14 mm. pig embryo.

applied to alteration of form and internal structure of developing cells without searching the immediate environment of the specializing cells or syncytium to ascertain whether or not these changes are attributable to forces outside of the differentiating zone. This applies particularly to the differentiation of bone and muscle tissue. If a cell changes in form successively through the spherical, ellipsoid, and spindle stages it undergoes a strain. A strain is usually due to an external force which elicits internal reacting stresses in the body acted upon. Cytological differentiation is frequently a manifestation of these internal reacting stresses to forces extrinsic to the differentiating zone.

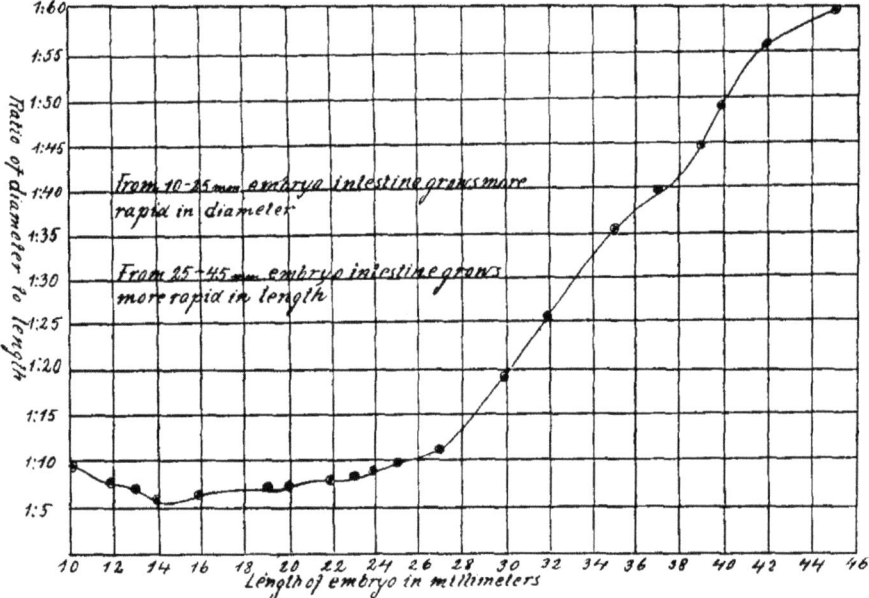

Fig. 11. Curve representing the ratio of the diameter to the length of the descending colon in embryos 10 to 45 mm. in length. Particular attention is directed to the fact that in embryos 10 to 25 mm. in length the intestine grows relatively more rapidly in diameter than in length. In embryos from 25 to 45 mm. in length the intestine grows more rapidly in length than in diameter. The inner circular smooth muscle is formed during the period of rapid growth of the intestine in diameter. During the period of the rapid growth of the intestine in length the histogenesis of the outer longitudinal smooth muscle coat is taking place.

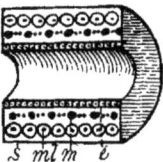

Fig. 12. Longitudinal section of intestine; *s*, peritoneal epithelium; *ml*, mesenchyme; *m*, circular muscle; *e*, epithelium.

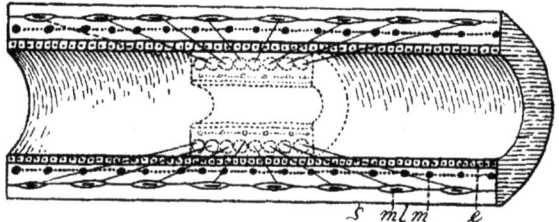

Fig. 13. Longitudinal section of intestine schematizing the elongation of the intestine represented in Fig. 12. Due to the resistance of the inner smooth muscle layer *m*, the intestinal epithelium grows in the longitudinal path of least resistance. This results in the elongation of the outer mesenchymal cells *ml* (Fig. 12) into the elliptical or spindle cells *ml* (Fig. 13).

A strain is produced in certain regions of the embryo by the expansion of a rapidly dividing group of cells against a less active or relatively passive group of cells. After their differentiation the relatively passive group of cells in their turn react upon the former. This action and reaction are objectively evident by a retardation or alteration of the rate of growth, or by a change produced in the external form or internal structure of the cells involved.

In this study, the initial zone of rapid growth is found in the epithelial tube. The rapid spiral expansion of the entodermal epithelial tube reacts against the surrounding splanchnic mesenchyme with the result that the less actively growing cells of the peripheral region of the intestinal wall are elongated. Later the elongated, differentiated mesenchymal cells cause a retardation of the growth in diameter of the epithelium. Immediately following this retardation of diametrical growth the period of rapid growth in length of the intestine takes place. In this development, therefore, the influence of unequal growth zones is definitely shown as furnishing a tensional stimulus for the differentiation of muscle.

This action is diagrammatically illustrated in Figs. 14 and 15. In Fig. 15 the growth in diameter of the intestine is schematized; the rapid increase in width is shown as due primarily to the increase in the lumen. This growth is due to rapid mitotic activity of the epithelium (e, Fig. 14, to e', Fig. 15). In the lumen of Fig. 15, the former is represented in a spiral manner, sg. In this growth the strain upon the surrounding mesenchymal cells m is illustrated. These cells are strained by the external applied forces of the progressively diverging radii. The internal reacting stresses are manifested by the changes in shape through spherical m, ellipsoidal m', and spindle m'' cellular phases in Fig. 15. In addition to the homogeneous strain to which the cells m, m' and m'' are subjected there is a definite pressure exerted by the epithelial cells lining the expanding lumen.

This force of expansion is represented by the arrow a–d. A resistance f, due to the peritoneal epithelium, is met. This causes a reaction d–a, With progressively increasing growth a zone of equilibrium of expansile and reacting forces is established in the middle, represented by the double arrows b–a, and d–c. This action and reaction of forces is another factor tending to compress the cells in the middle of the mesenchymal wall of the intestine. The action of the centripetal component of the spiral growth of epithelium in forming the rings of dense spindle-shaped muscle cells m'' is represented by the broken curved arrows. The spiral growth of the epithelial tube in a dexiotropic rotation exerts a centripetal force upon

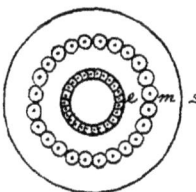

Fig. 14. Transverse section of intestine; e, epithelium; m, mesenchyme (spherical nucleus); s, peritoneal epithelium.

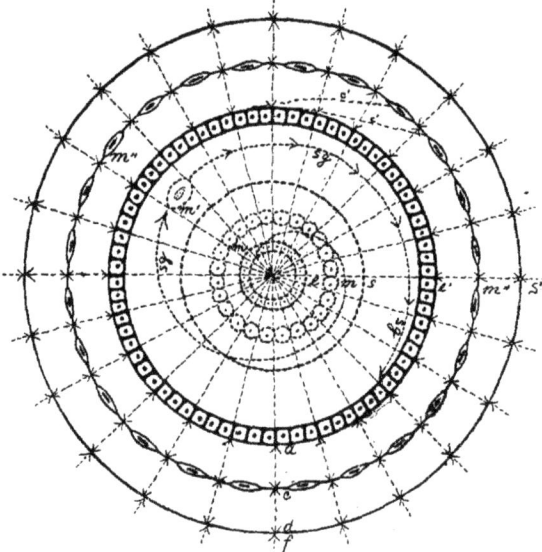

Fig. 15. Transverse section of intestine grown to three times the width represented in Fig. 14. Fig. 15 is represented in broken lines within the lumen. The spiral growth of the epithelium is represented by the broken lines *sg*. The tension, upon the mesenchyme, by the most rapidly growing epithelium, is shown in the elongated muscle cells *m″*. These cells are homogeneously strained in the centrifugal path *c′* due to the progressively diverging radii. Cells marked *m, m′,* and *m″* represent the progressive steps in the strain ellipsoid in the differentiation of a muscle from a mesenchyme cell. The expansile force of the epithelium is shown by the double arrow *a–d*; the reacting resistance of the serous membrane by the line *d–a*. Equilibrium is established in the middle of the mesenchyme and is graphically represented by the double arrows *a–b* and *d–c*. This is another factor in the tensional elongation of the middle cells. The smooth muscle ring exerts a centrifugal reaction to the applied centripetal force of the dexiotropic spiral rotation of the epithelial tube. The mitotic figures of the descending colon primarily follow the path of a left-handed helix.

the surrounding mesenchyme. The mesenchyme consequently exerts a simultaneous equal and opposite centrifugal force upon the epithelial tube. This growth is primarily in the form of a left-handed helix from the rectal to the ileocecal valvular regions of the large intestine. In Figs. 14 and 15 the right-handed helix is depicted.

Although by direct observation of serial sections no motion is seen, there is, however, objective evidence of homogeneous and ellipsoidal strains upon the surrounding mesenchyme. The mesenchyme is drawn out into concentric rings, the outermost of which are most viscid, by the spiral growth of the epithelial tube, roughly comparable to the increase of viscosity and concentric annular formation of egg albumin when subjected to an egg-beater.

Conclusion

The genesis and maintenance of muscle tissue represents a resultant or equilibration of converging factors which are active and formative during development. *One of these factors is the tensional stresses to which the mesenchyme is subjected by a force extrinsic to the differentiating zone*. In subsequent involution or degeneration of muscular tissue during the prenatal or postnatal periods, this equilibrium is upset by altering or destroying the tensional reacting stress.

Tension is developed when a muscle contracts. Contractility is a fundamental property of protoplasm and, when manifested, tension is developed. In both, the development and specific function of muscle tissue, tensional stresses are inseparably involved. The *Ameba* possesses the property of contractility in all possible directions. The function of contraction in one definite direction characterizes muscle tissue from that of undifferentiated protoplasm. What initiates the progressive series of physicochemical changes in the mesenchyme resulting in an alteration of its attribute from non-specificity to its specificity of direction of contractility? This question is answered as follows.

The mesenchyme before differentiating into muscle tissue must be subjected to a certain minimal homogeneous and ellipsoidal strain. This strain is objectively evident by an alteration of the form of the spherical nuclei, into the ellipsoidal and spindle conditions and by an elongation of the granular cytoplasm into parallel granular and continuous fibrillæ. The fibrillæ are arranged along lines of internal and reacting tensional stresses. The ends of the mesenchymal cells, in tension, must be attached to supports of which one, at least, is mobile. The tensional stresses are reactions to simultaneous forces extrinsic to the zone of myogenesis. The external forces are produced by a progressive divergence or separation of the mobile supports to which the mesenchymal cells are attached. Therefore, muscle tissue is not self-differentiating but is dependent upon an external dynamic stimulus. As regards smooth muscle this stimulus is the *tension of differential growth*.

Summary

1. The region of most active mitosis per mm. of cross-section in the intestine is the entodermal epithelial tube. The mitotic figures primarily follow the path of a right-handed helix. In one of the twenty embryos the mitotic figures describe the path of a right-handed helix.
2. The region of least active or relatively passive growth per mm. of cross-section is the mesenchyme, derived from the splanchnic mesoderm surrounding the epithelial tube.
3. The rapid expansion, due to epithelial growth in a rotating spiral manner, of the intestinal lumen is greater than the activity manifest in the surrounding mesenchyme. This causes a pressure in the latter resulting in a flattening and elongation of the mesenchymal cells. The successive changes in shape of these cells through the spherical, ellipsoidal, and spindle cellular phases are seen. The mesenchymal wall decreases in thickness, due to tension caused by epithelial tubular dilation.
4. The rotating spiral growth of the epithelial cells causes the formation of a series of mesenchymal cellular and fibrillar concentric rings due to the centripetal force of the former.

5. The circular, smooth muscle cells are differentiated in the outer, more condensed margins of the ring. At these points the developing tensional stresses are greater than within the ring.
6. The inner circular smooth muscle coat is the first one differentiated and is incident to the rapid growth of the epithelial tube in diameter. The former soon tends to restrict the growth of the epithelial tube in diameter. The tube, pursuing the lines of least resistance, grows in length. During the period of rapid growth in length the outer longitudinal muscle coat is in the process of formation.
7. The tensional stresses to which the elongated strained mesenchymal cells are subjected appear to be a dynamic stimulus to smooth muscle differentiation.
8. From this study of a closely graded and progressive series of sections of intestinal development, the conclusion is drawn that muscle tissue is not self-differentiating, in the strict sense of the term, but that the tension of differential growth acts as the stimulus to smooth muscle differentiation.

The writer wishes to express his indebtedness to Professor H. von W. Schulte for his interest and his valuable suggestions; to Madame Helen Ziska for the illustrations; and to his wife for her help in reading the proof.

Bibliography

Bardeen C R 1900 *Johns Hopkins Hosp. Rep.* ix 231 *Am. J. Anat.*, 1906–07, vi, 259

Boveri T *Sitzungsb. phys.-med. Ges. zu Wurzburg*, 1904, 16; Ergebnisse über die Konstitution der chromatischen Substanz des Zelikerns, Jena, 1904, 115

Carey E. J. *Anat. Rec.*, 1917, xi, 1; 1918, xiv, 30; 1919, xvi, 45, 114

Child C. M. Biological lectures from the Marine Biological Laboratory of Woods Hole, 1899, 231; Individuality in organisms, Chicago, 1915, 5

Conklin E G Biological lectures from the Marine Biological Laboratory of Woods Hole, 1896–97, 17; *J. Exp. Zool.*, 1905, ii, 145

Driesch H Analytische Theorie der organische Entwicklung, Leipsic, 1894, 97; *Biol. Centr.*, 1899, xix, 225; *Arch. Entwcklngsmechn. Organ.*, 1899, viii, 123

Harrison R G 1905 *Am. J. Anat.* iii 197

His W Unser Körperform und das physiologische Problem ihrer Enstehung, Leipsic, 1874, 165

Jackson C M 1909 *Anat. Rec.* iii 361

Keibel F and Elze C Normentafeln zur Entwicklun-geschichte der Wirbeltiere, Jena, 1908, viii, 1

Lewis W H *Am. J. Anat.*, 1904, iii, 505; *J. Exp. Zool.*, 1905, ii, 431; in Kiebel, F., and Mall, F. P., Manual of human embryology, Philadelphia and London, 1910, i, 456

Loeb J *Arch. Entwcklngsmechn. Organ.*, 1897, iv, 502; The organism as a whole, New York, 1916, 11, 152

Morgan T H Biological lectures from the Marine Biological Laboratory of Woods Hole, 1898, 196; *J. Morphol.*, 1895, x, 419

Roux W Der Kampf der Theile im Organismus, Leipsic, 1881, 152; in Merkel, and Bennet, Ergebnisse der Anatomie und Entwickelungsgeschichte, 1892, ii, 415; *Zool. Anz.*, 1893, 115

Whitman C O 1893 *J. Morphol.* **viii** 639

Wilson E B *J. Morphol.*, 1892, vi, 361; 1893, viii, 579; The cell in development and inheritance, New York, 1897, 23; *J. Exp. Zool.*, 1904, i, 1, 197

Zeleny C 1904 *J. Exp. Zool.* **i** 293

Case study 4: Depolarization Induces Neuron Division

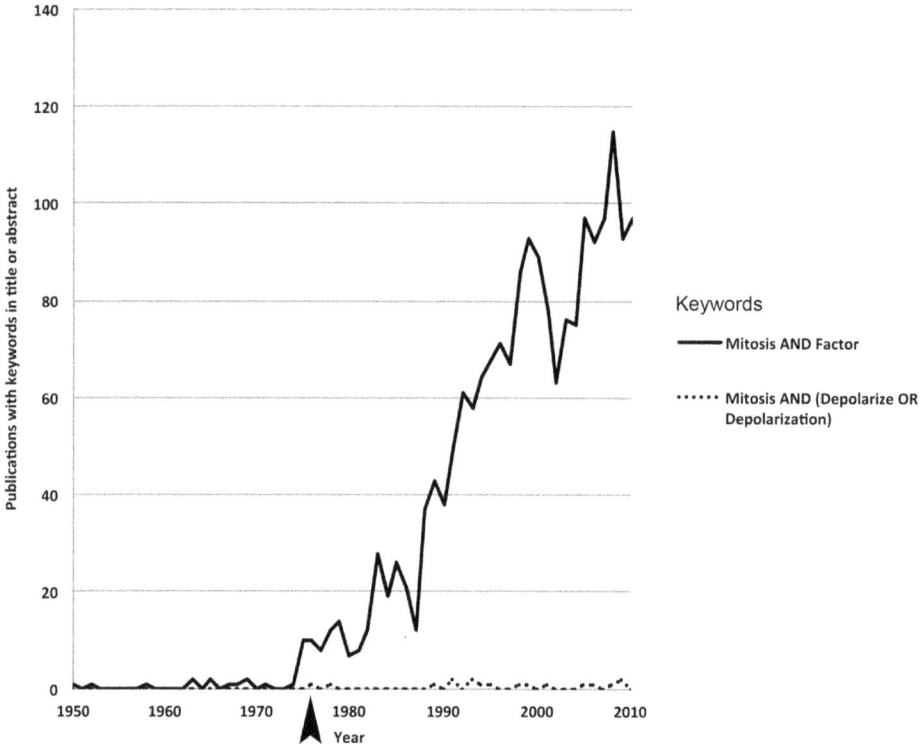

Cone and Cone (1976) Induction of mitosis in mature neurons in central nervous system by sustained depolarization.

Perspective on *Induction of mitosis in mature neurons in central nervous system by sustained depolarization*

Michael Levin, PhD
Allen Discovery Center,
Professor, Tufts University

'Electrical Potential: a new control knob for cell functions'

Salamanders can regenerate large portions of their brain after injury. A major limitation in the regenerative capacity of the brain in other species is the fact that post-mitotic neurons generally do not divide. Cone's work in the late 1960s and early 1970s showed that mitosis can be triggered by sustained depolarization of resting potential [1–8]. His group also extended it to the control of mitosis in cancer cells [9], formulating a theory of electrical control of cellular proliferation [10–12]. This work, though not widely recognized in subsequent decades, has major implications for a number of areas of biology and medicine. Its ramifications have begun to bear fruit in recent years [13], extending beyond the control of single cell behaviors toward a new aspect of developmental bioelectricity as a regulator of large-scale biological shape.

The resting potential (V_{mem})—the voltage difference between the inside and the outside of the cell surface—is established by ion channel and pump proteins in the cell membrane which facilitate the inward or outward movement of specific charged molecules. This electrical aspect of cell behavior is a crucial part of the living cell, being responsible for many aspects of its physiology and considered a hallmark of life [14]. Cone's work was perhaps the first to recognize the importance of this parameter for guiding key cell behaviors—proliferation and differentiation, beyond basic homeostasis. The study of endogenous, electrical control of cell functions other than excitable nerve and muscle is called developmental bioelectricity, and forms an exciting part of developmental, regenerative, and cancer biology today.

Over the last century, a number of waves of research studied the bioelectric aspects of life. Lund [15], Burr [16], Marsh and Beams [17], and many others in the 1960s–1980s studied the electric fields and ion currents that contribute to developmental pattern formation and regenerative repair [18–20]. The development of electrode strategies and breakthrough vibrating probe technology [21], which allowed accurate measurements of ion flux across membranes in vivo, led to an accumulating mass of data implicating endogenous electric fields in events like wound healing, guidance of migratory cell behavior, and regenerative response. Cone's work uniquely examined a different cellular parameter: the resting potential. While V_{mem} is related to ion flux and electric fields, it is a distinct property that must be studied separately. In the following paper, Cone and Cone used pharmacological depolarization and electrophysiology, and were able to show the instructive role that V_{mem} plays: depolarization is associated with a plastic, proliferative state (e.g. cancer cells), while hyperpolarization is associated with a post-mitotic, mature condition—a relationship that has been recently validated in a number of cell lines, including human stem cells [22, 23].

Cone's group's work served as an inspiration for the recent advances in developmental bioelectricity focused on resting potential. New methods using fluorescent voltage reporter dyes allow the electrical potentials of many cells to be imaged at once, without having to impale cells with electrodes. In this way, scientists detect the voltage potentials throughout the body, to observe the electrical conversations that cells are having with each other. Modern techniques of molecular biology (altering the expression of ion channels and pumps using genetic methods) have allowed us to manipulate those signals. By specifically altering the resting potentials of cells, we can re-write the natural electric patterns; recent work has shown that this has dramatic effects on gene expression and pattern formation. Bioelectric gradients, of the kind Cone studied in single cells, form patterns across tissues in vivo that are part of the way cells coordinate their activity towards constructing embryos and repairing/regenerating organs in adults [13, 24–29]. Thus, Cone's model of mitosis appears to be correct, but more than that, these voltage gradients control the types of organs that appear during embryogenesis and the geometric relationships between those structures. As conjectured by Burr, bioelectric gradients in vivo really do seem to be subtle prepatterns that guide growth and form. For example, recent work has taken advantage of the ability to control resting potentials in biological circuits in vivo to kickstart regeneration of spinal cord and muscle [30], to understand how mutations in ion channels induce craniofacial defects in embryogenesis [31, 32], and to re-specify the entire body plan [33], of model species like frog and flatworm.

Cone's conjectures about cancer were validated as well; recent work has shown that not only can tumors in vivo be recognized early by their aberrant electrical signature, but they can actually be prevented or reprogrammed into normal tissue by enforcing correct resting potentials [34–39]. Ion channel-targeting compounds are thus an important aspect of the search for new cancer drugs [40, 41], as well as an emerging class of 'electroceuticals' for applications in birth defects and regenerative medicine. C D Cone and his colleagues (including C M Cone, M Tongier, E F Stillwell) were way ahead of the community in recognizing the importance of resting potential as a key control parameter for cells, not just a housekeeping physiological side-effect. He would probably have been really pleased to see the exciting applications emerging in the modern study of developmental bioelectricity.

References

[1] Cone C D and Tongier M 1971 Control of somatic cell mitosis by simulated changes in the transmembrane potential level *Oncology* **25** 168–82

[2] Stillwell E F, Cone C M and Cone C D 1973 Stimulation of DNA synthesis in CNS neurones by sustained depolarisation *Nat. New Biol.* **246** 110–1

[3] Cone C D 1974 The role of the surface electrical transmembrane potential in normal and malignant mitogenesis *Ann. NY Acad. Sci.* **238** 420–35

[4] Cone C D and Cone C M 1978 Evidence of normal mitosis with complete cytokinesis in central nervous system neurons during sustained depolarization with ouabain *Exp. Neurol.* **60** 41–55
[5] Cone C D and Cone C M 1978 Blockage of depolarization-induced mitogenesis in CNS neurons by 5-fluoro-2′-deoxyuridine *Brain. Res.* **151** 545–59
[6] Cone C D, Tongier M and Cone C M 1977 DNA content of daughter nuclei from ouabain-induced nuclear divisions in central nervous system neurons *Exp. Neurol.* **57** 396–408
[7] Cone C D and Cone C M 1976 Induction of mitosis in mature neurons in central nervous system by sustained depolarization *Science* **192** 155–8
[8] Cone C D 1980 Ionically mediated induction of mitogenesis in CNS neurons *Ann. NY Acad. Sci.* **339** 115–31
[9] Cone C D 1969 Autosynchrony and self-induced mitosis in sarcoma cell networks *Acta. Cytol.* **13** 576–82
[10] Cone C D 1971 Unified theory on the basic mechanism of normal mitotic control and oncogenesis *J. Theor. Biol.* **30** 151–81
[11] Cone C D Jr. 1970 Variation of the transmembrane potential level as a basic mechanism of mitosis control *Oncology* **24** 438–70
[12] Cone C D and Tongier M 1973 Contact inhibition of division: involvement of the electrical transmembrane potential *J. Cell Physiol.* **82** 373–86
[13] Levin M 2014 Molecular bioelectricity: how endogenous voltage potentials control cell behavior and instruct pattern regulation in vivo *Mol. Biol. Cell.* **25** 3835–50
[14] Deloof A 1993 Schrodinger 50 Years Ago—What Is Life—the Ability to Communicate, a Plausible Reply *International Journal of Biochemistry* **25** 1715–21
[15] Lund E 1947 *Bioelectric fields and growth* (Austin: Univ. of Texas Press)
[16] Burr H S and Northrop F S C 1935 The electro-dynamic theory of life *Quarterly Review of Biology* **10** 322–33
[17] Marsh G and Beams H W 1952 Electrical control of morphogenesis in regenerating dugesia tigrina. I. Relation of axial polarity to field strength *J. Cell Comp. Physiol.* **39** 191–213
[18] Pullar C E 2011 *The physiology of bioelectricity in development, tissue regeneration, and cancer*Biological effects of electromagnetics series (Boca Raton: CRC Press)
[19] Borgens R, Robinson K, Vanable J and McGinnis M 1989 *Electric Fields in Vertebrate Repair* (New York: Alan R. Liss)
[20] Nuccitelli R, Robinson K and Jaffe L 1986 On electrical currents in development *Bioessays* **5** 292–4
[21] Jaffe L F and Nuccitelli R 1974 An ultrasensitive vibrating probe for measuring steady extracellular currents *J. Cell Biol.* **63** 614–28
[22] Sundelacruz S, Levin M and Kaplan D L 2009 Role of membrane potential in the regulation of cell proliferation and differentiation *Stem cell reviews and reports* **5** 231–46
[23] Sundelacruz S, Levin M and Kaplan D L 2013 Depolarization alters phenotype, maintains plasticity of predifferentiated mesenchymal stem cells *Tissue engineering. Part A* **19** 1889–908
[24] Bates E 2015 Ion Channels in Development and Cancer *Annu. Rev. Cell Dev. Biol.* **31** 231–47
[25] Tseng A and Levin M 2013 Cracking the bioelectric code: Probing endogenous ionic controls of pattern formation *Communicative & Integrative Biology* **6** 1–8
[26] Levin M 2013 Reprogramming cells and tissue patterning via bioelectrical pathways: molecular mechanisms and biomedical opportunities *Wiley Interdisciplinary Reviews: Systems Biology and Medicine* **5** 657–76

[27] Adams D S and Levin M 2013 Endogenous voltage gradients as mediators of cell-cell communication: strategies for investigating bioelectrical signals during pattern formation *Cell Tissue Res* **352** 95–122

[28] Mustard J and Levin M 2014 Bioelectrical Mechanisms for Programming Growth and Form: Taming Physiological Networks for Soft Body Robotics *Soft Robotics* **1** 169–91

[29] Levin M 2014 Endogenous bioelectrical networks store non-genetic patterning information during development and regeneration *The Journal of Physiology* **592** 2295–305

[30] Adams D S, Masi A and Levin M 2007 H^+ pump-dependent changes in membrane voltage are an early mechanism necessary and sufficient to induce Xenopus tail regeneration *Development* **134** 1323–35

[31] Adams D S *et al* 2016 Bioelectric signalling via potassium channels: a mechanism for craniofacial dysmorphogenesis in KCNJ2-associated Andersen-Tawil Syndrome *J. Physiol.*

[32] Vandenberg L N, Morrie R D and Adams D S 2011 V-ATPase-dependent ectodermal voltage and pH regionalization are required for craniofacial morphogenesis *Dev. Dyn.* **240** 1889–904

[33] Beane W S, Morokuma J, Adams D S and Levin M 2011 A Chemical genetics approach reveals H,K-ATPase-mediated membrane voltage is required for planarian head regeneration *Chemistry & Biology* **18** 77–89

[34] Chernet B T and Levin M 2013 Transmembrane voltage potential is an essential cellular parameter for the detection and control of tumor development in a Xenopus model *Disease models & mechanisms* **6** 595–607

[35] Chernet B and Levin M 2013 Endogenous Voltage Potentials and the Microenvironment: Bioelectric Signals that Reveal, Induce and Normalize Cancer *J. Clin. Exp. Oncol.* Suppl 1

[36] Lobikin M, Chernet B, Lobo D and Levin M 2012 Resting potential, oncogene-induced tumorigenesis, and metastasis: the bioelectric basis of cancer in vivo *Physical biology* **9** 065002

[37] Chernet B T, Adams D S, Lobikin M and Levin M 2016 Use of genetically encoded, light-gated ion translocators to control tumorigenesis *Oncotarget*

[38] Chernet B T, Fields C and Levin M 2015 Long-range gap junctional signaling controls oncogene-mediated tumorigenesis in Xenopus laevis embryos *Front Physiol* **5** 519

[39] Chernet B T and Levin M 2014 Transmembrane voltage potential of somatic cells controls oncogene-mediated tumorigenesis at long-range *Oncotarget* **5** 3287–306

[40] Arcangeli A 2011 Ion channels and transporters in cancer. 3. Ion channels in the tumor cell-microenvironment cross talk *American journal of physiology. Cell physiology* **301** C762–771

[41] Arcangeli A and Becchetti A 2010 New Trends in Cancer Therapy: Targeting Ion Channels and Transporters *Pharmaceuticals* **3** 1202

Induction of Mitosis in Mature Neurons in Central Nervous System by Sustained Depolarization

Abstract

DNA synthesis and mitosis have been induced in vitro in fully differentiated neurons from the central nervous system by depolarization with a variety of agents that produce a sustained rise in the intracellular sodium ion concentration and a decrease in the potassium ion concentration. Depolarization was followed in less than 1 hour by an increase in RNA synthesis and in 3 hours by initiation of DNA synthesis. Apparently normal nuclear mitosis ensued, but cytokinesis was not completed in most cells; this resulted in the formation of binucleate neurons. The daughter nuclei each contained the same amount of DNA as the diploid preinduction parental neurons; this implies that true mitogenic replication was induced.

The hypothesis has been advanced [1, 2] that intracellular cation levels associated with generation of electrical transmembrane potentials in somatic cells may be functionally involved in control of mitogenesis and, hence, of cell division. Results from studies with a variety of cell systems have supported this premise [3, 4]. As a corollary of the hypothesis, it was proposed [2] that mitogenesis might be activated in highly polarized, nondividing cells such as neurons in the central nervous system (CNS) and muscle by treatments that would produce and maintain a substantial increase in the level of intracellular Na^+ and a decrease in K^+. This proposition was recently investigated for the case of mature neurons from chick spinal cord depolarized with ouabain [5]; DNA synthesis was induced in a significant percentage of the neuron population by a range of ouabain concentrations. The question remained, however, as to whether the observed synthesis was truly mitogenic, rather than the result of an anomalous thymidine exchange or repair activity. We now report the induction of full DNA synthesis and subsequent mitosis in much larger percentages of chick spinal cord neurons by use of more favorable ouabain treatments, and an equally effective induction of mitotic activity in such neurons by other agents that depolarize by quite different mechanisms. The daughter nuclei from the induced mitoses each contain an amount of DNA identical to that of the mitotically quiescent G_1 (or G_0) neurons; this implies that true mitogenic replication has been induced by the imposed ionic changes.

The neurons were obtained by trypsin dissociation of spinal cords from chick embryos 7 to 10 days old and cultured as described [5]. Mature 16- to 20-day cover slip cultures of well dispersed, fully differentiated neurons were used in the experiments. The cultures contained numerous small neurons and similar-sized neuroglial cells in addition to the large, distinctive motoneurons, but primarily the latter were utilized in the present investigation to preclude any possible confusion with the glial cells; fibroblasts were also quite numerous. The motoneurons were readily identified by their large size, angular cytons with long axon and dendrites, extensively granular cytoplasm, distinctive vesicular nucleus with prominent nucleolus, and differential staining properties [6]. The glial population in the cultures increased continuously by mitotic division up to 14 days after plating but declined thereafter, although many glial mitoses were still observable; the ratio of glia (including the numerous microglia) to neurons at the time of culture use in experiments was approximately 10 to 1. In general, the glial cells tended to aggregate in clumps while the motoneurons were well dispersed. The cultured neurons had resting potentials of −35 to −60 mv, and were entirely devoid of mitotic activity after plating.

The ability of ouabain, veratridine, and the ionophore gramicidin to induce neuronal DNA synthesis and mitosis was determined over a range of concentrations for each agent. For DNA synthesis determinations, cultures were incubated for 6 hours with medium containing the test agent and [^{3}H]thymidine (0.1 µc/ml). The cover slips were rinsed, fixed, coated with Kodak NTB-3 nuclear track emulsion, and exposed for 14 days as described [5]. The time of onset of DNA synthesis following exposure to ouabain was determined by pulse labeling for 1 hour prior to serial fixations at 1 to 6 hours. Similar pulse labeling with [^{3}H]uridine (2.5 µc/ml) was used to study changes in RNA synthesis rate following ouabain treatment; the same autoradiographic procedures were used as with DNA, but with a 7-day exposure period. For mitotic activity determinations, cultures were incubated for 24 hours with medium containing the test agent, then fixed, stained, and examined for binucleate neurons [7]. Control cultures were assayed as were the test cultures, but were incubated in normal culture medium without addition of the depolarizing or other agents. Several supplementary time-lapse film studies were made of treated neurons to elucidate features of the mitotic process and the behavior of daughter cells. Relative amounts of DNA in the nuclei of control phase (G_1) neurons and in the daughter nuclei resulting from induced mitoses were determined by Feulgen microspectrophotometry of individual nuclei [8]. The nuclear DNA content of G_1 fibroblasts in the cultures was similarly determined by using an excess-thymidine blocking procedure for accumulating G_1 cells, to establish the basic diploid nuclear DNA level. The neuron cultures were Feulgen-stained by the procedure of DeCosse and Aiello [9], and the relative absorbency of each nucleus was determined at 550-nm wavelength by using a Zeiss fluorescence microscope integrated with a model 240 Gilford monochromater and digital photometer.

The percentages of neurons induced by the various agents to initiate DNA synthesis and to complete nuclear division are given in Table 1. All agents tested were effective in inducing both DNA synthesis and nuclear division. Since significant electroosmotic swelling [1] was found to accompany neuron depolarization in our initial studies with ouabain [5], an attempt was made to reduce the swelling (and to increase the Na$^+$ influx) by increasing the total medium osmolality 5 percent by addition of NaCl. This addition not only reduced the swelling but also lowered the ouabain concentration required for mitogenesis activation. Depolarization by the alkaloid veratridine produced less rapid swelling and detachment of neurons, and yielded the highest percentage of activated cells [10]. Like ouabain, the ionophore gramicidin produced substantial swelling and detachment upon depolarization, and was as effective as ouabain in its activation of neurons. Since both glial cells and fibroblasts showed spontaneous mitotic activity in control and test cultures, no quantitative assay was made of their mitotic activity in treated cultures. However, in some gramicidin-treated cultures a pronounced increase in the number of fibroblasts labeled with [^{3}H]thymidine and in the grain density of such labeling occurred in

Table 1. Percentages of CNS neurons induced to initiate DNA synthesis and nuclear division by various depolarizing agents. From 300 to 700 neurons were assayed at each test condition; four to six separate cultures were examined per condition. Grain density of labeled control cells (DNA synthesis) was one to two grains per nucleus, only slightly above background; that of labeled induced cells was severalfold higher (five to six grains per nucleus). The binucleate cells found in control cultures (mitotic activity) apparently resulted from neuroblasts in mitosis at the time of cord removal; the binucleate level in control cultures remained constant or decreased after 2 days in vitro. Data presented are for near-optimum concentrations of activating agent. A few cultures treated with 10^{-6}M ouabain plus NaCl had labeling frequencies as high as 58 percent.

Agent	Agent (M)	DNA synthesis (% of cells labeled)		Mitotic activity (% of binucleate cells)	
		Test	Control	Test	Control
Ouabain	10^{-5}	34	4	20	3
Ouabain	10^{-4}	34	2		
Ouabain + NaCl	10^{-6}	31	3	23	3
Veratridine	5×10^{-5}	63	3	30	8
Gramicidin	10^{-8}	32	6	29	6

portions of tightly confluent fibroblast monolayer areas in the neuron cultures. Numerous mitotic figures of fibroblasts and glia were seen in both control and test cultures, but very few binucleate cells of either type were found, which indicates completion of full cytokinesis in these cells. Phase-contrast photomicrographs of two binucleate neurons resulting from ouabain treatment are shown in Fig. 1.

Although the present experiments were concerned with assay of binucleate induction and the treatment time used was intended to permit full completion of nuclear division, a number of clearly discernible neuronal mitotic figures was found in the fixed test cultures. Metaphase, anaphase, and even prophase figures were clearly distinguished in neurons whose processes had not detached during treatment and which retained characteristic morphology and cytoplasmic staining properties. Also, a few neuronal mitotic figures on stained slides were independently identified by the periodic coordinate location procedure [7].

As shown in Fig. 2, DNA synthesis began in an appreciable fraction of the ouabain-treated cells within 3 hours. DNA synthesis was preceded by a significant rise in the RNA synthesis rate, which essentially reached its maximum level within 1 hour; this indicates that mitogenesis activation was induced quite rapidly with the ouabain concentrations used. A frequency distribution of the [^{3}H]uridine grain count per nucleus for control neurons and those treated with ouabain plus NaCl indicated that 78 percent of the treated cells had rates of RNA synthesis significantly greater than the mean level for the control neurons. In the time-lapse films of binucleate neuron formation, clear observation of individual chromosomal dispersion and movements was generally precluded because of the pronounced refractiveness of the rounded cyton, but the time of chromosome separation and formation of

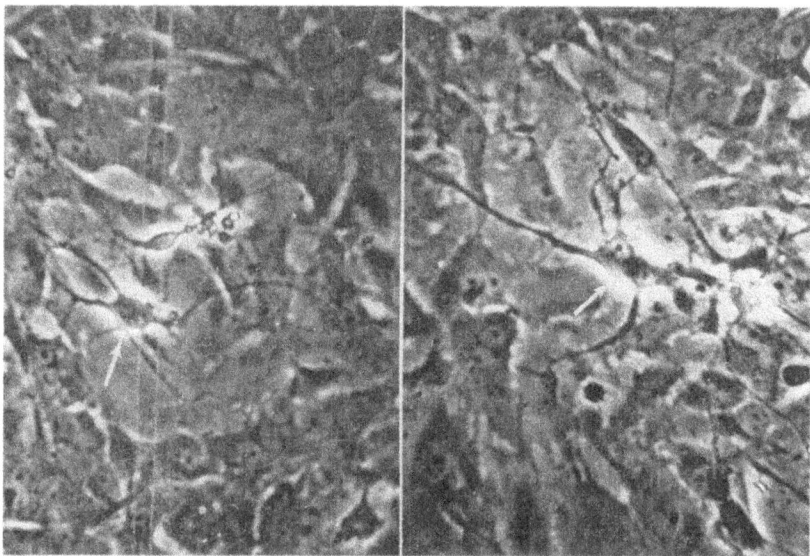

Fig. 1. Phase-contrast photomicrographs of two live binucleate neurons resulting from the ouabain treatment of 15-day test cultures. Ouabain was removed from the cultures after a 24 hour treatment. Several other binucleate neurons appear in these photographs, but are not so clearly discerned as those indicated by the arrows.

two nuclear masses could be discerned in several films. The elapsed time from beginning of treatment through completion of cytokinesis in one film of a gramicidin-treated neuron undergoing full cytokinesis was 12 hours. Although maximum swelling and "rounding up" [1] of the cell occurred within 2 to 3 hours after gramicidin addition, the actual onset of mitotic prophase did not begin until approximately 10 hours, and mitosis was completed by 12 hours. In cases where full rounding of the neuron took place with absorption of its processes, complete cytokinesis usually occurred; the resulting daughter cells were quite transient in shape and initially not morphologically identifiable as neurons.

Measurements of the DNA content of each of the daughter nuclei in 62 binucleate neurons resulting from ouabain-induced divisions revealed that each of the 124 daughters contained identical amounts of DNA. The DNA content of each daughter nucleus from the induced divisions was found to be identical also to that of the mononucleate neurons in control cultures. Since each of the 101 control neurons assayed contained the same amount of nuclear DNA as the G_1 (diploid) fibroblasts in the cultures, the control and test neurons were diploid cells prior to treatment.

The premise is generally accepted that mitotic activity ceases in fully differentiated neurons of the CNS and dorsal root ganglia, both in vivo and in vitro [11], although observations of divisions in such neurons in vitro have been reported [12]. In the present experiments, neurons that were apparently fully differentiated [13] were obtained by using cords in which neuronal mitotic activity had essentially ceased [5] and by maintaining the cultures for several days after attainment of full morphological and electrophysiological development before use in experiments. Neuronal mitotic activity was completely absent in pre-experimental cultures, as

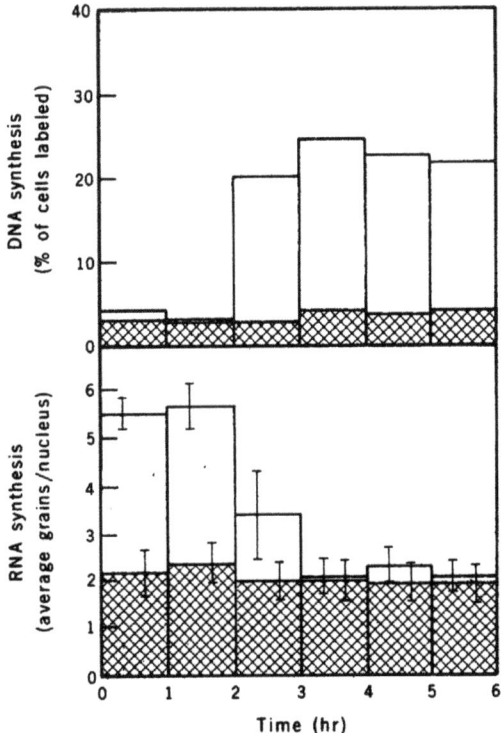

Fig. 2. Time sequence of changes in DNA and RNA synthesis induced in CNS neurons by treatment with 5×10^{-5} M ouabain plus NaCl, added to the culture medium at zero hour. Shaded bars refer to control cultures, and open bars to treated cultures. The labeling pulse was given 1 hour before culture fixation; approximately 350 neurons were assayed per time interval. Cells were considered to be synthesizing DNA only if their nuclear grain count was three times background (two to three grains per nucleus). Autoradiographic grain counts (RNA synthesis) are the mean per nucleus, averaged over 350 randomly chosen neurons in each time interval. The range of the mean at the 95 percent confidence level is indicated for each interval (the range for test neurons is indicated at left of interval bar, that for controls at right).

demonstrated by (i) constancy of the number of binucleate neurons after culture establishment (that is, beyond 2 days in culture; see Table 1 legend), and (ii) direct observations of some 200 individual neurons on a daily basis during 3 weeks of culture. Consequently, the mitoses reported herein were induced in neurons which had achieved the state of complete mitotic quiescence characteristic of the fully differentiated cells. An obvious and important extension of the present experiments, however, is determination of the mitotic response of neurons from adult animals and humans of various ages to these treatments.

The observation of neuronal mitotic figures in test cultures, along with the finding that daughter nuclei of binucleate neurons induced from diploid mononuclear parental neurons also contain the exact diploid level of DNA, strongly implies that the induction treatment activated true mitogenic replication of the neuronal nucleus. It is highly unlikely that anomalous nuclear fragmentation would result consistently in the production of daughter nuclei having equal amounts of DNA, corresponding exactly to the normal diploid level. Likewise, it is unlikely that the daughter nuclei resulted from nuclear amitosis, in view of the presence of neuronal mitotic figures in the test cultures and the dearth in observations of true amitosis in higher somatic cells. It is clear that the daughter nuclei did not result from the mitosis or fission of tetraploid parental nuclei already existing in the preexperimental cultures, since both the daughter and parental cell groups possessed the diploid level of DNA. Rather, the results indicate that the treatments induced an exact doubling of parental DNA followed by an exact halving to produce identical daughter nuclei, as in normal mitosis. The fact that most of the treated neurons retained some degree of process attachment and did not complete cytokinesis, while those few which fully rounded up underwent complete division [7], suggests the possible involvement of the neuronal microtubule system in preventing completion of cytokinesis, thus fostering binucleate rather than individual daughter neuron formation.

In regard to the validity of the basic hypothesis on ionic modulation of mitogenesis activity which the present experiments were primarily designed to test, it appears particularly significant that apparently normal nuclear mitosis has been induced in such highly differentiated and mitotically refractory cells as CNS neurons by three distinct agents whose only basic commonality of action is to effect an increase in cellular Na^+ and decrease in cellular K^+ [5, 14]. The present results strongly imply that the induced mitoses follow the usual sequence of mitogenic events and reinforce the previously advanced view that activation of mitogenic RNA synthesis by significant shifts in the intracellular Na^+ level or Na^+ to K^+ ratio (or both) may be a key event in the initiation of mitogenesis [1, 2, 4]. Timelapse film and direct visual observations have demonstrated that viable daughter cells result from such induced division when cytokinesis is completed. Whether the daughter cells of such neuron divisions still possess full neuronal phenotypic properties and functionality remains to be determined, but this question obviously constitutes a subject of considerable basic, and ultimately perhaps clinical, importance. In this regard, elucidation of those conditions that favor completion of cytokinesis, rather than binucleate neuron formation, is essential.

Clarence D. Cone, Jr.
Charlotte M. Cone
Cell and Molecular Biology Laboratory, Veterans Administration Center, Hampton, Virginia 23667

References and Notes

1. C. D. Cone, *Trans. N.Y. Acad. Sci. Ser. 2* **31**, 404 (1969).
2. ———, *J. Theor. Biol.* **30**, 151 (1971).
3. ——— and M. Tongier, *Oncology* **25**, 168 (1971); C. W. Orr, M. Yoshikawa-Fukada, J. D. Ebert, *Proc. Natl. Acad. Sci. U.S.A.* **69**, 243 (1972).
4. C. D. Cone and M. Tongier, *J. Cell Physiol.* **82**, 373 (1973).

5. E. S. Stillwell, C. M. Cone, C. D. Cone, *Nature (London) New Biol.* **246**, 110 (1973).
6. In addition to their distinctive nuclear and morphological characteristics the neurons were clearly differentiated by staining with methylene blue [O. Costero and C. M. Pomerat, *Am. J. Anat.* **89**, 405 (1951)], neuron polychrome [5], or methyl-green pyronin [L.- W. Chu, *J. Comp. Neurol.* **100**, 381 (1954)]. The latter stain was particularly useful with autoradiographic emulsions and for differentiating the larger ependy mal neuroglia.
7. In most of the dividing neurons some processes remained attached during mitosis. However, cytokinesis was not fully completed, and clearly discernible binucleate cells were thus produced. By determining the cover slip (location) coordinates of many individual neurons prior to treatment and then periodically relocating and observing these same cells during treatment and up to the time of fixation, it was found that substantial numbers of the neurons were lost during rinsing and fixation. Those few neurons that were observed to complete cytokinesis had rounded up fully during treatment, with detachment and absorption of their processes; the small daughter cells were also lost during fixation. In the present study, only clearly identifiable binucleate neurons (remaining on the fixed culture slides) were included in the assay of mitotic activity; the percentages of mitotic neurons given herein (Table 1) are thus minimums. The true percentages are possibly double those indicated, as estimated from the periodic observations. A similar situation existed in the case of the DNA synthesis determinations; detachment and loss of activated cells also occurred in substantial numbers, but with somewhat less frequency than in the mitotic cells, due perhaps to the shorter treatment period. The possibility that the binucleate cells observed arose from induced or spontaneous fusion of adjacent neurons, rather than from nuclear division, was precluded by use of cultures in which the individual test neurons were well dispersed.
8. C. D. Cone and M. Tongier, in preparation.
9. J. J. DeCosse and N. Aiello, *J. Histochem. Cytochem.* **14**, 601 (1966).
10. Although the possibility exists that the activated cells might comprise an immature neuronal sub-population specifically susceptible to induction, it appears more likely that all of the neurons are potentially capable of being activated; the non-activated fractions probably consist of neurons that had not adequately depolarized within the treatment periods used. Also, the high apparent percentages of nonactivated cells indicated by Table 1 may be misleading, due to the substantial loss of activated cells (but retention of the nonactivated population) during fixation [7].
11. M. J. Hogue, *Anat. Rec.* **108**, 457 (1950); *Am. J. Anat.* **93**, 397 (1953); C. E. Lumsden, in *The Pathology of Tumours of the Nervous System*, D. S. Russell, L. T. Ruben stein, C. E. Lumsden, Eds. (Arnold, London, 1959), pp. 272–309; J. Nakai and M. Okamoto, in *Morphology of Neuroglia*, J. Nakai, Ed. (Igaku Shoin, Tokyo, 1963), p. 65; H. A. Hansson and P. Sourander, *Z. Zellforsch. Mikrosk. Anat.* **62**, 26 (1964); M. R. Murray, in *Cells and Tissues in Culture*, E. N. Willmer, Ed. (Academic Press, New York, 1965), vol. 2, p. 373; P. Weiss, in

Mitogenesis, H. S. Ducoff and C. F. Ebret, Eds. (Univ. of Chicago Press, Chicago, 1956), p. 54.

12. M. R. Murray and A. P. Stout, *Am. J. Anat.* **80**, 225 (1947); R. S. Geiger, *Exp. Cell Res.* **14**, 541 (1958); M. R. Murray and H. H. Benitez, in *CIBA Symposium on Growth of the Nervous System*, G. E. W. Wolstenholme and M. O'Connor, Eds. (Churchill, London, 1968), p. 148.

13. Neurons obtained by culturing procedures identical to those used in the present experiments have demonstrated the capability of attaining such advanced developmental activities in vitro as formation of functional neuromuscular junctions [G. D. Fischbach, D. Fambrough, P. G. Nelson, *Fed. Proc. Fed. Am. Soc. Exp. Biol.* **32**, 1636(1973)].

14. R. C. Thomas, *J. Physiol. (London)* **220**, 55 (1972); M. Ohta, T. Narahashi, R. F. Keeler,7. *Pharmacol. Exp. Ther.* **184**, 143 (1973); B. C. Pressman, *Proc. Natl. Acad. Sci. U.S.A.* **53**, 1076 (1965).

15. We thank Burroughs-Welcome and E. R. Squibb and Sons for gifts of gramicidin. Dr. B. Chance originally suggested to us the use of ionophores as possible depolarizing agents for cells. Sponsored jointly by the Veterans Administration and the Eastern Virginia Medical Authority, Norfolk.

4 September 1975; revised 31 October 1975

Case Study 5: Cells Talk to Each Other via Electromagnetism

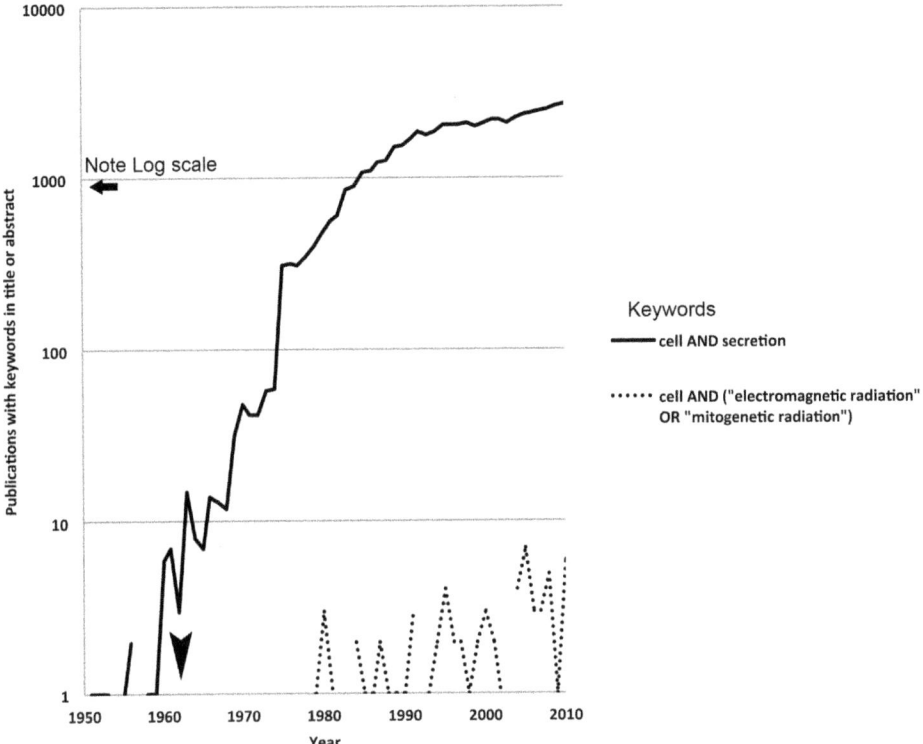

Gurvich *et al* (1961) Mitogenetic radiation of the neuromuscular system as a method of analyzing its molecular substrate.

Perspective on *Mitogenetic Radiation of the Neuromuscular System as a Method of Analysing its Molecular Substrate*

Lev V. Bellousov
Faculty of Biology
Moscow State University

Alexander Gavrilovich Gurwitsch (Gurvich in English transcription) is known for two main contributions: (1) a discovery (in 1923) of an ultraweak photon emission in UV range coming from a number of living objects and claimed to stimulate mitotic divisions at a certain distance, and (2) formulation and further extensive elaboration of a field concept, defined firstly as a 'force field' (1912), then 'embryonic field' (1922) and at last as a 'biological' or 'cellular' field (1944; posthumous edition 1991). The both had a dramatic and controversial fate. We start from the first one (for detailed account see Volodyaev and Beloussov 2015).

The 1923 crucial experiment was the result of G.'s prolonged interest to the causes of cell divisions. He came to the conclusion that at least one of these causes should be of a physical nature, able to 'resonate with cell membrane'. For detecting this factor, he brought into close vicinity two perpendicularly oriented onion roots, using one of them (oriented by his tip towards a side wall of another) as a sender of a physical signal while the second one as a receiver. As a result, he observed a statistically significant increase of the number of dividing cells in the receiver's area just opposed to the sender's tip. This effect was blocked by inserting between the both roots a glass plate (non-permeable to emission in the UV-range) but non-affected by inserting a UV-permeable quartz plate. In addition, the signal was shown to pass via the optical root of a UV spectrograph. Similar 'mitogenic' effects were reproduced in bacteria, yeasts (which became in G.'s labs the main every-day detector) and several kinds of animal cells.

This discovery quickly attracted a broad interest, going well beyond the borders of the former Soviet Union. Several groups in Germany, Netherlands, France, Italy, USA and other countries checked this effect and contrary to wide-spread opinion, most of them confirmed it. A properly balanced review of this period is given in Rahn's (1936) monograph. G. was several times nominated for a Nobel Prize. However, the only prize which he received was from the Soviet government. In the years 1930–48, G. was the head of large labs of a medical orientation in Leningrad (former and future St Petersburg) and Moscow, and used his discovery as a highly sensitive non-invasive tool for exploring a molecular basis of cells' and tissues' reactions to non-specific stresses (discovering what he called the 'degradation radiation'), different functional states of neuro-muscular system, and, most exciting, the cancer problem: it turned out that the blood of cancer patients stopped emitting UV photons at the very beginning of the malignant process (well before the appearance of any clinical symptoms), while the blood of healthy persons (or those suffering other diseases) permanently emitted photons. This discovery was successfully used for diagnostic and prognostic purposes not only in the Soviet Union in the 1930s and 1940s, but also in Germany before WWII.

This does not mean that the pathway of G. discovery was always triumphant. Its weak point was the capricious behavior of routinely used biodetectors unexpectedly changing their sensitivity. This paved the way for a huge paper by Hollander and Claus (1937) who claimed that G. results are at all irreproducible. At the same time the authors overtly claimed that they deviated from theexperimental protocols recommended by G. as mostly reliable ones. This paper played a fatal role, much reinforced by the events of WWII and subsequent Lysenko's campaign in the Soviet Union. As a result, to the end of the 1940s, the work on this direction was practically cancelled outside Russia, to be continued only in Moscow in a very small scale.

A new revival of interest in G.'s discovery was associated with technical progress in photo-registration techniques. Some of the most remarkable contributions were those by Quickenden and Tilbury (see Tilbury and Quickenden 1992) who fully confirmed the existence of UV radiation just of the same wave-length and emitted from the same sources (bacteria and budding yeasts) as described much earlier in G. labs. On the other hand, the authors did not confirm the mitogenic effect itself (stimulation of cell divisions). However, this could be expected, because similarly to Hollander and Claus, the authors deviated from G.'s protocols.

All in all, today's situation in this area of science looks paradoxical and controversial: if indeed the mitogenic effects of biogenic UV emission do not exist (which is still a dominating view) how could the previous investigations completely based upon the usage of biodetectors correctly predict the data confirmed later by physical measuring technique? In addition to Quickenden and Tilbury results several other examples of such correlations can be mentioned (see Volodyaev and Beloussov 2015). This urgently invites to revisit G. discovery using in a more adequate context the powerful modern technique.

The concept of a field arose in G.'s mind at the beginning of the second decade of the XXth century as an opposition to 'genocentric' ideas just emerging at that time and regarding a developing organism as a passive executor of a set of 'instructions' encoded in genes. Instead, the field concept tended to treat an embryo as a system of physical forces coordinated both in space and time and obeyed essentially wholistic laws, which G. stubbornly tried to express in mathematical terms. In the same years or slightly later, several other authors also used the term 'field' giving rise to the claims that G was not first in this topic. However, G.'s concept differs from all the others in making an accent on the temporal component of development ('our position different from the common one is that from an adequate analysis of a momentary state [of embryo] its passage to the next one should immediately follow'—G., published in 1991), rather than being restricted by a mere static registration of holistic effects. In other words, G.'s concept related to all the others as differential equations to algebraic ones. One may also say that G.'s concept, contrary to all the others, suggest certain generative rules aimed to reproduce the observed sequences of embryonic stages. In a number of his constructions G. intuitively previewed some principles of a self-organization theory. Although in spite of several reformulations, G. did not succeed in constructing a concise theory of embryonic morphogenesis (which was virtually impossible in his time due to poor knowledge of cellular and molecular basis of development), the fate of this direction of his work was probably happier than of that discussed beforehand. While being overshadowed in the midst of the XXth century by monopolistic domination of a synthetic theory of evolution, the field approach was rehabilitated in the framework of the contemporary Evo-Devo (Gilbert *et al* 1997) and G.'s contribution was properly acknowledged (see Commentary to Beloussov 1997).

References

Beloussov L V 1997 and additional commentary by J.M. Opitz and S.F. Gilbert Life of Alexander G. Gurwitsch and his relevant contribution to the theory of morphogenetic fields *Int.J.Dev Biol* **41** 771–9

Beloussov L V 2008 "Our Standpoint Different from Common…" (Scientific Heritage of Alexander Gurwitsch) *Russ J. Dev Biol* **38** 307–15

Gurwitsch A G 1991 *(in Russian) Principles of Analytical Biology and a Theory of Cellular Fields* (Moskva: Nauka)

Hollander A and Claus W D 1937 *An experimental study of the problem of mitogenetic radiation* (Washington, DC: Bull. Natl Res. Council)

Tilbury R N and Quickenden T I 1992 Luminescence from the yeast *Candida utilis* and comparisons across three genera *J.Biolumin. Chemilumin.* **7** 245–53

Rahn O 1936 *Invisible Radiations of Organisms* (Berlin: Gebruder Borntraeger)

Volodyaev I V and Beloussov L V 2015 Revisiting the mitogenetic effect of ultraweak photon emission *Front. Physiol.* **6** 241

MITOGENETIC RADIATION OF THE NEUROMUSCULAR SYSTEM AS A METHOD OF ANALYSING ITS MOLECULAR SUBSTRATE REPORT III. THE REGULATING EFFECT OF THE SPINAL CORD CENTERS ON THE MOLECULAR SUBSTRATE OF THE MUSCLES AT DIFFERENT AGE PERIODS OF THE ANIMALS AND THE SIGNIFICANCE OF THIS REGULATION FOR THE MUSCLE METABOLSIM

A. A. Gurvich, V. F. Eremeev, and M. A. Lipkind

From the Office of Mitogenesis (Head : A. A. Gurvich) of the Institute of Normal and Pathological Physiology (Director: Active Member of the Akad. Med. Nauk SSSR V. V. Parin) of the Akad. Med. Nauk SSSR, Moscow
(Presented by Active Member of the Akad. Med. Nauk SSSR V. V. Parin)
Translated from Byulleten 'Eksperimental' noi Biologii i Meditsiny, Vol. 51,No. 4, pp. 57-61, April,1961
Original article submitted April 25, 1960

In previous publications [2,3,6], data were presented on the antibathic processes (processes of a reverse nature) arising in spinal cord centers (presumably areas rich in cells—the motor cells of the anterior horns, the inter-neurons) and in muscle fibers (neuromuscular junctions).

The action of the muscles imparts a prepolymer character to the substrate, while the effect of the centers, on the other hand, results in a dispersed arrangement. The resultant state of the neural and muscular substrates is extremely labile, and may be characterized as a state of unequilibrated molecular regularity.[*]
The unequilibrated nature of the substrate in the muscle fibers has been demonstrated particularly clearly during the resting state of the muscles in experiments in situ.

Along with this, a previous report [3] introduced the concept of the constant regulation of the three-dimensional parameters in the molecular substrate of the muscles as a specific physiological phenomenon.

Further study of this regulation, serving as the goal of the present work, was carried out in the following directions.
1. We studied the evolution of the molecular substrate character in the muscles throughout the development of the animal, i.e. in the period when the full functional relationships are formed.
2. We studied the action of the centers on the muscle substrate as a factor limiting and, in this way, regulating the character of the metabolism. In this case we were thinking of mediators, specifically acetylcholine. Reducing the influence of the centers, we investigated the possibility of acetycholine arising under this condition in a resting muscle.

---

[*] The unequilibrated state of the molecular substrate in living systems is one of the basic concepts in the problem of mitogenesis [4], This state, of course, does not exclude the presence of myosine protein chains, etc.

METHOD

The first part of the work was performed on rabbits, the second, on frogs. We investigated the mitogenetic radiation of the gastrocnemius muscle in the resting state within the living animals. The adult rabbits were secured to a horizontally placed stand. The muscle was exposed, and the radiating area, 10–12 mm^2, was delineated in the upper portion. The detector was placed at a distance 10 mm from the muscle. The same conditions were observed in the work with the young rabbits, except that because they were so active it was necessary, in addition to the regular fastenings, to hold them in place manually.

The adult rabbits were also left in the horizontal positon for the spectral analysis of the radiation, directed into the objective aperture of the spectrograph[*] by reflection from a concave mirror. The young rabbits were placed in a the vertical position, directly in front of the spectrograph aperture.

The frogs were also centered in the vertical position, fastened to a cork disk.

Degradation radiation [3, 4, 6] was caused by irrigation of the muscle with chilled physiological saline(6-8° for the rabbits, and 1-2° for the frogs).

Separation of the narrow spectral bands was accomplished with the aid of a monochromatic aperture. However, the majority of the data was obtained by coarser separation of the spectrum into portions 100 A in width.

The biodetection method was used for the observation of the radiation. This method has been described in detail in a number of publications [4, 5, 6].

RESULTS

Just as in the case of the spectrum described for the corresponding muscle in the frog [3], the radiation spectrum of the gastrocnemius muscle in the adult rabbit is characterized by a single radiation maximum in the range of 2200-2300 A (Table 1). Only with finer separation was a small difference seen in the spectra: the greatest radiation intensity for the frog muscle lay in the range 2000-2260 A, while for the rabbit it was in the range 2210-2270 A.

The same type (as in the frog) and spectrum of degradation radiation was seen to arise in the rabbit muscle in association with its cooling.

Spectral analysis of the radiation encountered in the resting gastrocnemius muscle of the rabbits in the different age periods yielded results which are presented in the figure.

The clear evolution of the spectra toward reduction in the number of active bands makes these data completely conclusive, despite the small number of animals. The radiation spectrum of the muscle in the 16-day old rabbit is already comparable to the spectrum observed in the adult animals. It must be noted that at this age, i.e., after maturation of the animal, degradation radiation of the muscle also acquires the same spectral composition.

These data make it possible to postulate that the molecular substrate of the muscle fibers gradually acquires the character of an unequilibratedly regular system.

---

[*] Fuess Quartz Spectrograph, possessing adequate dispersion.

TABLE 1. Radiation Spectra of the Rabbits'. Gastrocnemius Muscle

Wave length (in A)	Resting radiation	Degradation radiation
	15-18 second exposure effects (in %)	6-8 second exposure
1 900–2 000	−3	3
2 000–2 100	4	−7
2 100–2 200	−1	5
2 200–2 300	52	45
2 300–2 400	4	0
2 400–2500	8	−3
2 500–2 600	6	5
2 600–2 700	−2	−
2 700–2 800	−8	13
2 800–2 900	10	−3
2 900–3 000	4	−3
3000–3 100	−1	5
3 100–3 200	−3	−

Note: Averages of 5 trials are presented in column 2; averages of 2 trials are presented in column 3.

TABLE 2. Intensity of the "Resting" and Degradation on Radiation of the Gastrocnemius Muscle in Rabbits at Various Age Periods

Age (in days)	Resting radiation		Degradation radiation	
	exposure, sec**	effect (in %)	exposure, sec	effect (in %)
1–2	2–3	40	4–5	5
			8	**36**
4–5	2–3	−3	2–3	−2
	4–5	**28**	4–5	**27**
9	2–3	10	2–3	−10
	4–5	**36**	4–5	**20**
16	4–5	2	2–3	**48**
	8	**24**		
Adults	4–5	−3	2–3	**39**
	8	**45**		

Note: Averages of the figures from 2 trials are presented. **The threshold exposures are indicated in boldface type.

The data were supported by the results obtained from comparing the intensity of "resting" and degradation muscle radiation in the young and adult rabbits (Table 2).

Intensity and threshold duration were inversely related: I.t = const. [5, 6] (I – intensity, t—threshold exposure). In these experiments the constant radiation area (10 mm^2) was strictly observed, as was the constant distance of the detector from the muscle, 10 mm.

Spectra of radiation encountered in the resting gastrocnemius muscle of the rabbits at various age periods:
1) 1-2 days; 2) 4-5 days; 3) 9 days; 4) 16 days; 5) adults.

Thus, we can speak of significant intensity of the resting radiation from the rabbit muscle at early ages of the animals, and of approximately the same elevation in the intensity of the degradation radiation, beginning with the rabbit's second weak of age.

The entire body of results permits drawing the following conclusions.

In the first days following the birth of the animal the radiation of the muscle in the resting state is directly related to metabolic processes, and must be very varied in that period. Later the radiation arises as a result of natural disturbances in the unequilibrated molecular substrate, i.e., appears as degradation radiation of a physiological type. Energy for the support of this unequilibrated state is supplied by the muscle metabolism. The three-dimensional regularities in the substrate's molecular order reproducible in general outline, are, as we already stated earlier, the product of an uninterrupted interaction of processes whose sources are the centers and the muscles.

Judging form the evolution of the spectra, the general steric regularities of the substrate gradually transform into more well defined structural parameters. Nevertheless the state of the substrate completely retains its unequilibrated character, i.e. requires a considerable expenditure of energy for its support.[*]

This latter conclusion leads to the concept, already advanced earlier, that the resting state of the neuromuscular system must be regarded as an active setting, and emphasizes that a high energy potential is specific for the muscle resting state.[**]

Judging from the low intensity of the "resting" radiation of the muscle, it is consumed very expeditiously and economically. The unequilibrated structural state of the substrate doubtlessly aids this.

---

[*] The concept that a more limited spectrum reflects greater uniformity of the substrate's structural organization has been completely substantiated.

[**] Especially manifest, apparently, for heavy muscles.

We considered it of interest to compare these data with the results of detailed investigations by I. A. Arshavskii and V. D. Rozanova [1, 8], demonstrating a low lability and excitability, and also low resistance of the neuromuscular apparatus in the early age periods of the animals, and elevation of these properties with development of the animal. The gradual formation of the unequilibrated structural arrangement corresponds to the rise in lability and excitability. The constant reproducibility of the structural arrangement formed may serve as one of the factors elevating the resistance of the muscle tissue.

But along with this the question may be raised as to whether such dynamic constancy of the structural arrangement does not serve as the kind of regulating factor which limits the accomplishment of various chemical activities.

In this case, reduction in the interaction between the spinal cord centers and the muscles could lead to "enrichment" of the spectrum for the resting radiation of the muscle and to inclusion of bands in it which are characteristic of metabolites not found under normal conditions. We settled on acetylcholine, whose spectrum is not observed during analysis of the resting radiation from the frog gastrocnemius muscle, and which arises during its excitation [7]. The experiments presented below were also carried out on frogs.

Reduction of the action of the centers was accomplished by cooling them. This, along with direct stimulating action, yields an increasing inhibitory effect on the enzymatic activity of the tissue, i.e. indirectly also on the functional activity of the centers.

TABLE 3. Radiation Spectrum of the Frog Gastrocnemius during Cooling of the Spinal Cord Centers (the acetylcholine line is shown in boldface type)

Wavelength (in A)	Exposure (in seconds)	Effect (in %)	Wavelength (in A)	Exposure (in seconds)	Effect (in %)
1 900–1 930	8–10	43	1 940–45	20–30	0
1 930–1 960	8–10	0	1 945–50	20–30	40
1 960–2 000	8–10	12	1 950–55	20–30	−10
2 000–2 100	8–10	3			
2 100–2 200	8–10	−3	2 330–35	20–30	6
2 200–2 300	8–10	−6	2 335–40	20–30	39
2 300–2 400	8–10	2	2 340–45	20–30	0
2 400–2 500	8–10	−8			
2 500–2 600	8–10	3			
2 600–2 700	8–10	12	2 370–75	20–30	3
2 700–2 800	8–10	6	2 375–80	20–30	45
2 800–2 900	8–10	−1	2 380–85	20–30	6
2 900–3 000	8–10	33			
3 000–3 100	8–10	20	2 610–20	20–30	0
3 100–3 200	8–10	16	2 620–30	20–30	60
			2 630–40	20–30	10

Note: The averages of two trials are presented in column 3, and the averages of 3 trials in column 6.

A flat piece of ice from frozen phsiological saline was affixed to the corresponding portion of the vertebrate, which had first been freed of the muscles covering it. After 5-8 min we began the spectral analysis of the radiation from the gastronemius muscle (Table 3).

We see, in this manner, that not only the general character of the spectrum changed, but weak acetylcholine lines also arose. The parallel occurrence of these two phenomena are of great interest from out point of view.

These data thus confirm the concept that regulation of the structuro-energy state in the molecular substrate of the muscle in association with the overall state of all systems involves the principle of limiting the various chemical processes, i.e., it determines, to a definite degree, the character of the metabolsim.

In line with this one must refer to the additional orientating data, showing that with adequate excitation of the muscles there occurs an "enrichment" of the total radiation spectrum in addition to the simultaneous genesis of acetylcholine,

SUMMARY

The authors present data which broadens the concept of the regulatory effect exerted by the spinal cord centers on the molecular substrate of muscles. Evolution of mitogenetic spectra of the resting rabbit gastrocnemius (connected with age) and the prevalence of mitogenetic radiation of degradational type in adult animals point to the gradual formation of the nonbalanced structural state of the substrate.

It was shown that the nature of metabolism is also connected with the substrate regulation. This follows from the altering of the mitogenetic spectrum lines of the resting frog gastrocnemius caused by reduced influence of the centers. In such conditions, the spectrum contains acetylcholine lines as well.

LITERATURE CITED

1. Arshavskii, I. A., Izv. Akad. Nauk SSSR, Seriya Biol., No. 1, p. 71 (1958).
2. Gurvich, A. A., Byull. Ekspt. Biol, i Med., No. 5, p. 67 (1960).
3. Gurvich, A. A., Byull. Ekspt. Biol, i Med., No. 10, p. 82 (1960).
4. Gruvich, A. G. and Gurvich, L. D., Mitogenetic Radiation [in Russian] (Moscow, 1945).
5. Gurvich, A. G., and Gurvich, L. D., Introduction to the Study of Mitogenesis [in Russian] (Moscow, 1948).
6. Gurwitsch, A., Die mitogenetische Strahlung, (Berlin, 1932).
7. Lipkind, M. A., Byull. Ekspt. Biol, i Med., No. 6, p. 65 (1958).
8. Rozanova, V. D., Fiziol. Zhurn. SSSR $\underline{25}$, p. 202(1938).
9. Rozanova, V. D., Fiziol. Zhurn. SSSR $\underline{30}$, No. 3, p. 346 (1941).

Case Study 6: Ion Currents Regulate Cell Polarity

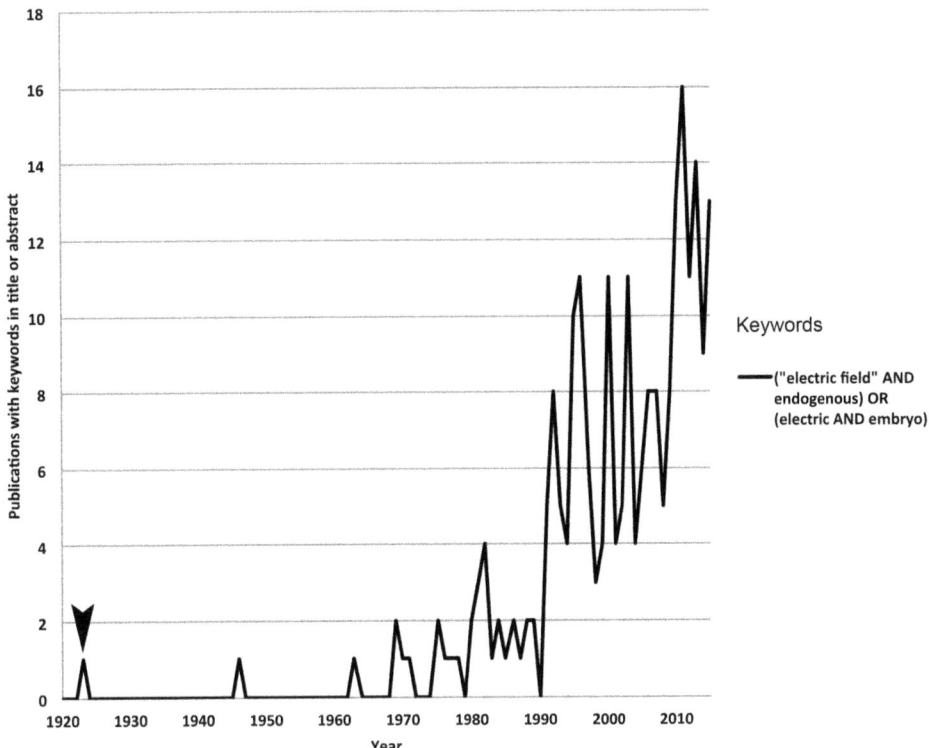

Lund (1923) Electrical control of organic polarity in the egg of Fucus. This paper is just one example of the range of ideas under study prior to the rise of molecular biology.

Perspective on *Electrical control of organic polarity in the egg of fucus*

Richard Nuccitelli, PhD
RPN Research and BioElectroMed Corp.

A true pioneer in the discovery of bioelectricity
Life as we know it relies extensively on electric fields and ionic currents. Not only is our main energy currency, ATP, generated by an electromotive force of 200 mV across our mitochondria membranes, but every sense that we have from vision to hearing utilizes electrical signals.

Consider our vision for example. Our retinal rods are highly polarized with Na^+ channels localized to the outer segment and K^+ channels localized to the inner segment. This polarization of ion channels generates a steady 'dark' current through these cells that is controlled by the concentration of cGMP that gates the Na^+ channels. When light hits the rhodopsin in the outer segment, it triggers a signal transduction cascade that results in the hydrolysis of cGMP and subsequent reduction in the Na^+ current. This leads to a hyperpolarization of the membrane potential of that rod and influences secretion to the next cell in the visual system to influence image formation in the brain. We are truly electric beings.

With this in mind, it is not surprising that our development utilizes electrical signaling as well. The first investigator to propose that natural electric fields might influence growth was Albert Mathews in 1903 [1]. He measured potential gradients along the surface of regenerating hydroids, indicating current entering the regenerating polyp end. He suggested that this current polarized the protoplasm or cells. However, the most significant and influential early investigations of developmental fields were surely those of Elmer Julius Lund [2, 3]. These studies began in 1921 and culminated in his book, *Bioelectric Fields and Growth* in 1947. His work stimulated a considerable number of contemporary investigators such as Spek [4] and Burr [5, 6] as well as later investigators such as Lionel Jaffe, Ken Robinson, Benjamin Peng, and myself. His paper included here describes a very simple method for applying an electric field to cells in sea water while measuring their polarized growth. He determined that rhizoid outgrowths of *Fucus inflatus* eggs would form towards the positive electrode if the eggs were exposed to an electric field of about 25 mV/egg throughout a period of about 10 h prior to rhizoid formation. These observations were both confirmed and improved upon by Ben Peng [7] and we now know from our more recent work [8, 9] that the direction of this imposed current is the same as the endogenous one that enters the egg in the rhizoid end and leaves the thallus. These eggs begin life completely spherically symmetrical with no preformed axis and can generate a rhizoid on any part of their surface. The selection of the optimal.

Growth axis can be guided by external cues such as light, pH or nearby neighboring eggs. Light is the most common guiding influence as they would normally grow away from the light to generate a holdfast on the intertidal rocks. Investigators stimulated by Lund's initial observations of fucoid egg sensitivity to imposed electric fields have determined that light triggers an influx if Ca^{2+} into the dark region of the egg surface [10] and an efflux of K^+ on the opposite end, and have measured the endogenous extracellular current flow resulting from these transmembrane ion movements [8]. We now know that these transcellular currents are found to be associated with cell function on most cells that have been investigated with the appropriate tools [11]. Thus, Lund's contributions 93 years ago have been seminal and inspired a great deal of investigation into the bioelectric properties of cells and tissues.

References

[1] Mathews AP Electrical polarity in the hydroids *American J Physiology* 19038 294–9
[2] Lund EJ 1923 Electrical control of organic polarity in the egg of Fucus *Bot. Gaz.* **76** 288–301
[3] Lund EJ 1947 *Bioelectric Fields and Growth* (Austin, Texas: Univ. Texas)
[4] Spek J 1930 Zustandsänderungen der Plasmakolloide bei Befruchtung und Entwicklung des Nereis-Eies *Protoplasma* **9** 370–427
[5] Burr HS 1941 Field Properties of the Developing Frog's Egg *Proc. Natl Acad. Sci. U S A* **27** (6) 276–81
[6] Burr HS and Bullock TH 1941 Steady State Potential Differences in the Early Development of Amblystoma *The Yale journal of biology and medicine* **14**(1) 51–7
[7] Peng HB and Jaffe LF 1976 Polarization of fucoid eggs by steady electrical fields *Dev. Biol.* **53** 277–84
[8] Nuccitelli R 1978 Ooplasmic segregation and secretion in the pelvetia-fastigiata egg is accompanied by a membrane generated electrical current *Dev. Biol.* 62
[9] Jaffe LF 1966 Electrical currents through the developing Fucus egg *PROC NATL ACAD SCI (USA)* **56** 1102–09
[10] Robinson KR and Jaffe LF 1975 Polarizing fucoid eggs drive a calcium current through themselves *Science (Washington D C)* **187** 70–2
[11] Nuccitelli R 1983 Transcellular ion currents: Signals and effectors of cell polarity *Modern Cell Biol.* **2** 451–81

ELECTRICAL CONTROL OF ORGANIC POLARITY IN THE EGG OF FUCUS

E. J. Lund

(with plate xxvii, and two figures)

Introduction

Organic polarity in animal and plant cells, organs, and whole organisms is one of the fundamental common properties of living material, and therefore it makes little difference whether the phenomena are considered in plant or in animal cells, the problem is probably, if not certainly, the same. This fact has long been recognized, but the problem of polarity is only a special (although in all probability primary) part of the general and more comprehensive problem of the origin and causes of symmetry in living material. An excellent treatment of certain aspects of the principle of symmetry in its relations to chemical and biological science is given by Jaeger (3).

The facts of physiology relating to growth, differentiation, and dedifferentiation have shown beyond question that the visible structure of the cell or organism is only a result of preceding physico-chemical processes, in the same sense that the formation of a precipitate is the result of a chemical or physical process, which in turn is always associated with an energy change. From this fundamental fact it follows that the energy changes, or forces which determine the polar or symmetrical arrangement of the material of living cells, organs, and whole organisms must to a greater or less degree be directed forces. Such energy changes or forces must have an axial or polar orientation, corresponding in degree to the axial or polar arrangement of the material laid down as structure. The kind of information desired, therefore, is facts relating to the type, magnitude, and distribution of the associated energy changes, and not primarily a knowledge of the quality or quantity of the materials laid down as structure. The primary problem, therefore, would appear to be the nature of the forces which bring about the characteristic orientation of the materials in the process of formation of the colloidal structure of cells.

It is well known that light is capable of determining the polarity of the germinating spores of many lower plants. This has been shown in the recent experiments by Miss Hurd (I) on the egg of *Fucus*. Effects of mechanical pressure and centrifugation of animal eggs in relation to orientation of cell cleavage have been described by many workers (see Kellicott 4), but so far as the writer is aware, the following is the first successful attempt at controlling the polarity of an egg and the future body axis by means of an electric current. The object in the present paper is (I) to show that by the application of a difference of electrical potential to the egg of *Fucus*, it is possible to determine the axis of symmetry of the egg and future plant body; and (2) to give a method by means of which it has been possible to calculate the magnitude of the potential difference necessary for this orientation.

Fucus inflatus is monoecious, and consequently the eggs are fertilized at the time of shedding. The first cleavage, which is equal, begins approximately sixteen hours after the shedding of the eggs, and is completed in a few hours at room temperature.

The eggs are perfectly spherical, and no axis of symmetry is evident before cleavage begins. As is well known, one of the cells of the first cleavage becomes the rhizoid or holdfast, while the other gives rise to the thallus, thus determining definitely the future polar axis of the organism.

Methods

The procedure in obtaining the eggs of *Fucus* was the same as that described by Miss Hurd. Plants with mature tips were collected from midtide level near the Friday Harbor laboratory, and wrapped in moist cloths for periods varying from ten to eighteen hours, and then exposed to air for about half an hour. The tips were then removed and floated in flat dishes containing fresh sea water. Within a few minutes or half an hour the tips began to shed the eggs, which then settled down upon the bottom of the dish and could be collected and removed with a pipette. The possibility of carrying out the experiments described depends entirely upon the fact that after fertilization the *Fucus* egg fixes itself permanently to the solid substratum (cover glass, for example) upon which it comes to rest. Permanent fixation takes place in about six to eight hours, or in some cases within a shorter time. All of the experiments with the electric current were carried out in total darkness, so that none of the results are attributable to light.[1]

The apparatus used is shown in text fig. I, which is about one-half the actual size. *A* is a general view of the arrangement, while *B* is a longitudinal section through the arm d of the trough shown in *A*. The V-shaped arrangement of the two arms *c* and *d* of the trough in *A* provided for a continuous flow of a stream of sea water, whose rate of flow was regulated by the stopcocks e and *f*. The stopcocks were connected by siphon to the same supply of fresh sea water kept in glass jars. The level of the water in the two arms of the bath was regulated by the adjustable outflow *g*. The apparatus was made from a glass plate having three holes bored through it. The two holes at the ends of the V-shaped trough were covered with porous clay diaphragms (*x*) sealed to the glass by means of de Kotinsky cement, as shown in *B*. Into the other hole (*h*) was fitted the outflow tube *i*, in which the zinc rod electrode was held by a special support not shown in the figure. The electric current was led into the arm (*d*) of the trough from the dish of sea water (*j*), in which the other zinc electrode was supported. This is shown best in the section *B*. The sea water in the dish *j* could be changed continually if necessary, thus removing all possibility of any products of electrolysis from entering the stream of water over the eggs. The products of electrolysis at the electrode *z* were washed away by the constant stream of sea water from the two arms of the trough, thus insuring for the eggs absolute freedom from contact with any foreign chemicals. As a result of this arrangement, the electrical resistance of the arm (*d*) of the trough was relatively low, so that a relatively large adjustable resistance (*r*) could be inserted, thereby making possible a complete control of the density of the electric current through the arm of the trough (*d*). The electric current from the battery *b* (about 18 volts) was passed through an accurately

---

[1] The writer is greatly indebted to his assistant, Miss Gladys Hamm, for the work connected with the preparation of the material for experiment.

calibrated milleammeter reading to 0.05 of a milleampere. The two arms of the trough provide for identical conditions except for the electric current passed through *d*. The trough was made from accurately fitted glass strips held together by de Kotinsky cement, and covered by glass plates (*l*, *m*), so as to provide a definite constant cross-sectional area of the stream of sea water in which the eggs were placed. Since the total electric current flowing through the arm *d* of the bath is known from the milleammeter reading, and the cross-sectional area of the column of sea water over the eggs is constant and known, then the current density (the number of microamperes per sq. mm. cross-section of the sea water) is known. Finally, if we can determine the resistance per unit length of the column of sea water under the

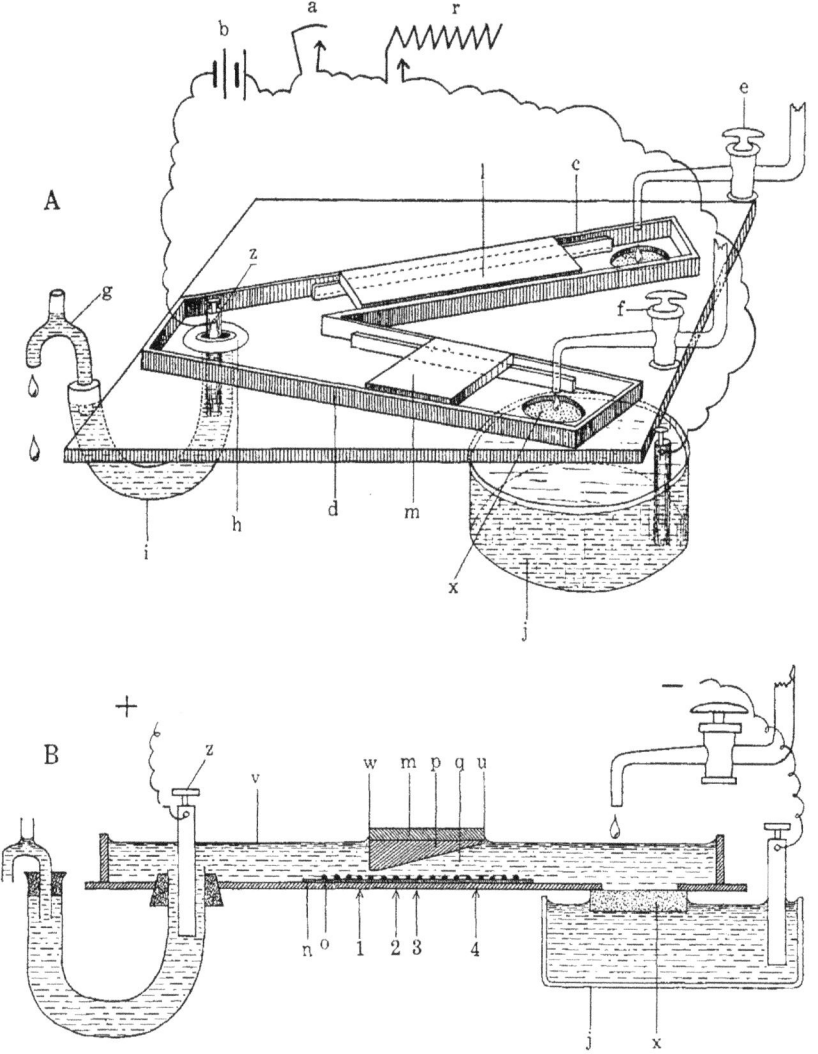

Fig. I. —*A*, general view of apparatus; *B*, longitudinal section of arm; explanation of details in text.

glass cover *m*, then of course the fall of electrical potential over any cross-section of water column can be calculated. Text fig. I*B*, which is a longitudinal section of the arm *d*, shows the essentials of the arrangement to better advantage. The cover glass (*n*) bearing the eggs (*o*) is fixed in place on the bottom of the trough by a thin film of vaseline. To the under side of the glass cover (*m*) is fixed a glass wedge (*p*), whose width is such that it fits snugly between the sides of the arm d of the trough. This provides for a sharply defined wedge of sea water (*q*) whose dimensions can be determined, and therefore provides for a gradient of known density of the electric current, and therefore a gradient of known fall of electrical potential over the eggs spread upon the cover glass. The electrical resistance of this wedge of sea water of known dimensions was determined in the ordinary way, using platinum electrodes which were inserted in the positions *u* and *w* covering completely the two ends of the water wedge. To check the accuracy of this method of determining the resistance of the wedge, one of the electrodes was placed in the position *u*, while the other was placed at some definite position *v*. The resistance of the sea water column was then determined with the electrodes at these positions. The electrode at *u* was then removed to the position at *w* and the resistance again determined. The difference between these values was the resistance of the water wedge *q*. Identical values (195 ohms) for the wedge of sea water were obtained by the two procedures.

Experimentation

Several preliminary experiments were undertaken to discover the appropriate current density at which effects of the electric current upon the eggs appeared. The use of the wedge facilitated the work; for, as was found in the writer's previous work on the electrical control of polarity in regenerating internodes of Obelia (6), a definite threshold of current density for orientation of the first cell cleavage exists in the *Fucus* egg.

In the first experiments the wedge was not used, but instead the current density was varied by placing the cover glasses holding the eggs upon glass squares of different thicknesses, which were then placed on the bottom of the trough under the glass covers (*m*, *l*). The results of a successful experiment, in which the appropriate current density for orientation was accidentally found in this way, is given in Plate XXVII, *a* and *b*, in which *a* is a typical field on the cover glass, the eggs upon which had been subjected to the electric current, while *b* is a field from the cover glass kept as a control in the other arm of the trough without the electric current. Other controls were also kept in finger bowls. These showed identical conditions of growth with that of the control *b*.

The photomicrographs *a* and *b* were taken seventy-one hours after seeding the eggs on the cover glasses, and seventy-two hours after shedding of the eggs by the plant tips. The electric current was turned on immediately after the eggs were placed on the cover glass, and at the same time the water current was turned on through the arm *c* of the trough, thus serving to wash the electrode *z*. The sea water in the dish *j* was changed several times during the experiment, thus insuring freedom of the eggs in *d* from chemicals, even though no water current was flowing through *d*. Since it takes about six to nine hours for the eggs to become firmly fixed, the current of sea water was not turned on in d until nine hours after seeding. The velocity of flow of sea water over the eggs in the arm c was slower than that over the eggs in arm d, because of the greater cross-sectional area of the water column in c; hence the control eggs in arm c were not in danger of being dislodged by the water current, even though this was started immediately after seeding the eggs upon the cover glass. This procedure permitted the eggs on the cover glass in d to become fixed firmly in relatively quiet water, while the electric current was being passed through the eggs from the beginning.

When the eggs in d had become fixed, the water current was turned on without danger of losing any of them. As time went on after seeding, the eggs became more and more securely fixed, until at the end of about twenty-four hours the swiftest current did not seem to be able to dislodge them.

In one of the preliminary experiments eggs were placed upon two glass squares of different thicknesses. These two glass squares were then placed on the bottom of the same trough d and the electric current passed through. The water was turned on in the arm c at the beginning, as in the experiment just described. Later the water current was turned on in d. The result was that the eggs upon the thicker glass square (higher current density) were oriented, while those on the thinner glass square were not.

A cover glass with control eggs previously kept in sea water in a finger bowl until they had become fixed was then put into the arm c, and the water current allowed to flow much more swiftly through the arm c than over the eggs in the arm d. No trace of any orienting effect of the water current upon the eggs in c could be seen. These and other similar preliminary experiments convinced the writer that the orienting effects observed in the electric current were not due to the flow of water, nor were they due to any reaction by the egg cells to chemicals from the electrodes. In figs, a and b no difference in the amount of growth of the eggs can be seen. The eggs in the control b appear slightly larger than those in a, but this is simply due to an accidental difference in magnification. Under the microscope no difference between the two lots of eggs, except that of orientation, could be seen. These experiments, together with others giving similar results, show that it is possible to orient the first cleavage of the egg, and therefore the polarity and future body axis by an electric current.

The following experiment is reported in full because the procedure enables us to determine the quantitative relations between the current density, or electrical potential, and the effects upon the eggs. The eggs were placed on cover glasses within one hour after being shed. Some of these cover glasses were kept as controls in finger bowls containing fresh sea water. One of the covers was placed in the arm d of the trough under the glass wedge exactly as shown in text fig. 1B. The water current was turned on through the arm c to wash the electrode z, and the electric current turned on in d immediately after seeding the cover glass. Twelve hours after the beginning of the experiment the water was turned on in d. Fourteen hours after the beginning of the experiment a cover glass containing control eggs was removed from the finger bowl and placed in the arm c of the trough. Forty hours after the beginning of the experiment the cover glass in d was removed for examination, and photomicrographs of typical microscopic fields were taken at the positions numbered 1, 2, 3, and 4 in text fig. I*B*. These figures are correspondingly numbered 1, 2, 3, and 4 in plate XXVII. It will be seen that the eggs shown in figs. 1 and 4, corresponding to positions 1 and 4 in text fig. 1*B*, do not show any orienting effects of the electric current.

Growth in these positions was not different from growth in the control kept in the arm *c*, or those in finger bowls. In all such lots of eggs there are always some eggs which do not divide, similar to what occurs for example in echinoderm eggs, so that the few eggs showing no growth have no special significance.

Fig. 2 from position 2 shows practically 100 per cent orientation. Examination of the eggs in position 2 showed that the planes of the first cleavage were very nearly at right angles to the direction of the electric current. This fact was verified by several members of the laboratory at the time.

Fig. 3 from position 3 in text fig. 1*B* shows that there is a very distinct threshold of current density (or *PD*) at which orientation occurs. The writer believes that this may prove to be an important fact for the significance of the phenomena of orientation as a whole. Similar well defined thresholds of electrical potential have already been described for *Obelia* (6). On the left side of fig. 3 most of the cells are oriented, while on the right side of the same field no orientation effects are evident. The attempt was made to show the region of the threshold in the middle of the photograph, but unfortunately manipulation of the camera resulted in a slight displacement, so that the orientation threshold lies at the left of the middle of the figure.

Under the glass wedge between the positions 1 and 2 a definite inhibition (delay) in cleavage occurred. Development occurred in most of these delayed eggs after removal of the cover glass to a finger bowl with sea water. As seen from the experiments (figs, *a* and *b*), orientation by the electric current does not necessarily involve inhibition (delay) of cleavage and growth. This is again evident in figs. 1,3, and 4, and consequently places the fact of orientation in a definite category of what might be called normal response comparable with the response to light as found by Miss Hurd (i).

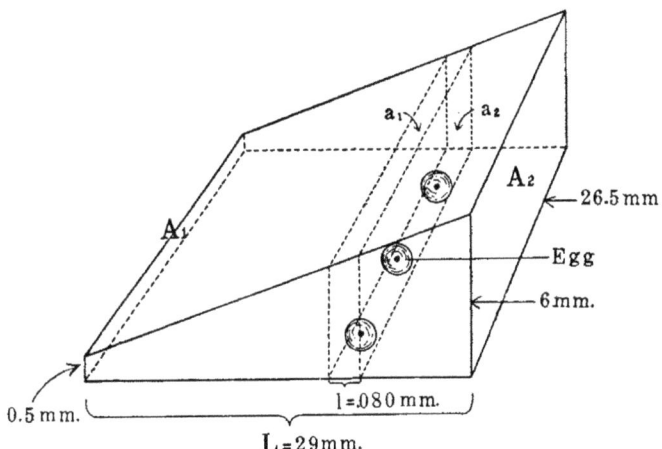

Fig. 2. —Diagram of water wedge q in fig. 1*B*; numbers give actual dimensions in millimeters.

Figs, *a,* 2, and 3 show that orientation of cells to the electric current is not perfect wherever the eggs lie in close groups. This may be interpreted simply as due to the fact that the electrical resistance of the group is high, and therefore the fall of electrical potential across an egg in such a group is less (that is, below the threshold value, other things being equal) than it would be if that same egg were isolated. An instance of group orientation as described by Miss Hurd is shown in fig. 4. This orientation by the cell to adjacent cells undoubtedly modifies the response to the electric current, similar in a way to the interference effect of a contact stimulus upon *Paramecium* when reacting to an electric current.

Calculation of threshold value of electrical potential for orientation

Let fig. 2 represent the shape of the water wedge q in text fig. IB. Since the total electric current (C) flowing through the wedge from face A_1 to face A_2 was known, and equal to 37.4 milliamperes in the experiment from which figs. I, 2, 3, and 4 were taken, and the total resistance (R) of the wedge was 195 ohms, then by Ohms' law, E, the total fall of electrical potential between the faces A_1 and A_2, was 7.29 volts. The P.D. causing the current to flow through any given egg is the P.D. between two planes parallel to A_1 and A_2, and separated by a distance equal to the diameter of an egg. Since the diameter of the *Fucus* egg is nearly 0.08 mm., the problem becomes one of calculating the magnitude of the P.D. between any such two planes 0.08 mm. apart, at any position along the length of the wedge, as for example a_1 and a_2 in fig. 2. Since the specific resistance K of the sea water (in this case the resistance of a millimeter cube of sea water) is given by the equation $K = R \frac{1}{L} \frac{(A_1 + A_2)}{2}$ (I), in which all the terms except K can be determined by direct measurement, and where $R = 195$ ohms (total resistance of water wedge); $L = 29$ mm. (total length of water wedge); and A_1 and $A_2 = 13.5$ and 159 sq. mm. respectively (areas of ends of water wedge); then $K = 580$ ohms.

The form of the equation for calculating the resistance of the wedge when K is known is therefore $R = \frac{KL2}{(A_1 + A_2)}$ (2). Since in equation (2) R is the total resistance of the water wedge, and since any cross-section of this wedge is similar to the whole, for any such section of the wedge as shown in fig. 2 between the parallel planes a_1 and a_2, we have $r = \frac{580 \times 2 \times l}{a_1 + a_2}$ (3); where r = resistance of wedge section between any two planes with areas equal to a_1 and a_2, and $l = 0.08$ mm., the diameter of the egg.

Position of photograph	$\frac{580 \times 2 \times l}{a_1 + a_2}$	r (ohms)	P.D. across egg (volts)	Current density (microamperes per sq. mm.)
I........	$\frac{580 \times 2 \times 0.08}{351.0}$	0.26	0.009	210
2........	$\frac{580 \times 2 \times 0.08}{76.6}$	1.21	0.036	970
3........	$\frac{580 \times 2 \times 0.08}{109.4}$	0.84	0.25	680
4........	$\frac{580 \times 2 \times 0.08}{111.3}$	0.41	0.012	330

The values for the areas of the planes a_1 and a_2 for each one of the positions 1, 2, 3, and 4 were determined by direct measurement from a plaster cast of the space below the glass wedge p in text fig. 1B. Since the distance between a_1 and a_2 is very small, the areas of a_1 and a_2 at any one position may be considered equal.

Applying equation (3), we can now determine r for these sections, and, since the total current is known, the P.D. between their faces, that is, the actual fall of electrical potential through an egg at these positions. Table I gives the values for the positions from which the photographs were taken. It will be seen from the table that the threshold P.D. for orientation is 0.025 volts. This value is several times greater than the electrical potentials which are effective in producing the results described by the writer for *Obelia* (**6**).

The difference between these thresholds of P.D. may be due in part to the fact that the *Fucus* egg possesses a tough membrane, which may have a high resistance to the electric current. It will be clear, of course, that the density of the electric current in the egg may be entirely different from that in the sea water, depending upon the ratio of the specific resistance of the egg protoplasm to that of the sea water. The last column in table I gives the densities of the electric current in the sea water at the positions of the photographs. The limits of accuracy of the method were tested by making calculations based on different sets of measurements of the water wedge. These showed, for example, that the different values for the threshold P.D. did not differ by more than 0.002 volt. Consequently it is believed that the values given lie within an error of 15 per cent of the actual values.

Discussion

Our present knowledge concerning the phenomena of orientation of the axis of symmetry of cells to an electric current is too limited to warrant any final conclusions as to its actual significance for morphogenesis, but attention may be called to certain facts related to the problem. Many free living cells orient to the electric current by movement of the cell as a whole, but in these experiments we are confronted with a process involving orientation of cell division and growth (synthesis of protoplasm). A close similarity appears to exist between the results of Ingvar (2) on the oriented growth of nerve cells in tissue cultures placed in an electric field, and the growth processes described here. The main difference is that in the Fucus egg we have cell division superposed, so to speak, upon the growth process (synthesis of protoplasm) as such. The orientation of Amoeba to an electric current is also of interest, since in this case morphological polarity is a fluctuating property, and in all probability may be considered identical with fluctuating functional polarity. A series of striking cases of what appear to be transitional conditions between functional polarity and morphological polarity and their reversal, in a cell (Bursaria), has been described by the writer elsewhere (**7**). In these cases it is clear that a highly specialized and apparently fixed visible morphological polarity in a single cell may disappear by a process of dedifferentiation (autolysis?), part of the material of the cell reappearing with a different polarity from what it had before. The facts cited show a very close interrelation between definitely directed energy transformations and the processes of orientation of the materials laid down as morphological structure (= colloidal crystallization?) in the process of cell morphogenesis, as stated in the first part of this paper. The experiments and suggestive speculations of Lillie (**5**) are to the point in this connection. If the physico chemical process in the cell, which determines morphological polarity, can be oriented by the application of a difference of electrical potential of external origin, does not this imply that the mechanism determining the morphological polarity possesses electrical properties?

The orientation by light presents no necessary difficulty, of course, for it is a well known fact that bioelectric currents are associated with photochemical changes, for example, plants, retina of the eye. The occurrence of constant electrical potentials in tissues and organs have long been known, for example, in epithelia, roots and shoots of seedlings, Muller-Hettlingen (**10**), hydroids, Mathews (**9**), Lund (**8**). These electrical potentials correspond to the polarity and symmetry properties of the tissues.

From these facts the following interesting question arises. May not such electrical differences of potential present in cells and tissues act upon adjacent cells and cell groups in such a way as to bring about orientation effects such as appear in phenomena of cell correlation, where cells in contact with one another behave in morphogenesis as an ordered whole? The writer believes that such an assumption may prove a valuable guide to the investigation of the nature of the phenomena of cell correlation. Experimental facts bearing upon this question will be presented elsewhere.

Summary

1. Polarity of the egg of *Fucus* can definitely be determined by applying a difference of electrical potential to the egg. Apparently this is the first egg, among either animals or plants, the polarity of which has been controlled by means of an electric current. The response of the *Fucus* egg to the electric current in total darkness is similar to its response to light.
2. A method is given by which it is possible to measure the P.D. across a single egg which is placed in an electric current flowing in a stream of sea water of constant cross-section.
3. The magnitude of the electrical potential applied to opposite sides of an egg, which is just able to cause orientation (threshold P.D.) of the axis of symmetry of the protoplasm of the *Fucus* egg, is 0.025 ± 0.002 volts.
4. This threshold P.D. is definite and does not appear to involve any marked inhibitory effect (delay) upon growth. A P.D. of 0.036 volts applied to the egg does show an inhibitory effect (delay) upon cleavage.
5. The establishment of an electrochemical polarity of some sort may possibly be a fundamentally associated condition for the development of morphological polarity, because the physiological mechanism in the *Fucus* egg which determines morphological polarity can be controlled and directed by the application to the egg of a difference of electrical potential of external origin.

University of Minnesota
Minneapolis, Minn.

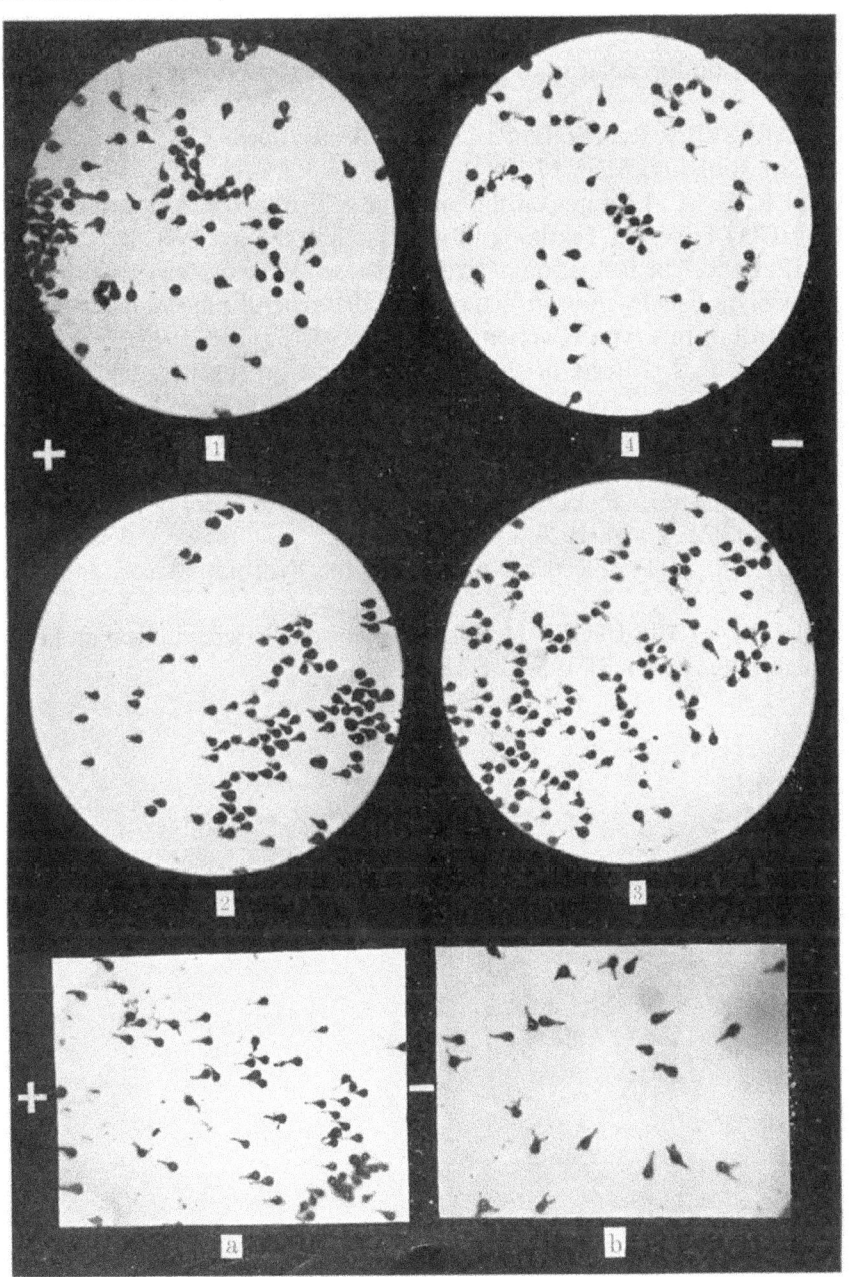

LUND on FUCUS

LITERATURE CITED

1. HURD, ANNIE M., Effect of unilateral monochromatic light and group orientation on the polarity of germinating *Fucus spores*. BOT. GAZ. **70**:25–50. 1920.
2. INGVAR, SVEN, Reactions of cells to the electric current in tissue cultures. Proc. Soc. Exp. Biol, and Med. **17**:198–199. 1920.
3. JAEGER, F. M., Lectures on the principle of symmetry. Amsterdam. 1920.
4. KELLICOTT, W. E., Textbook of general embryology. 1913.
5. LILLIE, R. S., The formation of structures resembling organic growths by means of electrolytic local action in metals, and the general physiological significance and control of this type of action. Biol. Bull. **33**:135–186. 1917.
6. LUND, E. J., I. Effects of the electric current on regenerating internodes of *Obelia*. Jour. Exp. Zool. **34**:471–494. 1921.
7. ———, Reversibility of morphogenetic processes in *Bursaria*. Jour. Exp. Zool. **24**:1–33. 1917.
8. ———, II. The normal electrical polarity of *Obelia*. A proof of its existence. Jour. Exp. Zool. **36**:477–494. 1922.
9. MATHEWS, A. P., Electrical polarity in the hydroids. Amer. Jour. Physiol. **8**:294–299. 1903.
10. MULLER-HETTLINGEN, J., Über galvanische Erscheinungen an keimenden Samen. Pflüg. Arch. **31**:193–214. 1883.

Chapter 3

Inheritance

Case Study 7: Induced Eye-Defects Can Be Passed On to Future Generations

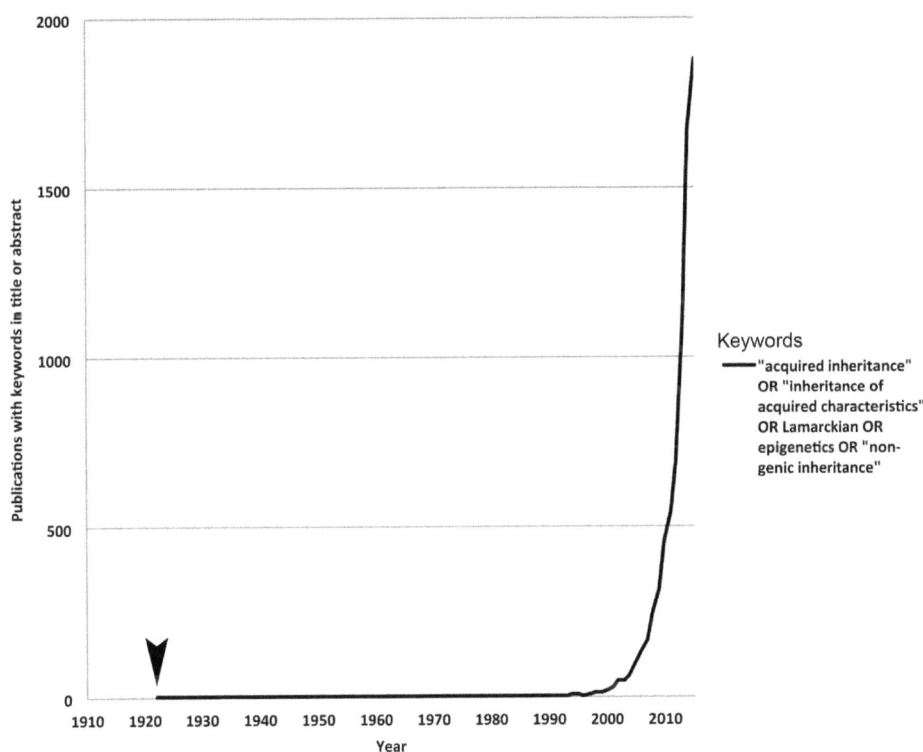

Guyer (1922) The production and transmission of certain eye defects.

Perspective on *The Production and Transmission of Certain Eye Defects*

Author: Edward J Steele PhD
Affiliation: CYO Foundation, 24 Genomics Rise, Piara Waters, WA 6112 Australia.

Correspondence: Edward J Steele PhD, CYO'Connor ERADE Village Foundation Inc., 24 Genomics Rise, Piara Waters, WA 6112 Australia.

Origin of Congenital Defects: Stable inheritance through the male line via maternal antibodies specific for eye lens antigens inducing autoimmune eye defects in developing rabbits *in utero*

In April 1922 Michael F Guyer, Professor of Zoology at the University of Wisconsin, stood and addressed the International Congress of Othalmology in Washington D.C. He presented the results of a remarkable series of 'acquired inheritance' experiments which he and Elizabeth A Smith had been conducting at the Zoological Laboratory since 1916 (Guyer 1922). He also had with him live rabbits bearing defective eyes so that the gathered opthalmologists could examine some of the evidence for themselves. Guyer and Smith showed that specific antibodies for eye lens antigens crossed the placenta in rabbits producing offspring with severe eye defects. By itself this is perhaps not surprising. They then showed that these induced eye defects were passed on to future generations arising spontaneously in progeny rabbits via the male or female lines without any further immunization or antiserum treatment of subsequent parental rabbits with eye lens antibodies (Guyer and Smith 1918, 1920, 1924). In my view these transgenerational genetic data are on a historical par with Mendel's landmark publication on inheritance of genetic factors affecting characteristics of pea plants 60 years earlier.

I first wrote on this topic almost 40 years ago (Steele 1979 p. 59–63) yet the title of this perspective today has a contemporary ring about it—or at least I would like to think so. In fact I would like to think that in a few years it will be the type of title appearing on *many* biomedical and developmental biology papers particularly in relation to the origins of numerous congenital diseases afflicting mankind not just induced-eye defects. Thus Rachel Caspi (2010) of the National Eye Institute opened her recent review thus: 'Autoimmune and inflammatory uveitis are a group of potentially blinding intraocular inflammatory diseases that arise without a known infectious trigger and are often associated with immunological responses to unique retinal proteins' (also see Perez and Caspi 2015).

However there are other settings where the approach to 'acquired inheritance' we will discuss here are likely to be relevant viz. the origins of multiple sclerosis and its potential genetic transmission as a risk factor (van den Elsen 2014, Zager *et al* 2015) through the widespread use of 'EAE animal models', which use the Guyer and Smith immunization strategy to create experimental MS disease (experimental autoimmune encephalitis, EAE; Dendrou *et al* 2015).

The tissues of certain key vertebrate organs such as the eye, central nervous system, and the gonads are relatively immunologically privileged. Whilst the self antigens in these tissues are immunogenic in the host they are not normally exposed to the adult immune system because of various blood–organ barriers (Blood–Brain Barrier; Blood–Retinal Barrier). Yet following injury to the tissue (disease, physical lesion) immune cells may gain access (antigen presenting cells and lymphocytes such as antibody producing B cells, and cytotoxic and helper/regulatory T cells). Quite vigorous immune responses can then ensue often becoming chronic as a waxing and waning autoimmune condition of increasing destructive severity in the individual and patient takes hold (indicative of an ongoing tugger war between the host tissue and its immune system with the latter usually winning). These diseases can be debilitating unless brought under control. It is this type of autoimmune disease which is our topic under discussion here described in the accompanying classic paper(s) published by MF Guyer and EA Smith between 1918 and 1924.

Thus autoimmunity affecting the eyes and central nervous system has to first breach the blood–retinal and blood–brain barriers. However there is another 'barrier' of equal importance breached in the Guyer–Smith experiments. This barrier is still surprisingly less well known to modern geneticists and biomedical scientists, because its existence is tacit and ingrained in our education system.

This is the Weismann Barrier, the tissue barrier protecting the germ cells from any type of somatic genetic influence from the body's somatic cell populations of circulating white cells in blood and lymph, as well as the cells in the fixed tissues and organs. Thus Weismann's Barrier traditionally *forbids* the flow of genetic information (DNA/RNA) from somatic cells to germ cells (whereupon it could be integrated into germline DNA thus affecting a possible 'directional' genetic change in the offspring, Lindley 2010). Accordingly the Weismann Barrier forbids the inheritance of somatically acquired characteristics, such as somatically hypermutated (Steele 2016) antibody V genes (Steele and Lloyd 2015)—it forbids what is known as 'Lamarckian modes of inheritance'. The reader can Google the significance of this barrier to theories of biological evolution—but the Weismann Barrier is *the* central pillar of mainstream genetics particularly the neo-Darwinian New Synthesis fashioned in the 1930s and 1940s giving rise to modern Population Genetics (Steele 1979, Steele *et al* 1984, Steele *et al* 1998, Steele *et al* 2002, Lindley 2010, Steele and Lloyd 2015).

Discussing or researching this topic openly has been a highly hazardous exercise for one's academic and scientific career. Therefore *many* scientists *do not entertain doing such research.* The present author nevertheless crossed that red line over 40 years ago (Steele 1979, Gorczynski and Steele 1980, 1981, Steele *et al* 1984, Steele *et al* 1998, Blanden *et al* 1998, Steele *et al* 2002, Lindley 2010, Steele and Lloyd 2015). In my view this area of genetics is of the utmost importance to explore, rationalise and explain, *particularly now* in this era of whole genome sequencing and personalised genomic medicine.

The heat and intensity of this battle has diminished some what in the past 20 years – but not the conviction among mainstream geneticists that the 'acquired inheritance' proposition is both wrong if not absurd and a contradiction in terms. Indeed Max Planck's observation at the turn of the 20th Century that one needs to wait for the vanguard of 'dogmatic orthodoxy' to literally die out certainly seems to apply in this case. No doubt the publication of numerous papers describing controlled and induced transgenerational phenomena particularly via the male has forced mainstream science to take the ideas behind such work far more seriously today (Campbell and Perkins 1988, Lindley 2010).

Thus in recent years there has been a flowering of reports on so called Lamarckian 'Acquired Inheritance' or 'Epigenetic' phenomena in many animal and plant systems where the claim is clearly made either overtly or tacitly, that Weismann' Barrier has been breached and is clearly permeable—but only up to a point (Jablonka and Lamb 1995, Rassoulzadegan *et al* 2006, Whitelaw and Whitelaw 2008, Grossniklaus *et al* 2013, Sharma 2013, Devanapally *et al* 2015, Sharma *et al* 2016). If the fine print is read carefully you will find the disclaimer: 'Epigenetics, the transgenerational transfer of phenotypic characters without modification of gene sequence' (e.g. Ho and Buggren 2010). So the inheritance is 'soft' and thus 'regulatory' and reversible in nature, involving DNA methylation-type effects and miRNA inhibitory effects on protein synthesis (translation), quantitatively controlling the magnitude and intensity of the 'transmitted' phenotype in the progeny.

The present author has spent 40 years studying the conditions for the transmission of 'hard' Lamarckian inheritance effects in the immune system of mammals (mice, rabbits, humans) whereby the outcome *must be* the integration of the selected somatic gene mutation into the germline DNA (Steele 1979, Gorczynski and Steele 1980, 1981, Steele *et al* 1984, Steele *et al* 1998, Blanden *et al* 1998, Steele *et al* 2002, Steele and Lloyd 2015). So there is still a type of stand-off here—although in recent tumor systems in mice by the Spadafora group (Cossetti *et al* 2014) and in microscopic worms (*C. elegans*, Devanapally *et al* 2015) forms of clear cut soma-to-germline transmission have been demonstrated. And 10 years ago the very clear demonstration in mice that somatic regulatory microRNAs (miRNA) can been shown to be transmitted in a non-Mendelian manner to progeny via spermatozoa (Rassoulzadegan *et al* 2006).

The problem is often the long term nature as well as the level of technical and logistical difficulty in conducting a soma-to-germline experiment in the current prejudicial research funding environment. Further, in the case of the immune system the key genes, the variable region genes (V) involved in foreign antigen recognition by B and T cells exist not as single genes but as *highly similar* genes in repertoire clusters spanning at least a megabase in the germline (where $N = 50 - 150$ V elements). These loci are very difficult to analyse, particularly in the case of human antibody V clusters. It is therefore a challenge to accurately and easily DNA sequence on the scale required for animal experiments or inheritance studies on human subjects (Steele and Lloyd 2015). Strategies based on informative three-generation pedigrees in humans have been proposed and are likely to be taken seriously for the heavy chain IGHV gene cluster on human chromosome 14 at band 14q32.33, but it will still require significant technical innovation in long-read DNA sequencing (Steele and Lloyd 2015).

Let us now summarise in more detail the results obtained in the experiments of Guyer and Smith. From a perspective of 100 years these are remarkable and revolutionary experiments. In my opinion these data provide the *first clear-cut controlled studies* unequivocally showing the germline transmission of a specific antibody-induced somatic character thus demonstrating the inheritance of an acquired autoimmune disease (the reader is advised to ignore the negative comments in Wiki and read Guyer–Smith very carefully and then consult Steele 1979 p. 59–63 and references therein, and this paper).

Here is a dot-point summary of some of their key findings, but the reader is urged to carefully read all the papers, particularly the final 1924 paper:

- In the first experimental line of rabbits bearing transmitted induced eye-defects specific antiserum to eye-lens and associated proteins was raised in fowls after grinding an emulsion of rabbit eye-lens tissue into normal saline. This antiserum was injected in multiple doses to pregnant does during gestation. The lethal toxicity of this manoeuvre on both the embryo (spontaneous abortions) and the occasional death of a mother was apparent—as would be expected. However from about 15 pregnant mothers treated with anti-eye lens antiserum 62 offspring were produced and 11 had clear eye defects. These defects were propagated through to the ninth generation either via the male and/or female line.
- The control serum and other vaccination exposures are very significant, as such mothers produced progeny with *no eye defects*:
    * Normal fowl serum or fowl anti-rabbit testis (50 progeny with no eye defects)
    * A large control group of progeny arising from immunised rabbit ($N > 500$) where mothers (and males in some cases) were hyperimmunised with different types of antigens. It is informative to quote Guyer and Smith directly at this juncture:

> *'...we, together with others in our laboratory, have been doing many experiments in which shortly before or during pregnancy female rabbits have been injected with typhoid fever vaccines or living typhoid germs, or with various kinds of foreign serum or serum immunized against proteins other than lens, and in not a single one of more than five hundred young born after such treatments has there been a case of eye defect. But even should the defect have originally been a general rather than a specific one, it is obvious that the germinal condition sooner or later must have become specific since the anomaly reappears generation after generation without any recognizable accompanying malformations of other parts of the body.' (Guyer and Smith 1924)*

These results clearly show that the effects induced were specific for eye lens and associated eye proteins, and do not normally appear in rabbit populations at the frequency observed by Guyer and Smith in their test experiments.

- The Transmission of the defects by male progeny with defective eyes rules out ongoing direct maternal antibody effects down many (up to 9) generations of breeding and thus suggests stable genetic inheritance associated presumably (in modern parlance) with chromosomal DNA sequences.

- Backcrossing experiments to show that the transmitted defects, which may disappear in one or two generations, can reappear again suggesting a type of Mendelian recessive inheritance. Although there are clearly several specific genetic factors involved as they point out.
- Establishing a separate line via the defective-eyed male progeny where they directly induced the production of anti-eye lens antibodies in the pregnant mothers by multiple vaccine doses of rabbit eye lens emulsion in normal saline. This demonstration is very important and remarkably modern.

The current animal models for induced autoimmune disease for Uveitis (EAU) and Multiple Sclerosis (EAE) depends *precisely* on this type of autoimmunization strategy first used by Guyer and Smith (Caspi 2010, Perez and Caspi 2015, Dendrou *et al* 2015). The modern twist would be for the experimenter to set up EAU or EAE disease in pregnant mothers (or males for that matter cf Gorczynski and Steele 1980, 1981, Steele 1984a, 1984b, Steele *et al* 1984) and then breed them to see if tissue-specific defects appear in progeny—with or without possible co-factor manoeuvres which might involve mild or graded injury to target tissues in progeny (to allow easier lymphoid and white blood cell access to targets and speed up the assay read-out).

An important rider I would add, after much experience of animal (mouse) breeding programs (Steele 1984a, 1984b, Steele *et al* 1984, 1986) is this: sufficient breeding pairs should be set up because not every autoimmune or antigen-treated mother (or father) will produce affected progeny (cf Gorczynski and Steele 1980, Figure 1)—Guyer and Smith also noted this. But the frequency is still high—a rule of thumb is that the expectation should be about 10% of treated breeding pairs will produce defective progeny at high frequency—and that upwards of 10% of progeny are likely to be affected (in the case of Guyer and Smith up to 15%–20% progeny were overtly affected).

The assay methods of 100 years ago were visual inspection backed by the standard ophthalmology techniques of the day. Note that ophthalmology experts examining the Guyer–Smith rabbits found *many more* eye defects than visually recorded so the eye-defect incidence in offspring is far higher than the 15%–20% reported. Modern Ophthalmology molecular assay and physical technical methods and whole genome DNA sequencing can be used today.

In wrapping up my interpretation of the Guyer–Smith data which I published in 1979 I stated the following:

'All these features are consistent with the following model (a) abnormal development of a target organ previously inaccessible to the immune system, (b) some of the "new" genetic information specifying this abnormal development pattern is transferred to and laid down in the offspring's germline, (c) this information is passed onto the next generation. Paralleling this now genetically specified organ defect is a chronic autoimmune response causing further damage to the target organ, and therefore favouring the germline fixation of high binding affinity, auto-specific V-region genes in the germline.' (Steele 1979 p. 62).

I cannot improve on this summary. So I urge the interested reader, and budding young scientist, to read Guyer–Smith very carefully. Their classic papers still inform and speak directly to us 100 years later in easily understood scientific prose—and they are an instructive outline of careful objective scientific method.

Finally, there are now published results of ongoing current experiments and clinical observations on the etiology of human children suffering from Austism Spectrum Disorders (ASD) which are now even more compelling in the light of the Guyer–Smith experiments. These contemporary findings force all of us take the issues discovered by Guyer and Smith *very seriously* (Martin *et al* 2008, Bauman *et al* 2013; the whole field reviewed in Estes and McAllister 2015). In short these data, and their careful behavioural observations, make the compelling claim that maternal antibodies (IgG) for foetal brain antigens cross the placenta and affect long term behavioural damage causing the autism ASD syndrome in their children. These findings are so important we can quote the summary of the key findings from the group of David G Amaral (at the University of California, Davis) in their ground breaking 2008 paper using rhesus monkeys as the experimental vehicle:

'*.... four rhesus monkeys were exposed prenatally to human IgG collected from mothers of multiple children diagnosed with ASD. Four control rhesus monkeys were exposed to human IgG collected from mothers of multiple typically developing children. Five additional monkeys were untreated controls. Monkeys were observed in a variety of behavioral paradigms involving unique social situations. Behaviors were scored by trained observers and overall activity was monitored with actimeters. Rhesus monkeys gestationally exposed to IgG class antibodies from mothers of children with ASD consistently demonstrated increased whole-body stereotypies across multiple testing paradigms. These monkeys were also hyperactive compared to controls. Treatment with IgG purified from mothers of typically developing children did not induce stereotypical or hyperactive behaviors. These findings support the potential for an autoimmune etiology in a subgroup of patients with neurodevelopmental disorders.' (Martin et al 2008).*

Thus Step I of the Guyer–Smith experiment has already been achieved in human subjects. When these ASD children grow up and produce their own offspring, and so forth down the generations, we will have secured Step II.

References

Bauman M D *et al* 2013 Maternal antibodies from mothers of children with autism alter brain growth and social behavior development in the rhesus monkey *Transl. Psychiatry* **3** e278

Blanden R V, Rothenfluh H S, Zylstra P, Weiller G F and Steele E J 1998 The signature of somatic hypermutation appears to be written into the germline IgV segment repertoire *Immunol. Rev.* **162** 117–32

Campbell J H and Perkins P 1988 Transgenerational effects of drug and hormonal treatments in mammals: a review of observations and ideas *Progress in Brain Research* **Vol 73**; Boer G J, Feenstra M G P, Mirmiran M, Swaab D F and van Haaren F pp 535–53

Caspi R R 2010 A look at autoimmunity and inflammation in the eye *J Clin. Invest.* **120** 3073–83

Cossetti C *et al* 2014 Soma-to-germline transmission of RNA in mice xenografted with human tumour cells: possible transport by exosomes *PLoS ONE* **9** e101629

Dendrou C A, Fugger L and Friese M A 2015 Immunopathology of multiple sclerosis *Nat. Rev. Immunol.* **15** 545–58

Devanapally S, Ravikuma S and Jose A M 2015 Double-stranded RNA made in *C. elegans* neurons can enter the germline and cause transgenerational gene silencing *Proc. Natl Acad. Sci. USA* **112** 2133–8

Estes M L and McAllister A K 2015 Immune mediators in the brain and peripheral tissues in autism spectrum disorder *Nat. Rev. Neurosci.* **16** 469–86

van den Elsen P J, van Eggermond M C J A, Puentes F, van der Valk P, Baker D and Sandra Amor S 2014 The epigenetics of multiple sclerosis and other related disorders *Multiple Sclerosis and Related Disorders* 2014 **3** 163–75

Gorczynski R M and Steele E J 1980 Inheritance of acquired immunologic tolerance to foreign histocompatibility antigens in mice *Proc. Natl. Acad. Sci. USA* **77** 2871–5

Gorczynski R M and Steele E J 1981 Simultaneous yet independent inheritance of somatically acquired tolerance to two distinct H-2 antigenic haplotype determinants in mice *Nature* **289** 678–81

Grossniklaus U, Kelly W G, Ferguson-Smith A C, Pembrey M and Lindquist S 2013 Transgenerational epigenetic inheritance: how important is it? *Nat. Rev. Genet.* **14** 228–35

Guyer M F 1922 *The production and transmission of certain eye defects* (Washington D.C: Transactions of an International Congress of Ophthalmology) April 25–28, 1922

Guyer M F and Smith E A 1918 Studies on cytolysins I Some prenatal effects of lens antibodies *J. Expt. Zool.* **26** 65–82

Guyer M F and Smith E A 1920 Studies on cytolysins II Transmission of induced eye-defects *J. Expt. Zool.* **31** 171–216

Guyer M F and Smith E A 1924 Further studies on inheritance of eye defects induced in rabbits *J. Expt. Zool* **38** 449–75

Ho D H and Burggren W W 2010 Epigenetics and transgenerational transfer: a physiological perspective *J. Exp. Biol* **213** 3–16

Jablonka E and Lamb M J 1995 *Epigenetic Inheritance and Evolution: The Lamarckian Dimension* (Oxford: Oxford University Press)

Lindley R 2010 *The Soma: How Our Genes Really Work and Why That Changes Everything!* ISBN1451525648, POD book Amazon.com CYO Foundation)

Martin L A *et al* 2008 Stereotypies and hyperactivity in rhesus monkeys exposed to IgG from mothers of children with autism *Brain Behav. Immun.* **22** 806–16

Perez V L and Caspi R R 2015 Immune mechanisms in inflammatory and degenerative eye disease *Trends Immunol.* **36** 354–63

Rassoulzadegan M, Grandiean V, Gounon P, Vincent S, Gillot I and Cuzin F 2006 RNA-mediated non-Mendelian inheritance of an epigenetic change in the mouse *Nature* **441** 469–74

Sharma A 2013 Transgenerational epigenetic inheritance: focus on soma to germline information transfer *Prog. Biophys. Mol. Biol.* **113** 439–46

Sharma U *et al* 2016 Biogenesis and function of tRNA fragments during sperm maturation and fertilization in mammals *Science* **351** 391–6

Steele E J 1979 *Somatic Selection and Adaptive Evolution: On the Inheritance of Acquired Characters* 1st Edn. (Toronto: Williams-Wallace) 1979; 2nd Edn. Chicago, IL: University of Chicago Press. 1981

Steele E J 1984a Acquired paternal influence in mice. Improved reproductive performance of CBA/H males immunized with rat blood cells. *Aust. J. Exp. Biol. Med. Sci* **62** 155–66 (now Immunology and Cell Biology)

Steele E J 1984b Acquired paternal influence in mice. Altered serum antibody response in the progeny population of immunized CBA/H males *Aust. J. Exp. Biol. Med. Sci.* **62** 253–68

Steele E J 1986 Idiotypes, allotypes and a paradox of inheritance *Paradoxes in Immunology* ed G W Hoffman, J G Levy and G T Nepom (CRC Press Inc) pp 243–52

Steele E J 2016 Somatic hypermutation in immunity and cancer: Critical analysis of strand-biased and codon-context mutation signatures *DNA Repair* **45** 1–24

Steele E J, Hapel A J and Blanden R V 2002 How can DNA patterns of somatically acquired immunity be imprinted on the germline of immunoglobulin variable (V) genes? *IUBMB Life* **54** 305–7

Steele E J, Gorczynski R M and Pollard J W 1984 The somatic selection of acquired characters *Evolutionary Theory: Paths into the Future* ed J W Pollard (London: John Wiley) 217–37

Steele E J, Lindley R A and Blanden R V 1998 *Lamarck's Signature: How Retrogenes are Changing Darwin's Natural Selection Paradigm* (Reading, MT: Addison-Wesley-Longman)

Steele E J and Lloyd S S 2015 Soma-to-germline feedback is implied by the extreme polymorphism at IGHV relative to MHC *Bioessays* **37** 557–69

Whitelaw N C and Whitelaw E 2008 Transgenerational epigenetic inheritance in health and disease *Curr. Opin. Genet. Dev.* **18** 273–9

Zager A, Peron J P, Mennecier G, Rodrigues S C, Aloia T P and Palermo-Neto J 2015 Maternal immune activation in late gestation increases neuroinflammation and aggravates experimental autoimmune encephalomyelitis in the offspring *Brain, Behavior, and Immunity* **43** 159–71

Reprinted from the Transactions of AN INTERNATIONAL CONGRESS OF OPHTHALMOLOGY, *held in Washington, D. C., April 25 to 28, 1922*

THE PRODUCTION AND TRANSMISSION OF CERTAIN EYE DEFECTS[1]

PROF. M. F. GUYER
University of Wisconsin

By way of introduction to the discussion of eye defects, I wish to review briefly some points in the embryologic development of the eye. Although it will prove to be an old story to ophthalmologists, I feel, nevertheless, that by so doing I can get before you most effectively the materials I have to present.

Cleavage of the fertilized ovum, formation of the three fundamental germinal layers, and general embryogeny in the rabbit do not differ in any important ways from these same processes in other mammalian forms. Through the successive divisions which begin shortly after penetration of the ovum by the spermatozoon, a mulberry-like mass of cells enclosed by the *zona radiata* is built up. Some of the cells divide more rapidly than others, so that the resulting spherical mass comes to consist of a central group of larger, more granular cells surrounded by a superficial layer of smaller, clearer elements. Soon fluid appears between the central cells and the peripheral layer except at one side. As the liquid accumulates the entire mass becomes transformed into a fluid-filled vesicle consisting of a single layer of small transparent cells with the original central mass projecting from one side into the cavity. The outer layer, termed the *trophoblast,* is concerned only with the establishment of relations between the developing organism and the uterine mucosa. The inner mass is the part out of which the embryo is formed. At this stage the developing ovum is commonly termed the *blastodermic vesicle* or *blastocyst.*

Seen from without, the *germinal area* appears as a circular disc at the upper pole of the blastocyst. Within this disc the cells are rapidly shaping up into the two primitive germinal layers—*ectoderm,* and *entoderm.* By unequal growth the disc soon becomes oval, then more or less pear-shaped. At its smaller end a median denser streak, formed by a keel-like thickening of the ectoderm, appears. This is serologist has put important tools and ideas into the possession of the experimental biologist which may be utilized in new attacks upon certain fundamental biological problems.

The hemolysins, for example, discovered by Bordet in 1895, are now known to be special members of a general class of substances termed cytotoxins or cytolysins. For just as alien red blood-cells lead to the production of specific hemolysins, so various other materials, as leukocytes, nervous tissues, spermatozoa and crystalline lens—any foreign protein, in fact—when injected into the blood-stream of an unrelated species, will cause the formation of lytic substances more or less specific for the antigen used in the immunizing process. All cytolytic sera so far studied have been found to be more or less hemolytic, and it is probable that none acts exclusively

---

[1] Illustrated by lantern-slides and living animals.

upon its own antigen. While a particular cytolytic serum may affect some other tissues, it attacks the special tissue used as antigen much more vigorously.

Although presumably distinct from one another, the various classes of the so-called *antibodies*—precipitins, agglutinins, bacteriolysins, cytolysins or cytotoxins, etc.—seem to have many points of similarity, as, for instance, their method of origin, their reaction to heat, and, in some cases, their mode of operation. Chemically their natures are still unknown. Considerable evidence of their close association in some way with the euglobulin constituent of the blood is appearing in various recent researches.

To the biologist viewing this fascinating field, many questions arise. If, for example, it is possible to originate in living organisms antibodies which will destroy particular tissue-elements, is it not possible to secure similar selective action on certain parts of the developing embryo? May not serologic methods enable us to make a new attack upon the long-standing problem of the inheritance of somatic modifications, or that of provoking specific modifications in the germ through direct operation of external agents? If a special serum can be developed which will single out and destroy a certain element of the adult, is it not possible that there is sufficient constitutional identity between the mature substance of such a part and one or the other of its material antecedents in the germ, that the latter may also be influenced specifically by the serum in question? If external influences *can* be transmitted to the germ-cell, it is clear that in higher animals the one obvious means of conveyance is the blood.

In an attempt to find answers to certain questions of this kind I and my research associate, Dr. E. A. Smith, began various experiments some six years ago which we are still continuing. Among other things we undertook, by means of cytolysins, to produce antenatal effects in fetuses. Our main work in this direction has been on rabbits with fowl-serum immunized against rabbit-lens, although we have also experimented somewhat with mice and with guinea-pigs. I shall confine my discussion largely to certain eye-abnormalities we secured in fetal rabbits, and to the inheritance of such defects.

In our first experiments[2] the lenses of newly killed young rabbits were pulped thoroughly in a mortar and diluted with normal salt solution. About four cubic centimeters of this emulsion was then injected intravenously or intraperitoneally into each of several fowls. Four or five weekly treatments with such lens-emulsions were given. A week or ten days after the last injection the blood-serum of the fowls was ready for use. The rabbits had been so bred as to have their young advanced to about the tenth day of pregnancy, since from the tenth to the thirteenth day seems to be a particularly important period in the development of the lens. As we saw in reviewing the embryology of the eye, the lens is then growing rapidly and is surrounded by a rich vascular network that later disappears. From four to seven cubic centimeters of the immunized fowl-serum were injected intravenously into the pregnant rabbits at intervals of two or three days for from ten days to two weeks.

---

[2] Guyer and Smith, 1918, 1920.

A number of the rabbits died from the treatment and many young were killed in utero. Of sixty-one surviving young from mothers thus treated, four had one or both eyes conspicuously defective and five others had eyes that were clearly abnormal. It is possible that still others were more or less affected, as we judged only by conditions easily visible. In some of the descendants of this stock, indeed, ophthalmologists who have examined the eyes more thoroughly have pointed out defects which we had overlooked, and occasionally rabbits, that in their earlier months passed for normal, have later manifested defects in the lens or in other parts of the eye.

The commonest abnormality seen in both the original subjects and in their numerous descendants was partial or complete opacity of the lens (Plate I, Fig. 4), usually accompanied by reduction in size of the eye (Plate I, Fig. 2). In a few of our later strains in a different experiment, however, we have had several cases of enlargement of the eye, or *buphthalmia* (Plate I, Fig. 3). Among the rabbits I brought with me for demonstration there is one of this type which I shall be glad to have you examine. Other common defects which have appeared are cleft-iris, displacement of the lens, persistent hyaloid artery, bluish or silvery color instead of the characteristic pink of the albino eye, microphthalmia, and even almost complete disappearance of the eyeball. The cases of cleft-iris, or *coloboma,* range all the way from a narrow slit in the lower edge of the iris to a broad wedge- or U-shaped opening which amounts practically to the absence of the entire lower part of the iris. The cleft may be confined to the iris or it may extend back deeper into the eye. When one takes into account the early embryology of the eye, it is easy to see how such clefts result from failure of the choroidal fissure to close as it should do normally. The bluish or silvery color, I am told by ophthalmologists who have examined the rabbits, is due mainly to detachment of the retina. Here again, when one recalls the loose embryologic connection between the retinal layers of the eye and the outer coats, even in the

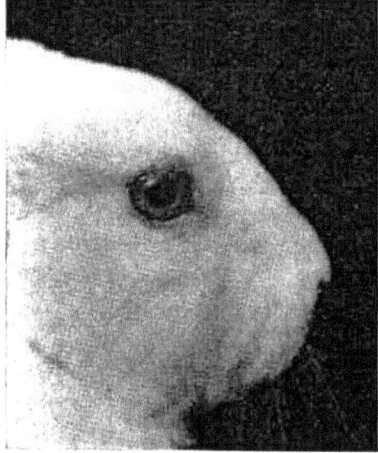

Fig. 1. Showing appearance of normal eye.

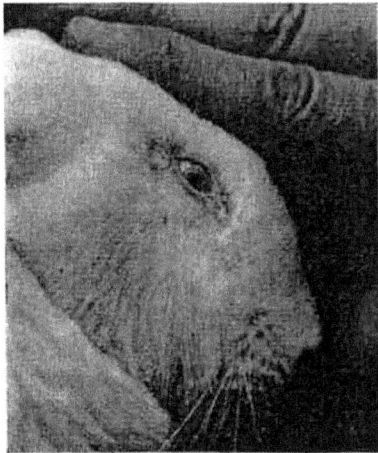

Fig. 2. Microphthalmia eye with cleft iris and opaque lens; eyeball rotated downward somewhat.

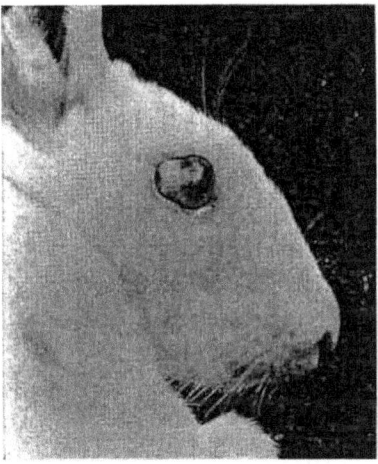

Fig. 3. Buphthalmic eye with staphylomatous sclera. The eyeball is so rotated backward that the edge of the cornea, is just visible at the upper outer angle of the lids; the lenses in both eyes are opaque.

normal eye, it is easy to see how almost any distortion of the eyeball, unevenness of growth, or accumulation of fluid might bring about such detachment.

Many of the eyes take abnormal postures (Plate I, Fig, 3). This is particularly true in some of our later strains. One or both eyes are likely to be strongly rotated downward or backward. The backward- rotation is carried to such an extreme in some cases that the cornea is visible only when the eyelids are drawn back at the outer corner (Plate I, Fig. 3), or occasionally when the animal attempts to roll its eyeball forward. In such eyes the exposed sclera in front usually bulges (*staphyloma*) and becomes transparent, simulating a cornea. When we first came across this

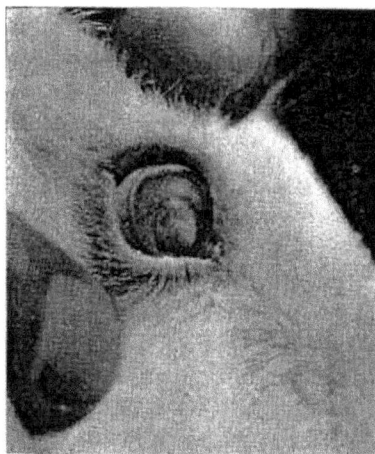

Fig. 4. Showing opaque lens and coloboma of the iris.

anomaly, in fact, we thought that we had a rabbit with a double eye on each side. I have brought one such individual with me for demonstration.

Taking into account the method of embryologic development—the relations of lens, optic cup and choroidal fissure—the defects, except those of the muscular attachment, are practically all such as might reasonably be attributed to arrests of development based upon early lens-defect. It is possible, to be sure, that we have developed antibodies against other eye-tissues as well as against the lens, since undoubtedly more or less of the aqueous humor and the vitreous body adhered to the lenses when we removed and pulped them for the original injections. Moreover, if proteins from other parts of the eye are ever in solution in the humors, they too may have been present in the antigen. Each individual protein, of course, has the capacity for engendering antibodies specific for itself. Even the lens is composed of at least four proteins: albuminoid (constituting the lens-fibers), alpha-crystallin, beta-crystallin and albumin. According to Jess and Reiss (Jess, 1920), in their study of the chemical changes which take place in cataract, alpha- and beta-crystallin, both soluble in water, make up the greater part of the lens of the young animal. These gradually decrease in quantity with age, accompanied by sclerosis—a process even more in evidence in cataractous lenses.

In some of our animals we find that an eye defective at birth, particularly if microphthalmic, may undergo further degeneration, characterized by collapse of the eyeball and resorption, so that the eyeball may eventually disappear entirely. The eyes of the mothers originally injected have always remained apparently unaffected. This is probably due to the fact that the lens-tissue of the adult is largely avascular, and that, therefore, the injected antibodies did not come into contact with it.

That the changes in the eyes of the fetuses resulted from the specific action of lens-antibodies is indicated by the fact that in the original experiment, in not one of the forty-eight controls obtained from mothers which had been treated with pure fowl-serum or with fowl-serum immunized to rabbit-tissues other than lens, was there any evidence of eye-defects. I may add that since then, among over five hundred young obtained from mothers which are being experimented upon for other purposes with various types of sera or protein-extracts, or with typhoid bacilli, just before or during pregnancy, not a single case of eye-defect has appeared. To one familiar with the results obtained by the experimental embryologist, which show how susceptible the eye is in early embryogeny to any kind of harmful influence, the natural inclination is to regard such abnormalities as due merely to a general poisonous or inhibitive effect, rather than to specific antibodies in the blood-serum. That lens-defects may be produced by general chemical or physical means is undeniable. I know of no case yet, however, where they have become inheritable. Bagg (1922), for example, has recently found that as a result of exposure of rats to radium emanation (gamma-ray radiation) during late pregnancy, some of the young, after birth, developed eye-defects. In his paper he gives photographs of an adult in which both lenses have become opaque and the left eyelids nearly closed. As a rule, such fetally irradiated young showed other marked defects, particularly of the nervous system, and were usually sterile.

Regarding our own rabbits I can only repeat that we have never obtained the defects in question except with serum carrying specific antibodies. In any event, should the effect have originally been a general rather than a specific one, it is obvious that, germinally considered, it must sooner or later have become specific, since the anomalous eye-condition appears generation after generation without any recognizable accompanying malformations of other parts of the body.

Before passing on to the question of inheritance, I may say that by way of control, for genetical studies, in addition to what we have termed our 3A1 line, we developed another line from wholly unrelated stock, our so-called 16A1 line. Moreover, we have established still a third strain, the 84 line, which was started, not by means of fowl-serum immunized to rabbit-lens, but by the use of pulped rabbit-lens intravenously injected directly into rabbits just before or during their pregnancy. In this last case the rabbit must herself have developed antibodies against the invading lens-material. Out of eleven different females so treated, in twenty-three matings, only one individual gave us young with abnormal eyes. These defects are of the same general nature as those secured by means of fowl-serum immunized to rabbit-lens, and they behave similarly in inheritance.

As already indicated, once the defect is secured, it may be transmitted to subsequent generations through breeding (Fig. 1). So far, in the 3A1 line, we have succeeded in passing it down through nine generations. There is no reason apparent why it will not go on indefinitely, since the imperfections tend to become worse in successive generations, and also to occur in a proportionately greater number of young. The same genetical conditions hold for the other lines, although because of their more recent origin, we have manifestly not been able to carry them through so many generations.

The transmission is not infrequently of an irregular unilateral type (Fig. 1), sometimes only the right, at others only the left, eye showing the defect. In this respect it resembles genetically such anomalies as brachydactyly or Polydactyly in

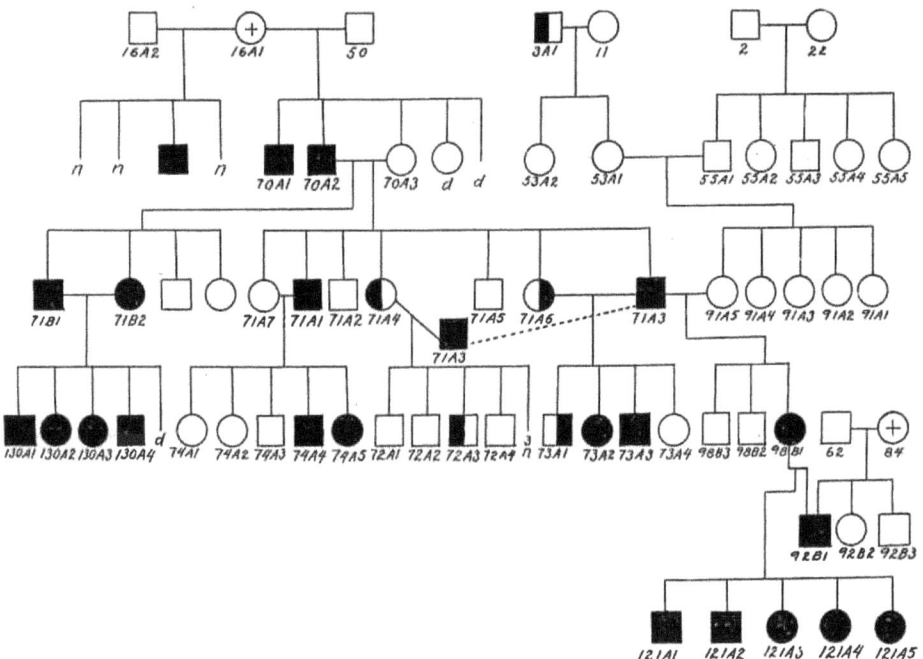

Fig. 1. Chart showing pedigree of some of the defective-eyed individuals. Only a few of the numerous matings are shown. The circle with the + sign in it indicates the female treated with lens-immunized fowl-serum (16A1) or directly with pulped rabbit-lens (84). The mother of 3A1 was treated with lens-immunized fowl-serum. Squares indicate males; circles, females; symbol all black, both eyes defective; right half black, right eye defective; left half black, left eye defective; unshaded, presumably normal-eyed; *d*, died; *n*, normal. Males 16A2, 50 and 2 (upper row) were of normal untreated stock, as were females 11 and 22. Male 62 (fourth row) was of normal untreated stock, while female 84 was injected directly with pulped lens.

man. In later generations there has been an increasing number of young with both eyes affected.

Though not analyzed completely as to its exact mode of inheritance, the abnormal condition has in general the characteristics of a Mendelian recessive. When either defective-eyed males or females are bred to normal-eyed individuals from other strains, for instance, only normal-eyed progeny result in the first generation, but the abnormal condition may be made to reappear in subsequent generations if appropriate matings are made. If we were dealing with a pair of simple Mendelian characters, the young from two individuals with the same recessive trait should all show this trait. Two of our defective-eyed rabbits, however, when bred together, are likely to produce some normal-eyed young. If, therefore, this inheritance is to be interpreted in terms of Mendelism, there is probably more than one pair of unit-factors involved.

To meet the objection that we are not getting instances of true inheritance but merely placental transmissions of antibodies or related substances from the bloodstream of the mother in each successive generation, we have established the descent

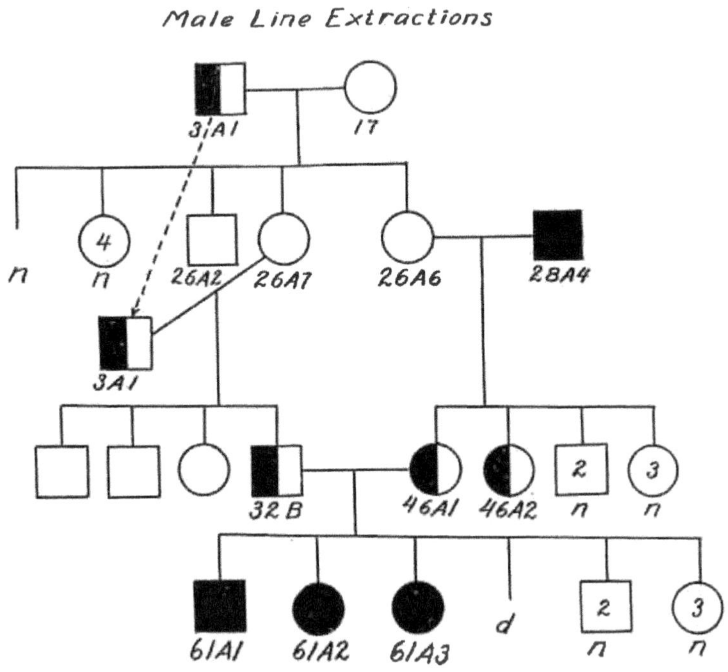

Fig. 2. Inheritance of the defects through the male line. It is plain that individuals of the 32B, 46A and 61A aeries could have derived their defects only from male ancestry originally, since female 17 was of normal and unrelated stock. Symbols same as in Fig. 1.

through the male line in a number of cases, one of which is represented in Fig. 2. To do this, females from strains of rabbits unrelated to our treated stock were mated to defective-eyed males. The first generation produced in this way was invariably normal-eyed; that is, the defective condition was recessive to normal condition. When, however, females of this generation were mated to defective-eyed males, or to normal-eyed males of similar derivation to themselves, the defects reappeared in some of the progeny, somewhat after the manner of an extracted Mendelian recessive. It is obvious that the normal condition could have been introduced into these new strains only through the germ-cells of the males, and that its transmission is, therefore, an example of true inheritance.

I feel that in establishing and developing from unrelated stock three different strains of defective-eyed rabbits—two (3A1 line and 16A1 line) by the use of fowl-serum immunized to rabbit-lens, the other (84 line) by direct injection of rabbit-lens into a pregnant rabbit—we have placed our results beyond the bounds of coincidence or chance. We can also cite further the production recently of similar lens-defect in the young of the guinea-pig, if need be, although we are not yet ready to report on this latter series of experiments.

To the biologist, perhaps the most interesting fact brought to light in these researches is the possibility of directly or indirectly inducing germinal changes by means of antibodies developed in an animal's own body against tissues taken from individuals of the same species. Such a result together with another I have obtained in inducing the male rabbit to develop spermatotoxins against its own spermatozoa (Guyer, 1922a), lend support to the idea that an animal can build antibodies against its own tissues when these are misplaced, altered or injured, and that such antibodies may so affect the germ-cells as to induce germinal changes. Since I have discussed this point rather fully in recent papers (Guyer, 1921, 1922b, 1922c), I need not enter into it here.

LITERATURE CITED

Child, C. M. Individuality in Organisms. University of Chicago Press, 1915. The Origin and Development of the Nervous System. University of Chicago Press, 1921.

Cole, W. H. The Transplantation of Skin in Frog Tadpoles, with Special Reference to the Adjustment of Grafts over Eyes and to Local Specificity of Integument. Jour. Exp. Zool., v. 35, no. 4, May, 1922.

Fischel, A. Weitere Mitteilungen über die Regeneration der Linse. Arch. f. Entw.-Mech., v. 15, 1902. Uber rückläufige Entwicklung, Arch. f. Entw.- Mech., v. 42, 1916.

Guyer, M. F. Immune Sera and Certain Biological Problems. Am. Nat., v. 55, Mar.-Apr., 1921. Studies on Cytolyains: Experiments with Spermatotoxins. Jour. Exp. ZooL, v. 35, No. 2, Feb., 1922. Serological Reactions as a Probable Cause of Variation, Am. Nat., v. 56, Jan.-Feb., 1922. Orthogenesis and Serological Phenomena. Am. Nat., v. 56, Mar.–Apr., 1922.

Guyer, M. F., and Smith, E. A. Studies on Cytolysins: Some Prenatal Effects of Lens Antibodies. Jour. Exp. Zool., v. 26, No. 1, May, 1918. Studies on Cytolysins: Transmission of Induced Eye Defects. Jour. Exp. Zool., v. 31, No. 2, Aug., 1920.

Herbst, C. Formative Reize in der tierischen Ontogenese, 1901.

Jess, A. Die Monoaminosauren der Linsenproteine. Ztschr. f. physiol. Chem., 110, 266, 1920.

King, H. D. Experimental Studies on the Eye of the Frog Embryo. Arch. f. Entw.-Mech., v. 19, 1905.

Lewis, W. H. Experimental Studies on the Development of the Eye in Amphibia. Am. Jour. Anat., v. 3, 1904. Experimental Studies, etc. ... On the Origin and Differentiation of the Lens. Am. Jour. Anat., v. 6, 1907. Lens Formation from Strange Ectoderm in R. Sylvatica. Am. Jour. Anat., v. 7, 1907.

Speman, H. Über Correlationen in der Entwickelung dea Auges. Verh. Anat. Ges., Anat. Anz., v. 19, Erganzungsbd., 1901. Über Linsenbildung bei defekter Augenblase. Anat. Anz., v. 23, 1903. Neue Versuche zur Entwicklung des Wirbeltierauges. Verh. d. deutsch. Zool. Ges. (Stuttgart), 1908. Zur Entwicklung des Wirbeltierauges. Zool. Jahrb., Abt. f. allg. Zool. u. Physiol., v. 32, 1912.

Stockard, C. R. The Development of Artificially Produced Cyclopean Fish. Jour. Exp. Zool., v. 6, 1909. The Independent Origin and Development of the

Crystalline Lens. Am. Jour. Anat., v. 10, 1910. An Experimental Study of the Optic Anlage in Amblystoma punctatum, with a Discussion of Certain Eye Defects. Am. Jour. Anat., v. 15, 1913. The Artificial Production of Structural Arrests and Racial Degeneration. Proc. N. Y. Path. Soc., N.S., v. 13, 1914. Developmental Rate and Structural Expression: an Experimental Study of Twins, Double Monsters and Single Deformities, etc. Am. Jour. Anat., v. 28, 1921.

Wachs, H. Neue Versuche zur Wolff'schen Linsenregeneration. Arch. f. Entw. -Mech., v. 39, 1914.

Werber, E. I. Critical Notes on the Present Status of the Lens Problem. Biol. Bull., v. 34, No. 4, April, 1918.

Case Study 8: Heritability Via the Cytoskeleton

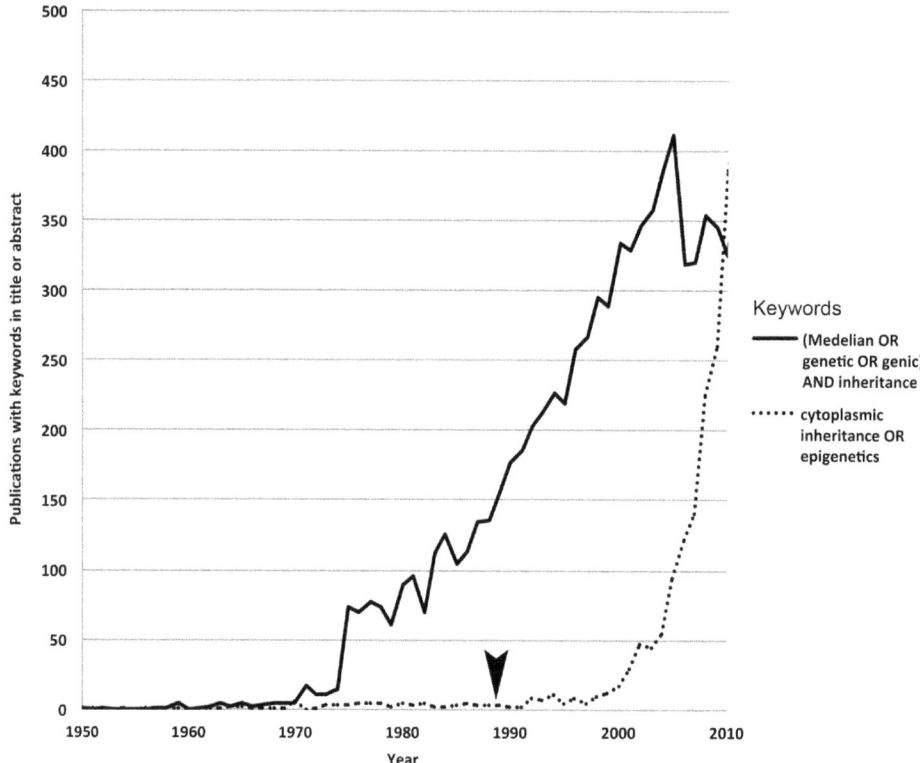

Non-genic inheritance of cellular handedness [46].

Perspective on *Non-Genic Inheritance of Cellular Handedness*

Jack Tuszynski
Professor
University of Alberta

Cystoskeleton as an Information-Processing and Memory Storage Structure in Cells

One of the key problems in biology is how cells store and process information, during their lifetimes and during propagation across organismal generations. The cytoskeleton is a central player in this process, contributing to information storage and processing in parallel to DNA, both in the process of cellular inheritance and the function of the brain.

How the brain works is still one of the great mysteries of modern science. The Human Brain Project, with strong EU funding aims to computationally simulate a brain, with each computer processor simulating millions of neurons. However, computer simulations of an individual neuron, let alone its intricate intra-cellular structure, are still out of reach. Similarly, the US has launched the Brain Research through Advancing Innovative Neurotechnologies (BRAIN) initiative to understand how the brain works. These projects focus on the brain above the level of the neuron leaving deeper questions still unresolved. Currently accepted theories in neuroscience represent information processing as signals propagating from neuron to neuron by synaptic connection. However, single-celled organisms demonstrate information processing at an intra-cellular level. It is, therefore, reasonable to assume that each neuron, and indeed each living cell, is as a plastic entity with its own complex network architecture that rewires itself due to input signal activities [1]. Indeed, single-celled organisms such as the paramecium have shown intelligent behaviour [2], pointing to the possibility of information processing within a single cell. Previous research indicates that the cytoskeleton, in general and the dendritic cytoskeleton of the neuron in particular, functions as a computational device [2]. Also, increasing evidence supports electrical and computational properties of the cytoskeleton [3], which is summarized below.

GMicrotubules (MTs) are unique among cytoskeletal structures in that they are cylindrical, highly dynamic in conformation and assembly, and supermolecular. The cylindrical walls of the tubulin dimers that form the MTs also display the seams of protofilaments, the physiological role of which is still largely unknown. The MT nanopores periodically found on the MT. surface provide an open passage to the MT lumen and are relevant to the action of clinically important drugs. For example, the nanopores are a molecular target for cyclostreptin and are important for Taxol to access its binding site in the MT lumen. It has been previously shown that MTs behave like biomolecular transistors arranged in nonlinear transmission lines, which are capable of processing and amplifying electrical signals, and also demonstrated that the presence of local Ca^{2+} ions modifies the amplifying capacity of MTs [4]. MTs provide shape to cells and act as conveyor belts, ferrying molecular cargo throughout cells. They are among the most rigid cellular structures and their unique property of dynamic instability, whereby they are able to grow and shrink rapidly, enables them to affect cellular shape changes and perform chromosome segregation in cell division. It has been recently hypothesized that MTs are involved in computation and memory storage [5, 6] by postulating a plausible mechanism for encoding synaptic memory at a subcellular level in MTs [6]. The standard experimental model for neuronal memory involves long-term potentiation where intense, pre-synaptic excitation results in prolonged post-synaptic sensitivity. The enzyme calcium/calmodulin-dependent protein kinase II (CaMKII) is a key player in this process. Calcium ions entering post-synaptic neurons activate CaMKII and cause it to phosphorylate a substrate, and thus encode one 'bit' of synaptic information. Computer modelling has shown that each CaMKII can encode 6 bits of information in an MT lattice. This allows for an enormous information storage capacity of the brain at extremely low energy cost, vastly superior to the most powerful supercomputers today. This mechanism explains how MTs can influence axonal firings, regulate synapses, and affect differentiation and various functions of cells. This could provide a real-time, interactive code in contrast to the static genetic information stored in DNA. Decoding and interfacing to MT bits could enable therapeutic intervention in a host of pathological processes, such as Alzheimer's disease, in which MT disruption plays a key role, as well as brain injury, with respect to which MTs can repair and regenerate neurons and synapses [7]. A recent review paper summarized the numerous attempts to characterize the conductivity of MTs and actin filaments (AFs) [3]. Since MTs are highly electronegative, they can amplify ionic waves support electrical signals through the outer counterionic cloud surrounding the MT, which can propagate through the lumen and through the protein lattice itself. Recent work supports the possibility that, at certain electromagnetic excitation frequencies, MTs act as semiconductors [8, 9], a finding which indicates that MTs may act in a coherent manner to support cognition and consciousness [10]. Although all MTs have very similar morphology, they are marked on their outside surface with a variety of chemical groups via PTMs, mainly to their C-termini tails. These markers impact a cell's activity either by changing the stability of MTs, thus affecting cell shape, or by repositioning molecular cargo travelling on the MTs. The tubulin subunits in an MT lattice may vary in genetic isoform, as well as in PTMs. MT lattices are thus mosaics of tubulin states, which are based on both genetic and learned information ('nature and nurture') [11, 12]. On this pre-programmed background, molecular automata and computation in MT lattices may occur.

Regarding memory storage, dendritic and cell body MTs are far more stable than either synaptic membrane components or MTs in axons or non-neuronal cells (because of PTMs and specialized proteins such as MAPs, which prevent disassembly), and are thus capable of storing at least short-term memory. Long-term memory may require further encoding, e.g., into extremely stable neurofilaments. One of the signatures of damaged cells in blunt trauma and neurodegenerative disease is a change in the pattern of post-translational markers [7, 13, 14]. Each PTM of the MT lattice essentially alters the electromagnetic profile of the polymer, and therefore the electrostatic profile of the surface. It is well known that both MTs and AFs have fields around them, including those due to ion potential concentrations [5], which can affect stability, orientation, and interaction with motor proteins. Additionally, flow or hopping of electrons through the material will also be altered. Such electron movement can act as a signal transmission and may be used to write a bit of memory either by electro-reduction of a disulfide or redox action on a metal ion embedded within the lattice. These same electron currents can be used to perform logical operations and process information.

The fundamental questions of how a single neuron gains a specific structure for a specific function and what are the neuronal computational capabilities are still unresolved. MTs, being essential for cellular polarization, migration and growth, exhibit unique structural and physical properties. The neuronal MTs form dense parallel arrays in axons and dendrites that are required for the growth and maintenance of these neurites. MT organization differs between axons and dendrites. First, axonal MTs have uniform orientation, with their plus ends facing the axon tip, whereas dendritic MTs have mixed orientation, with their plus ends facing either the cell body or the dendritic tip [15]. Second, MTs differ in their complement of MAPs: for example, MAP2 is found mostly in dendrites and tau is found mainly in axons [16]. Modulation of MTs structure is important for differentiation of immature processes into axons or dendrites. MTs serve as tracks for directed transport and as transducers of force generated by molecular motors, which control neuronal morphology and function [17–20].

Neurons in development and growth react to the extracellular environment, which in turn may trigger electrical activity and morphological formation. It has been shown that external stimulation—electrical, mechanical or chemical—can alter neuronal organization [21–24]. A key mechanism in neuronal growth is the ability of the sensory-motile growth cones at the tips of growing processes to measure long- and short-range environmental cues and use them to develop accordingly [25–27]. A potent environmental cue is physical. A recently developed paradigm, 'substrate-cytoskeletal coupling' model, links MTs to the response of neurons to the external cues as well [28, 29]. According to this model, growth cones can move forward if they are capable of coupling intracellular motility signals to a fixed extracellular translocation substrate via cell surface adhesion receptors [29]. This makes topography a key factor in neuronal growth and an important method for stimulating neurons [30–37]. Intracellular investigations of down-stream effectors revealed a large group of cell adhesion molecules that are involved in axonal guidance such as focal adhesion kinase (FAK) [38] and several MT-associated genes such as Mtap1b, Mtap2b, Tubulin beta 2B, MPA MT stabilizers and cross-linking factors. The importance of MT structure is highlighted by identification of mutations in tubulin genes and genes encoding MT-related proteins in patients showing brain defects or disorders [17–20].

Neuronal growth is also critically influenced by the level of neuronal connectivity and activity [37]. Dr. Shefi's group has shown that neurons in culture significantly change their growth strategy following recognition and targeting of neighbouring cells. Isolated, non-contacted, neurons tend to elaborate their dendritic tree. In contrast, following the establishment of contact between neurons, branching strategy is changed, favouring higher efficiency in neurite volume [36, 37, 39]. Extracellular recordings of cultured neuronal populations have demonstrated, that the pattern of neuronal activity is also self-regulated following establishment of interconnections [40]. However, the mechanism underlying the regulation of growth strategy is still unresolved.

We hypothesize that the nanopores of the MT wall are indeed 'ion channel-like' structures [41], the function of which may be altered by the presence of local ionic species, and/or MAPs that modify surface charges, and/or the stability of MTs would modulate electrical transmission through MTs, and consequently regulate their computational capabilities. It has been demonstrated that the nanopore dimensions (10 × 40 Å) and an electric potential gradient (approx. 100 mV) are similar to membrane IC characteristics. This provides grounds to assume that they perform useful roles in the cell, but thus far no empirical knowledge has been gained with respect to this connection. The MTs arrangement within axon and dendrites reflects (1) the morphology of neurons and distribution of adhesion sites, affected also by topography and (2) responds to the level of activity as MTs act as 'ion channel-like' structures. The MTs could modulate electrical transmission through MTs, and consequently regulate their computational capabilities with correlation to the geometry. We should also take into account the ability of electrical fields to direct neuronal growth.

Most current models of axonal elongation suggest that MTs, which are a major component of the cytoskeleton, are important for axonal guidance [16, 42, 43], and that MTs assembly in the growth cone is critical for axonal lengthening [16, 44, 45]. Electrical stimulation has been previously shown to affect neuronal growth in culture. Axon and dendrites are promoted and directed according to the electrical field direction. Neuronal connectivity, which usually is followed by functional interconnection is also important, affecting neuronal growth strategy with a significant change of morphology following neuron–neuron contact. The MTs and AFs form the cytoskeleton of the neuron along development and regeneration. Correlating the cytoskeleton elements organization with the electrical stimulation may advance our understanding of neuronal response to the stimulation and may link neuronal morphology and function. It can also be used to explain a numer of intriguing biological phenomena. For example, ciliates exhibit an asymmetry in arrangement of surface structures around the cell termed handedness [46]. The usual organization of cellular structures is defined to be 'right-handed' (RH) while the reversed organization is called 'left-handed' (LH). An example of LH forms are those produced in *Tetrahymena thermophila* and such forms were genetically analyzed for nuclear genetic changes [46] indicating non-genetic origin of the change in handedness. The authors of this study [46] pointed to direct cortical perpetuation as the most likely cause of this effect implicating the cytoskeleton as an information-bearing subcellular structure which is consistent with the arguments present in our paper.

Moreover, embryogenesis and stem cell differentiation are known to utilize epigenetic mechanisms to give rise to the fully developed organism but the spatiotemporal coordination of cells during morphogenesis is still a major scientific enigma. The cell state splitter and differentiation wave model can shed light on experimental findings for embryonic stem cell, pluripotency and differentiation [47] by introducing a bistable cytoskeletal mechanism which once again points to an important role in both the mechanics and information processing (decision making in this case) played by MTs and AFs.

References

[1] Abbott A 2013 Neuroscience: solving the brain *Nature* **499** 272–4
[2] Armus H L, Montgomery A R and Gurney R L 2006 Discrimination learning and extinction in paramecia (P. caudatum) *Psychol. Rep.* **98** 705–11
[3] Friesen D E, Craddock T J A, Kalra A P and Tuszynski J A 2015 Biological wires, communication systems, and implications for disease *BioSystems* **127** 14–27
[4] Priel A, Ramos A J, Tuszynski J A and Cantiello H F 2006 A biopolymer transistor: electrical amplification by microtubules *Biophys J* **90** 4639–43
[5] Woolf N J, Priel A and Tuszynski J A 2009 *Nanoneuroscience: structural and functional roles of the neuronal cytoskeleton in health and disease.* (New York: Springer)
[6] Craddock T J A, Tuszynski J A and Hameroff S 2012 Cytoskeletal signaling: is memory encoded in microtubule lattices by CaMKII phosphorylation? *PLoS Comput Biol* **8** e1002421
[7] Craddock T J A, Tuszynski J A, Chopra D, Casey N, Goldstein L E, Hameroff S R and Tanzi R E 2012 The zinc dyshomeostasis hypothesis of Alzheimer's disease *PLoS ONE* **7** e33552
[8] Sahu S, Ghosh S, Hirata K, Fujita D and Bandyopadhyay A 2013 Multi-level memory-switching properties of a single brain microtubule *Appl. Phys. Lett.* **102** 123701
[9] Sahu S, Ghosh S, Ghosh B, Aswani K, Hirata K, Fujita D and Bandyopadhyay A 2013 Atomic water channel controlling remarkable properties of a single brain microtubule: correlating single protein to its supramolecular assembly *Biosens Bioelectron* **47** 141–8
[10] Hameroff S and Penrose R 2014 Consciousness in the universe: a review of the "Orch OR" theory *Phys. Life Rev.* **11** 39–78
[11] Hameroff S R 1987 *Ultimate computing: biomolecular consciousness and nanotechnology* (North-Holland: Elsevier Science)
[12] Verhey K J and Gaertig J 2007 The tubulin code *Cell Cycle* **6** 2152–60
[13] Janke C and Chloë Bulinski J 2011 Post-translational regulation of the microtubule cytoskeleton: mechanisms and functions *Nat. Rev. Mol. Cell Biol.* **12** 773–86
[14] Garnham C P, Vemu A, Wilson-Kubalek E M, Yu I, Szyk A, Lander G C, Milligan R A and Roll-Mecak A 2015 Multivalent Microtubule Recognition by Tubulin Tyrosine Ligase-like Family Glutamylases *Cell* **161** 1112–23
[15] Baas P W, Deitch J S, Black M M and Banker G A 1988 Polarity orientation of microtubules in hippocampal neurons: uniformity in the axon and nonuniformity in the dendrite *Proc Natl Acad Sci USA* **85** 8335–9
[16] Conde C and Cáceres A 2009 Microtubule assembly, organization and dynamics in axons and dendrites *Nat. Rev. Neurosci.* **10** 319–32
[17] Keays D A *et al* 2007 Mutations in alpha-tubulin cause abnormal neuronal migration in mice and lissencephaly in humans *Cell* **128** 45–57
[18] Jaglin X H *et al* 2009 Mutations in the beta-tubulin gene TUBB2B result in asymmetrical polymicrogyria *Nat. Genet.* **41** 746–52

[19] Jaglin X H and Chelly J 2009 Tubulin-related cortical dysgeneses: microtubule dysfunction underlying neuronal migration defects *Trends Genet.* **25** 555–66
[20] Tischfield M A, Cederquist G Y, Gupta M L Jr and Engle E C 2011 Phenotypic spectrum of the tubulin-related disorders and functional implications of disease-causing mutations *Current Opinion in Genetics & Development* **21** 286–94
[21] Wood M and Willits R K 2006 Short-duration, DC electrical stimulation increases chick embryo DRG neurite outgrowth *Bioelectromagnetics* **27** 328–31
[22] Park J S, Park K, Moon H T, Woo D G, Yang H N and Park K-H 2009 Electrical Pulsed Stimulation of Surfaces Homogeneously Coated with Gold Nanoparticles to Induce Neurite Outgrowth of PC12 Cells *Langmuir* **25** 451–7
[23] Lowery L A and Van Vactor D 2009 The trip of the tip: understanding the growth cone machinery *Nat Rev Mol Cell Biol.* **10** 332–43
[24] Koppes A N, Nordberg A L, Paolillo G M, Goodsell N M, Darwish H A, Zhang L and Thompson D M 2014 Electrical stimulation of schwann cells promotes sustained increases in neurite outgrowth *Tissue Eng. Part A* **20** 494–506
[25] Whitington P M 1993 Axon guidance factors in invertebrate development *Pharmacol. Ther.* **58** 263–99
[26] Tessier-Lavigne M and Goodman C S 1996 The molecular biology of axon guidance *Science* **274** 1123–33
[27] Huber A B, Kolodkin A L, Ginty D D and Cloutier J-F 2003 Signaling at the growth cone: ligand-receptor complexes and the control of axon growth and guidance *Annu. Rev. Neurosci.* **26** 509–63
[28] Rajnicek A and McCaig C 1997 Guidance of CNS growth cones by substratum grooves and ridges: effects of inhibitors of the cytoskeleton, calcium channels and signal transduction pathways *J. Cell Sci.* **110** Pt 23 2915–24
[29] Suter D M and Forscher P 2000 Substrate-cytoskeletal coupling as a mechanism for the regulation of growth cone motility and guidance *J. Neurobiol.* **44** 97–113
[30] Mahoney M J, Chen R R, Tan J and Saltzman W M 2005 The influence of microchannels on neurite growth and architecture *Biomaterials* **26** 771–8
[31] Lee J W, Lee K S, Cho N, Ju B K, Lee K B and Lee S H 2007 Topographical guidance of mouse neuronal cell on SiO2 microtracks *Sensors and Actuators B: Chemical.* **128** 252–7
[32] Hwang H, Kang G, Yeon J H, Nam Y and Park J-K 2009 Direct rapid prototyping of PDMS from a photomask film for micropatterning of biomolecules and cells *Lab. Chip.* **9** 167–70
[33] Fozdar D Y, Lee J Y, Schmidt C E and Chen S 2010 Hippocampal neurons respond uniquely to topographies of various sizes and shapes *Biofabrication* **2** 035005
[34] Xie C, Hanson L, Xie W, Lin Z, Cui B and Cui Y 2010 Noninvasive neuron pinning with nanopillar arrays *Nano. Lett.* **10** 4020–4
[35] Fricke R, Zentis P D, Rajappa L T, Hofmann B, Banzet M, Offenhäusser A and Meffert S H 2011 Axon guidance of rat cortical neurons by microcontact printed gradients *Biomaterials* **32** 2070–6
[36] Baranes K, Chejanovsky N, Alon N, Sharoni A and Shefi O 2012 Topographic cues of nano-scale height direct neuronal growth pattern *Biotechnol. Bioeng.* **109** 1791–7
[37] Baranes K, Kollmar D, Chejanovsky N, Sharoni A and Shefi O 2012 Interactions of neurons with topographic nano cues affect branching morphology mimicking neuron–neuron interactions *J. Mol. Histol.* **43** 437–47

[38] Rico B, Beggs H E, Schahin-Reed D, Kimes N, Schmidt A and Reichardt L F 2004 Control of axonal branching and synapse formation by focal adhesion kinase *Nat. Neurosci.* **7** 1059–69
[39] Shefi O, Golding I, Segev R, Ben-Jacob E and Ayali A 2002 Morphological characterization of in vitro neuronal networks *Phys. Rev. E Stat. Nonlin. Soft Matter. Phys.* **66** 021905
[40] Ayali A, Fuchs E, Zilberstein Y, Robinson A, Shefi O, Hulata E, Baruchi I and Ben-Jacob E 2004 Contextual regularity and complexity of neuronal activity: From stand-alone cultures to task-performing animals *Complexity* **9** 25–32
[41] Freedman H, Rezania V, Priel A, Carpenter E, Noskov S Y and Tuszynski J A 2010 Model of ionic currents through microtubule nanopores and the lumen *Phys. Rev. E* **81** 051912
[42] Tanaka E M and Kirschner M W 1991 Microtubule behavior in the growth cones of living neurons during axon elongation *J. Cell Biol.* **115** 345–63
[43] Sabry J H, O'Connor T P, Evans L, Toroian-Raymond A, Kirschner M and Bentley D 1991 Microtubule behavior during guidance of pioneer neuron growth cones in situ *J. Cell Biol.* **115** 381–95
[44] Dent E W and Gertler F B 2003 Cytoskeletal dynamics and transport in growth cone motility and axon guidance *Neuron.* **40** 209–27
[45] Xie C, Lin Z, Hanson L, Cui Y and Cui B 2012 Intracellular recording of action potentials by nanopillar electroporation *Nat Nanotechnol.* **7** 185–90
[46] Nelsen E M, Frankel J and Jenkins L M 1989 Non-genic inheritance of cellular handedness *Development* **105** 447–56
[47] Lu K, Cao T and Gordon R 2012 A cell state splitter and differentiation wave working-model for embryonic stem cell development and somatic cell epigenetic reprogramming *BioSystems* **109** 390–6

Non-genic inheritance of cellular handedness

E. MARLO NELSEN, JOSEPH FRANKEL and LESLIE M. JENKINS

Department of Biology, University of Iowa, Iowa City, Iowa 52242, USA

Summary

Ciliates exhibit an asymmetry in arrangement of surface structures around the cell which could be termed handedness. If the usual order of placement of structures defines a 'right-handed' (RH) cell, then a cell with this order reversed would be 'left-handed' (LH). Such LH forms appear to be produced in Tetrahymena thermophila through aberrant reorganization of homopolar doublets back to the singlet condition. Four clones of LH forms were selected and subjected to genetic analysis to test whether this drastic phenotypic alteration resulted from a nuclear genetic change. The results of this analysis indicate that the change in handedness is not due to a genetic change in either the micronucleus or macronucleus. The LH form can, under certain circumstances, revert to the RH form, but typically it propagates itself across both vegetative and sexual generations with similar fidelity.

While this analysis does not formally rule out certain possibilities of nuclear genic control involving regulatory elements transmitted through the cytoplasm, when the circumstances of origin and propagation of the LH condition are taken into account direct cortical perpetuation seems far more likely. Here we outline a conceptual framework centred on the idea of longitudinally propagated positional information; the positive evidence supporting this idea as well as further application of the idea itself are presented in the accompanying paper.

Key words: pattern formation, inheritance, cytoplasmic, ciliates, patterning.

Introduction

Cytoplasmic localizations are of great importance in embryonic development. These localizations typically are a consequence of the action of the maternal genome in the oocyte or newly fertilized egg. A classic example of such maternal predetermination is the control of the direction of asymmetry of the body and shell of gastropod molluscs (Sturtevant, 1923; Freeman & Lundelius, 1982). In this case, the effect is exerted by way of a stored cytoplasmic product that functions shortly after fertilization to bring about a reversal in the asymmetry of early cleavage that in turn affects the subsequent macroscopic asymmetry (see Freeman & Lundelius, 1982, and references cited therein). In *Drosophila,* a reversal of polarity of up to one-half of the egg can be brought about by mutants at one of several *bicaudal* loci which involve a maternal predetermination that functions before fertilization (Nüsslein-Volhard, 1977; Mohler & Wieschaus, 1986).

In contrast to the maternal predetermination characteristic of embryonic organization, ciliates have been among the classic exemplars of true cytoplasmic inheritance (reviews: Nanney, 1983; Sapp, 1987). The majority of these cases have turned out to be effects of nucleic acids packaged within symbionts or organelles, or to selfsustaining feedback loops (Nanney, 1983; Preer, 1988). There remains, however, a residuum of cytoplasmically inherited conditions in ciliates that has not yet yielded to the hegemony of nucleic acids. This involves the inheritance of patterns of structural organization. The two classic examples are the inheritance of 180°-rotated (inverted) ciliary rows in *Paramecium* (Beisson & Sonneborn, 1965), which was subsequently observed in *Tetrahymena* (Ng & Frankel, 1977), and the inheritance of the homopolar-doublet condition (Sonneborn, 1963). These examples all involve inherited differences in the number or spatial orientation of structures (or sets of structures) that show normal internal organization.

Recently, investigators in several laboratories have found ciliate forms in which a large part of the cell surface exhibits an arrangement of structures that is a mirror image of the normal, much as in *bicaudal Drosophila*. These include mirror-image *janus* forms in *Tetrahymena* (Jerka-Dziadosz & Frankel, 1979) and mirror-image doublets in *Stylonychia* and its close relatives (Tchang *et al.* 1964). In these forms, only the global arrangement of structures is reversed, while the local architecture within each structure is not reversed (Grimes *et al.* 1980; Jerka-Dziadosz, 1981, 1983; Frankel *et al.* 1984). However, there is no apparent uniformity in the origins of these mirror-image forms: while the *janus* condition in *Tetrahymena* results from the action of recessive mutations at particular gene loci (Frankel & Jenkins, 1979; Frankel & Nelsen, 1986*b*; Frankel *et al.* 1987), mirror-image doublets in *Stylonychia* are generated by microsurgical operations carried out on normal wild-type cells (Tchang *et al.* 1964; Shi, personal communication).

Still more recently, phenocopies of the *janus* configuration were discovered in *Tetrahymena* cells regulating from a homopolar-doublet to a normal singlet condition (Frankel & Nelsen, 1986*a*). In the course of that investigation, occasional *reverse* singlets were observed (Nelsen & Frankel, 1986), presumably resulting from preservation of the 'wrong' handedness during regulation from a *janus*-like condition to a singlet state. Contrary to early expectations (Frankel, 1984), such 'left-handed' cells could feed and grow, allowing selection and maintenance of clones of these cells. At about the same time, Suhama generated a clone of reverse singlet cells in *Glaucoma* (a close taxonomic relative of *Tetrahymena*) by longitudinal transection of mirror-image doublets (Suhama, 1985). In that case, the 'left- handed' singlet was clearly of non-genic origin, but the origin of the preceding mirror-image doublet clone was unknown (Suhama, 1982).

The capacity to clone and maintain cells with a reversed handedness in arrangement of structures in *Tetrahymena thermophila,* a genetically domesticated ciliate, provided an opportunity for analysis of the inheritance of these differences. This paper is a report of such a genetic analysis. Here we present evidence for the conclusion that the difference in cellular handedness is not due to a difference in nuclear genes. Since the emphasis of this paper is genetic, the reversed phenotype itself will be described schematically without supporting documentation, and only the general concept underlying our developmental model will be presented in the Discussion. The accompanying paper (Nelsen & Frankel, 1989) will provide the details of the phenotypic differences between 'right-handed' and 'left-handed' cells, the consequences of these differences, and a reconstruction of the detailed course of events during changes in cellular handedness.

Materials and methods

Stocks, media, general procedures

Tetrahymena thermophila was used in this study. Wild-type cultures were from the 20th inbred generation, established in 1979 (B2079). Mutant stocks used were IA-104, IA-121, IA- 267 and IA-330. IA-104 and IA-121 are homozygous for recessive, temperature-sensitive mutations, *cdaA1* and *cdaC2,* respectively, in which cell division is arrested at restrictive temperatures (Frankel *et al* 1976). IA-267 and IA- 330 both are 'homozygous functional heterokaryon' (Bruns & Brussard, 1974) stocks, homozygous for cycloheximide resistance (dominant) in the micronucleus and cycloheximide sensitivity in the macronucleus [*ChxA2/ChxA2* (cycl.-s)], and of mating types III and V, respectively. In addition, a B* (B-star)-VII stock (Doerder & Berkowitz, 1987), with a defective micronucleus that does not contribute a pronucleus during conjugation, was obtained from Dr F. P. Doerder.

Stocks were routinely maintained axenically at 19 °C in 5 ml tubes containing 1 % proteose peptone plus 0.1 % bacto yeast extract (1% PPY). Homopolar doublets (see below) were maintained in 2 % proteose peptone plus 0.5 % bacto yeast extract (2% PPY). A richer medium (PPYGFe, described in Nelsen *et al.* 1981) was used to screen for LH (reverse) cells and to maintain them after selection. Dryl's salt solution, made up as described by Nelsen & DeBault (1978), was used to prepare cells for conjugation.

The standard technique for carrying out conjugation involved washing cells three times and resuspending them overnight in Dryl's salt solution before mixing with appropriate mating types in non-shaking cultures at 30°C. Cell densities for conjugation were 8–20×10^4 cells ml^{-1}. Isolated conjugating pairs were kept at room temperature. These basic procedures were modified for specific purposes, as indicated below.

Assessment of cell-surface geometry was carried out using two silver-staining procedures: Chatton-Lwoff silver impregnation using the procedure of Frankel & Heckmann (1968) as modified by Nelsen & DeBault (1978), and protein-silver (protargol) staining following the general methods of Ng & Nelsen (1977) with the improvement of Aufderheide (1982) except that a thick albumin film was used.

Procedures for selecting LH clones

All LH clones were obtained from homopolar doublets (RH cells fused side by side). Homopolar doublet stocks were constructed using three techniques. The first two have been described previously (Nelsen & Frankel, 1986). In brief, the doublets from which the LH1 clone was obtained were selected from a vegetative culture of stock IA-104 [*cdaA1/cdaA1* (II)] after subjecting it to restrictive conditions (36 °C for 2h) which arrest cleavage; the doublets that served as the foundation for the LH2 clone were selected from wild-type cells [B2079 IV×VI, locally constructed stocks which are isogenic but of different mating types] fused during conjugation using immobilization antiserum. The doublets used to select the LH3 and LH4 clones were obtained from the homozygous functional heterokaryon stocks IA-267 and IA-330 using a new procedure. Cells of these two stocks were allowed to conjugate for 5 h in Dryl's' salt solution at 30 °C. Conjugation then was blocked by subjecting the culture to 40 °C for 1 h and 39 °C overnight in Dryl's solution. The cells were fed with 2 % PPY the next day and 'stuck' pairs were isolated after 3 to 4 h of growth at room temperature. Many of these yielded doublet cultures. Subcultures of the doublet clones selected were tested for fertility and retention of their heterokaryon character. All doublet clones were originally isolated by visual selection, and subsequently subcloned for weeks to months by selection of doublets every 2 to 3 days. Details of this procedure were reported previously (Nelsen & Frankel, 1986).

To obtain LH singlets, the 'Poisson lottery' techniques of Orias & Bruns (1976) were employed to dispense samples of doublet subclones into microtitre plates in PPYGFe medium containing antibiotics (1·4 g penicillin G potassium and 2·2 g streptomycin sulphate per litre). After growth for 4 to 7 days at room temperature or 3 days at 25–27°C, plates were screened to select for clones showing slow growth. Slow growth was expected because LH cells commonly have defective oral structures. The slow-growing clones were subcloned and samples stained to check for LH (reverse) geometry. Growth for 3 days at 25–27°C proved to be optimal for finding these cells. Most of the slow-growing clones were not reversed; however, with experience one learns to recognize LH cells by their swimming behaviours: they often 'hesitate' and/or exhibit a 'twisty' spiralling during forward swimming.

LH clones were maintained by weekly subcloning in an 'Edgegard' (Baker Co., Sanford, Maine, USA) laminar flow hood into tubes containing 5 ml of sterile PPYGFe (without antibiotics) kept at room temperature. For long-term maintenance, stocks were kept in liquid nitrogen using the procedures of Simon & Flacks with minor modifications (1975).

Crosses involving LH clones
Crosses using stocks carrying temperature-sensitive cell-division arrest mutations (IA-104 and LA-121) were carried out using standard procedures (see above). In these LH×RH and control RH×RH crosses, numerous pairs were seen in all cultures one day after mixing. These cultures were maintained at 30°C, and, after three days, nutrients (2% PPY) were added. Subsequently, cultures were shifted to 39°C, and scored for the presence of growing cells 3 days later.

Crosses between LH1 and LH2 stocks were carried out using standard procedures with one exception. To promote mating (costimulation) in the LH cells, a small proportion (1:10) of a third cell culture (B*VII) was added. These are RH cells that lack a functional micronucleus. In these crosses, pairs were isolated, grown in PPYGFe at room temperature and tested for capacity to mate with mating type testers within 15 fissions after conjugation. All 'mature' conjugating lines were presumed to have retained their old macronuclei following an abortive conjugation (Allen & Gibson, 1973).

Crosses between the LH3 and LH4 stocks were carried out by the same general procedures used in the LH1×LH2 crosses. Two subclones of each of the LH parents were prepared for mating and B* costimulators were added to the mating mixture. However, instead of isolating pairs, cyclo-heximide dissolved in nutrient (2% PPY) medium (final concentration of 30 μg ml^{-1}) was added 32 h after mixing. Surviving LH cells were isolated, grown in PPYGFe and tested as illustrated in Fig. 3 and described in the Results.

Results

Selection of LH clones

Four clones of singlet cells displaying an apparent right-left reversal of global asymmetry were isolated from doublet cultures using a screening protocol which selects for slow growth. The first clone of reverse singlets (designated LH1) was isolated in 1983 from a doublet culture produced through a temporary fission arrest in cells homozygous for the temperature-sensitive cell-division-arrest mutation *cdaA1*. This conditional mutation, under permissive conditions, has no known effect

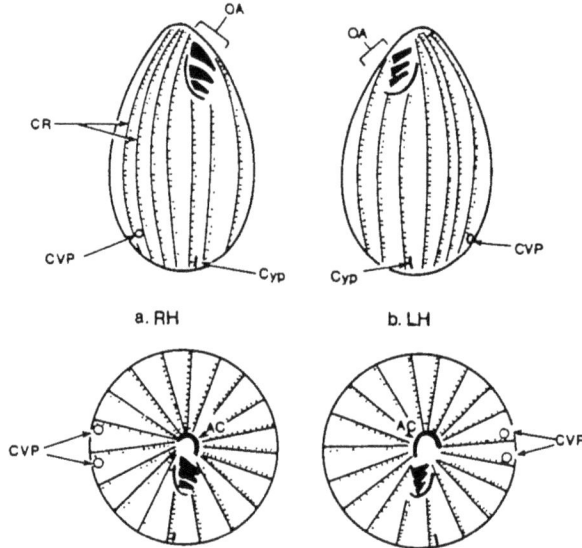

Fig. 1. Diagrams of the cell-surface organization of (a) 'right-handed' (RH) and (b) 'left-handed' (LH) cells of *Tetrahymena thermophila*. Schematic ventral views are shown in the top row, polar projections in the bottom row. In the polar projections, the anterior pole is at the centre, the posterior pole at the margin. The diagrams show the ciliary rows (CR), with longitudinal microtubule bands (lines) always to the cell's right (viewer's left) of the basal bodies (dots). Both RH and LH cells have two postoral ciliary rows, located posterior to the oral apparatus (OA). The cytoproct (Cyp, cell anus) is found along the right postoral ciliary row in RH cells, and along the left postoral row in LH cells, but always to the cell's *left* of the respective postoral ciliary rows. Similarly, the CVPs are in mirror-image positions relative to the OAs in RH and LH cells, but always just to the cell's left of ciliary rows. The apical crown (AC) of basal body doublets, drawn in a highly schematic form in the polar projections, also is in a mirror-image configuration in RH and LH cells. The typical arrangement of structures of the OA is shown for the RH cells, while only one of several arrangements found in LH cells is shown. The individual structural elements within the OA are rotational permutations, not true mirror images, of the corresponding normal structures, (see Nelsen & Frankel, 1989 for a general account; Nelsen *et al.* 1989 for details).

on the arrangement of cell structures. LH1 is obviously vegetative in its origin and retains the mating type II of the parental stock.

The second clone of LH phenotype (LH2) was isolated in 1984 from a doublet clone produced by fusion of conjugating wild-type cells after treatment with homologous immobilization antiserum. The LH2 cells retain the mating type of one of the 'parental' clones, indicating a probable vegetative origin.

The LH3 and LH4 clones were isolated in 1987 from doublet cultures produced by heat treatment of conjugating functional heterokaryon stocks. These two LH clones were, as expected for heat-treated conjugating cells (Scholnick & Bruns, 1982a,b), vegetative 'progeny' of abortive conjugation, retaining both the drug-sensitivity (cycloheximide-sensitive) and the mating types (III and V, respectively) determined by the macronuclei of the parental cells.

The feature of crucial importance for this investigation is the vegetative origin of all of the LH clones. This vegetative origin can be considered certain for clones LH1, LH3 and LH4, and probable for clone LH2.

The LH phenotype

Some major differences between the typical 'right-handed' (RH) phenotype and the reverse 'left-handed' (LH) phenotype are presented schematically in Fig. 1. The evidence for these and other differences is presented in the accompanying paper (Nelsen & Frankel, 1989). The arrangement of structures in LH cells is the reverse of

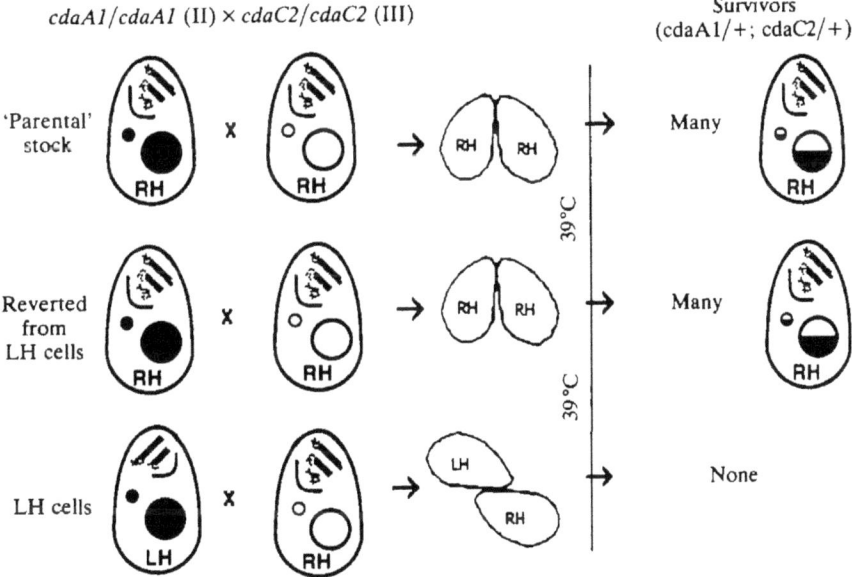

Fig. 2. Crosses testing the capacity of LH and RH cells to produce progeny. Roman numerals indicate mating types. The shading of nuclei in these diagrams indicates the genotype at the *cda* loci, with *cdaA1* nuclei shaded and *cdaC2* nuclei unshaded. Nuclei heterozygous at both loci are half-shaded. Only such double heterozygotes can propagate at 39°C ('survivors' in right column). Some cells from both RH×RH crosses survived the nonpermissive conditions, indicating successful conjugation, while the cells from the LH×RH cross all died, indicating that they did not mate successfully. For further explanation, see the text.

that in RH cells. The positions of the contractile vacuole pore (CVP) and cytoproct (Cyp) and the pattern of basal body couplets at the anterior end of the cell all reflect this reversal. At the local level, however, the polarity within the ciliary rows, positions of microtubule bands, kinetodesmal fibres and cortical mitochondria are the same in LH and RH cells.

The reversal of large-scale asymmetry has serious consequences for development of the OA and hence for cell multiplication (Nelsen & Frankel, 1989). LH cells are therefore at a large selective disadvantage relative to RH cells. Hence, to maintain LH cultures one must periodically subclone the LH cells, since occasional RH 'revertants' will overgrow unselected cultures.

Breeding analysis
LH×RH crosses
Attempts to cross LH with RH cells proved unsuccessful. A breeding scheme utilizing two recessive temperature-sensitive cell-division-arrest (*cda*) mutations was devised to select for even rare sexual success in an LH×RH cross (Fig. 2). All pairs that had gone through the complete conjugation process would have macro-nuclei as well as micronuclei heterozygous for both of the recessive mutations (cf. Fig. 4), and hence would continue to divide even at a temperature restrictive for these mutations. Cells that had paired but subsequently aborted the conjugation process and retained the old macronuclei, as well as cells that did not pair, would undergo division arrest under restrictive conditions.

LH cells form heteropolar pairs with RH cells rather than the homopolar pairs normally seen during conjugation. While the mating stimulus sufficed to trigger meiosis and even the development of macronuclear anlagen in some LH×RH pairs (Nelsen & Frankel, 1989), putative exconjugants did not divide under restrictive conditions, indicating a prior failure to produce genuine progeny. Many RH 'parental' cells from which the LH1 clone was derived, and RH 'revertants' from the LH clones, survived the restrictive conditions following conjugation with RH cells, and hence were successful in producing progeny.

Successful mating of RH×RH pairs, while LH×RH pairs of the same genetic stock failed, suggests that the anterior–posterior mismatch of the bonding surface of the heteropolar LH×RH pairs might account for failure to produce progeny (Nelsen & Frankel, 1989). This implies that an LH×LH cross might be successful.

LH×LH crosses
An LH×LH cross is sufficient to test the hypothesis of a nuclear genie origin for the LH condition. The reason for this lies in the vegetative origin of the LH clones, described above. Since it has been demonstrated that when the micronucleus and macronucleus have different genotypes the micronucleus has no effect on the cellular phenotype (*Tetrahymena*: Nanney & Dubert, 1960; Bruns & Brussard, 1974; Mayo & Orias, 1985; *Paramecium*: Sonneborn 1947 p. 270; Tam & Ng, 1987), a mutation present in the micronucleus alone would not be expressed so long as the macronucleus retained the original wild-type genes specifying the RH condition. It therefore follows that if the origin of the LH condition is based on a nuclear

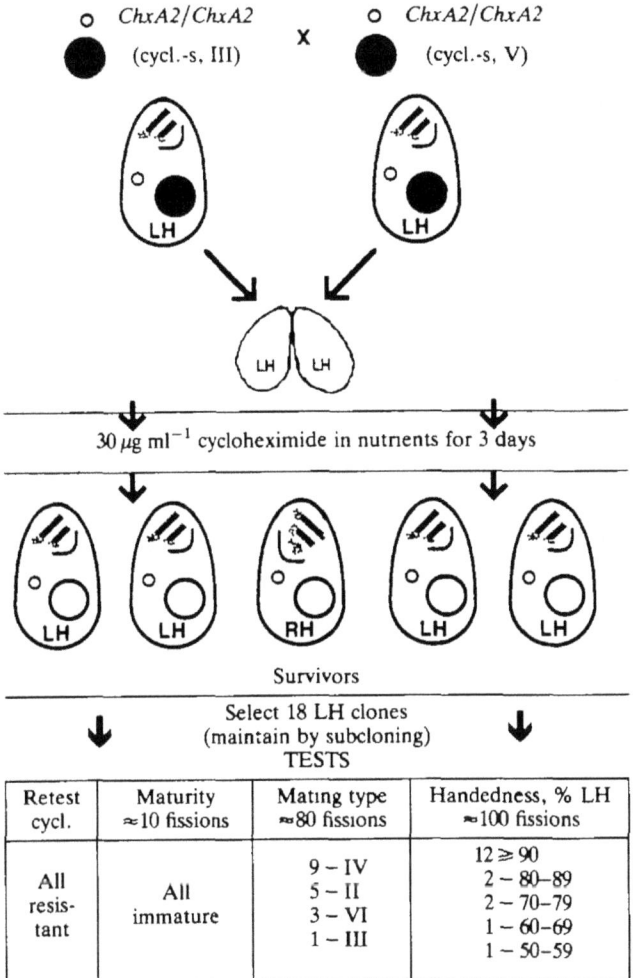

Fig. 3. LH×LH crosses to test the possible mutational origin of the LH phenotype. Homozygous functional heterokaryons, *ChxA2/ChxA2* (cycl.-sens) [cycloheximide resistant micronucleus (sensitive macronucleus)] of two mating types (indicated by Roman numerals) were utilized (the B* cells used as costimulators are not shown). The shading of nuclei indicates the genotype at the *ChxA2* locus, with the genotype conferring sensitivity shaded and genotype conferring resistance unshaded. Only cells that undergo complete conjugation (Fig. 4), forming new macronuclei from micronuclear sources, can survive the cycloheximide poisoning. The survivors included many LH cells, which after subcloning maintained their LH character. This result refutes the hypothesis that the LH phenotype is based on a mutation that originated and was maintained in the macronucleus. For further explanation, see the text.

mutation, it must be in the macronucleus. A dominant macronuclear mutation might come to expression quickly, while a recessive macronuclear mutation could only be expressed after phenotypic assortment (Nanney, 1964; Orias & Flacks, 1975; Doerder *et al.* 1977). The final possibility, of simultaneous independent mutations in *both* the micro-nucleus and macronucleus, is extraordinarily unlikely on probabilistic grounds. It is important to note that there is excellent genetic evidence that macronuclear *genetic changes in T. thermophila* do not secondarily affect micronuclei during vegetative growth (Nanney & Dubert, 1960; Bruns, 1986), although in certain unusual cases they may affect new macronuclei during conjugation (Doerder & Berkowitz, 1987).

The logic of the test for a nuclear genie origin of the LH condition is shown in Fig. 3. Since any putative LH mutation must be in the macronucleus but not the micronucleus, all true progeny of this cross, which have formed new macronuclei from a zygotic micronucleus (cf. Fig. 4), should have the wild-type (RH/RH) genotype, and hence should revert to the RH form.

Preliminary tests with two LH clones (LH1 and LH2) of different mating types yielded few pairs, which tended to fall apart on attempted isolation. These failures presumably resulted from difficulty in pairing due to a short preoral suture in LH cells (Nelsen & Frankel, 1989). To help overcome this problem, RH cells of a third mating type were added as costimulators (Bruns & Palestine, 1975; Suganuma *et al.* 1984) to promote the development of mating surfaces. This third mating type was a B* (B-star) (VII) clone that lacked a functional micronucleus, and would ordinarily yield no progeny except through an alternative form of conjugation known as 'genomic exclusion' (Allen, 1967), which involves two successive rounds of mating before a new macronucleus can be produced. In these LH×LH crosses, no genuine progeny were obtained following pair isolations, but assessment of protargol-stained mass cultures showed that some LH–LH pairs were proceeding through the nuclear stages diagnostic of normal conjugation. However, by the time this assessment was made the LH2 clone was showing indications of general decline (probably independent of the LH condition), so new LH clones were selected.

The definitive test then was carried out with the newly constructed LH3 and LH4 clones. These clones were 'homozygous functional heterokaryons' (Bruns & Brussard, 1974), which were homozygous for a dominant allele (*ChxA2*) conferring cycloheximide resistance in the micronucleus and the recessive allele conferring cycloheximide sensitivity in the macronucleus (Fig. 3). In this scheme, only cells that develop new macronuclei can survive the cycloheximide poisoning (Bruns & Brussard, 1974). Since the B* costimulators are cycloheximide-sensitive, they could survive cycloheximide poisoning only after conjugating with RH 'revertants' (recall that RH×LH crosses do not yield progeny); they would ordinarily have to do so twice in succession to produce cells with new macronuclei [except possibly in rare cases of 'short-circuit genomic exclusion' (Bruns *et al.* 1975)]. Thus their contribution to the population of survivors after cycloheximide poisoning probably is negligible.

Four crosses between the LH3 and LH4 clones were carried out. In each of these crosses, both LH and RH survivors were observed after 3 days of growth. The surviving RH cells may have resulted from the mating of 'reverted' RH cells of the starting cultures with each other or possibly from rare cases of matings with the defective costimulators. Nine surviving putative LH cells were isolated for each of the four crosses. Of these 36 cells, four died, and four clones appeared (by inspection) to be RH in character and were discarded. The 28 remaining clones were immature, cycloheximide resistant and LH in character. Ten of these clones

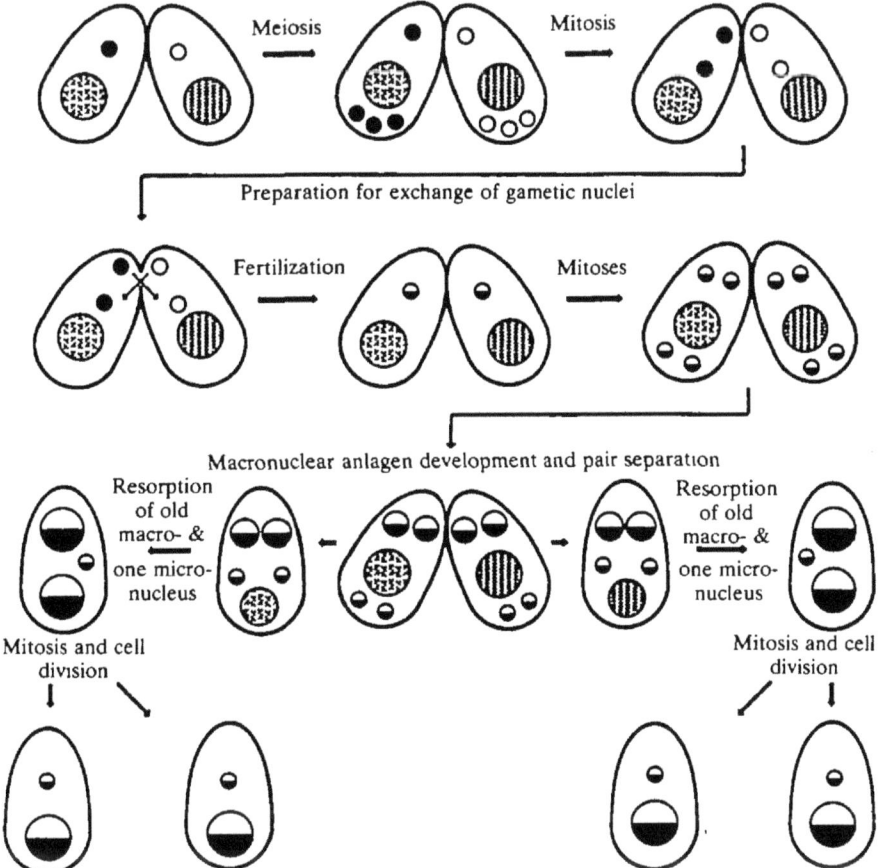

Fig. 4. A schematic diagram of conjugation between *Tetrahymena* cells with differing micronuclear and macronuclear genotypes. Here, *all four* of the nuclei involved are of differing genotypes. The micronuclei differ in alleles at one or more loci unrelated to the LH phenotype, while the macronuclei are assumed to have acquired mutations that specify the LH phenotype. The micronuclei of the two conjugating cells enter meiosis synchronously. Three of the four products of meiosis are resorbed, while the remaining one divides mitotically to produce two gametic nuclei. These are exchanged reciprocally. Each gametic nucleus from one cell fuses with a gametic nucleus from the other cell, to form zygotic nuclei that are genically identical in the two cells. The zygotic nuclei then undergo two postzygotic divisions. Of the resulting four nuclei in each conjugant, two form macronuclear anlagen that develop into new macronuclei, one forms the new micronucleus, and one is resorbed. The old macronuclei also are resorbed. The two new macronuclei of each exconjugant are segregated to two daughter cells. The new micronuclei and macronuclei are heterozygous for the allelic differences between the micronuclei of the preceding sexual generation; however, they do not carry any of the macronuclear mutations from the previous sexual generation, including the putative mutations presumed to give rise to the LH phenotype. For a detailed description of conjugation in *Tetrahymena*, see Orias (1986).

died during the subsequent subcloning, while 18 were carried through all of the tests indicated in Fig. 3.

The demonstration of cycloheximide resistance, immaturity (initial inability to mate) and new mating types upon subsequent maturity indicates that new macronuclei *were* formed in these cells. These nuclei were formed either by normal exchange of gametic nuclei between partners or by fusion of gametic nuclei without exchange (cytogamy) (Orias *et al.* 1979). In either case, all macronuclear mutations should have been eliminated, thereby removing any nuclear genetic basis for the LH form. Yet LH forms continued to predominate at the time of quantitative assessment 100 generations after conjugation in all of the 18 clones (Fig. 3) and persisted at 140 fissions when most of the clones were discarded and samples of three clones were selected for cryopreservation. The LH condition was also maintained in these three 'stock' clones for 250 to 300 generations postconjugation. These results indicate that the LH condition is not a simple consequence of a nuclear mutation.

Discussion

A non-genic basis for cellular handedness
In this study, we have demonstrated that the propagated difference between two mirror-image arrangements of cell structures on the surface of *Tetrahymena thermophila*, the 'right-handed' and 'left-handed' arrangements, are not based on differences in nuclear genes. This demonstration required a somewhat unconventional procedure. A standard cross-breeding analysis in ciliates is based on three aspects of conjugation: temporary union of conjugating cells, reciprocal fertilization of gametic nuclei derived from micronuclei, and development of new macronuclei from mitotic products of the zygotic (micro) nuclei (Fig. 4). Thus the standard way to initiate a genetic analysis is to cross cells of different micronuclear genotypes, as shown in Fig. 4. A cross-breeding analysis of this type was not possible in our analysis of the difference between RH and LH clones, in part because while RH and LH cells could conjugate they did not yield any viable progeny. We were nonetheless able to carry out a genetic analysis of the LH phenotype by relying on two other features of ciliate genetics: the control of the cellular phenotype by the macronucleus and the loss of old macronuclei during conjugation. Since all of our LH clones originally arose vegetatively, i.e. *without* the formation of new macronuclei from micronuclei, we know that if the LH condition had arisen by mutation, that mutation would have originated in the macronucleus. This is illustrated schematically in Fig. 4 by the hatching of the macronuclei. Since macronuclei are lost during conjugation (Fig. 4), if the LH phenotype depended on a (macro)nuclear mutation, that phenotype should have been lost in *any* cross involving LH cells, including an LH×LH cross. It was not lost, and therefore it could not have arisen from a nuclear mutation.

There is one apparent and one real ambiguity in this demonstration. The apparent ambiguity centres around the instability of the LH phenotype which is evident both before and after the critical LH×LH cross. RH 'revertants' appear sufficiently often that LH clones can be maintained only with frequent selection of

LH cells. Does this phenotypic instability have a genic basis? In the case of the *PEP4* gene in *Saccharomyces cerevisiae,* a new (hydrolase-minus) phenotype of recessive *pep4* haploid yeasts derived from *PEP4/pep4* diploids comes to expression after a phenomic lag of 80 generations or more, with sectoring indicative of phenotypic instability (Zubenko *et al.* 1982). However, the LH condition in *Tetrahymena* persists for a longer time after the critical cross—at least 300 generations—than was demonstrated for the hydrolase-plus phenotype in *pep4* yeast segregants. Even more important, unlike the hydrolase-plus phenotype of yeast carrying the wild-type *PEP4* allele in yeast, the LH phenotype of *Tetrahymena never* is truly stable, either before or after the cross. The precise reversion rate per generation is not readily computable from the data in Fig. 3 (the percentage of LH cells was assessed in tube cultures started by 6 cells each), but appears roughly comparable to that of vegetative LH clones. Thus, unlike the yeast example, there is no indication of a major phenotypic difference before and after the cross, and hence no support for the hypothesis of a corresponding nuclear genic difference.

The second and genuine ambiguity concerns the possibility of a *genetic* rather than a phenotypic feedback system: Specifically, while ciliate micronuclei and macronuclei are sealed off from each other during vegetative growth, they are not during macronuclear development. There is a long history of studies in *Paramecium* which demonstrate that pre-existing macronuclei can influence the characteristics of new macro-nuclei that are formed during conjugation (Sonneborn, 1977); recent evidence shows that this influence affects the genetic structure of the developing macronucleus (Epstein & Forney, 1984; Harumoto, 1986). A similar genetic feedback has recently been discovered in *Tetrahymena* (Doerder & Berkowitz, 1987), although it is not effective in the genetic situation comparable to that of the cross shown in Fig. 3. One could postulate that the LH macronuclear genotype specifies the LH phenotype and also specifies a cytoplasmic state that causes wild-type micronuclei to lose RH-determining base sequences during their development into macronuclei. However, the LH phenotype differs in its persistent vegetative instability from the cases of genetic feedback, in which phenotypic stability is the rule after the macronucleus has become fully differentiated (Sonneborn, 1977).

While we cannot conclusively rule out the possibility that such a genetic feedback system may control the maintenance of the LH condition, both the circumstances of origin and conditions of maintenance of LH clones argue against this alternative. All LH cells were derived vegetatively from RH cells through regulation of homopolar doublets. Furthermore, LH cells probably regulate back to the RH phenotype *via* a homopolar-doublet pathway: the frequency of such reverse-regulation is greatly increased under conditions that promote doublet formation through blockage of cell division (Nelsen & Frankel, 1989). The close connection between the degree of vegetative instability and the pre-existing cellular organization is not characteristic of phenotypes controlled by genes; it suggests a different conceptual model, whose essence will be presented here and whose details will be elaborated in the accompanying paper.

The clonal cylinder and the propagation of large-scale asymmetry
As first pointed out by Tartar (1962), a ciliate clone can be thought of as a cylinder that is growing longitudinally and periodically is segmenting (dividing) transversely (Fig. 5). This geometry lends itself to the inheritance of any cellular feature that is capable of being propagated longitudinally. The best known example, the inheritance of the spatial orientation of ciliary rows (Beisson & Sonneborn, 1965; Ng & Frankel, 1977), can be thought of as a selfperpetuating supramolecular scaffold. The imagery of a selfperpetuating scaffold does not, however, provide a clear explanation for why a structure such as the newly formed oral apparatus, which develops at a considerable distance from any preexisting structure of the same kind (Fig. 5), should normally be formed along the same cell longitude as the pre-existing structure. Ciliates appear capable of perpetuating not only microscopically visible structural ensembles but also the as-yet-invisible instructions that determine where visible structures such as the oral apparatus and the contractile vacuoles may be formed.

These invisible instructions can be represented as a grid of positional latitudes and longitudes (Fig. 6). The values of the latitudes (letters in Fig. 6) must be respecified in every division (segmentation) cycle, but the longitudes (numbers in Fig. 6) could in principle be perpetuated indefinitely. Further, a single complete set of longitudes can be wound only in two ways around the cell, as shown in Fig. 6a and b, respectively. If longitudes were propagated as the cylinder grows, there is no way in which either type of winding could be converted to the other without cutting into the circle and substituting subsets of longitudes.

If we imagine that the different numbers represent positional values that are 'interpreted' to promote formation of different structures (such as the oral apparatus at 5, the CVP-set at 7), the cylindrical topology of ciliate growth would allow for a propagation of these positional values. The direction of winding of the ensemble of positional values would then be the basis for the difference between the RH and LH

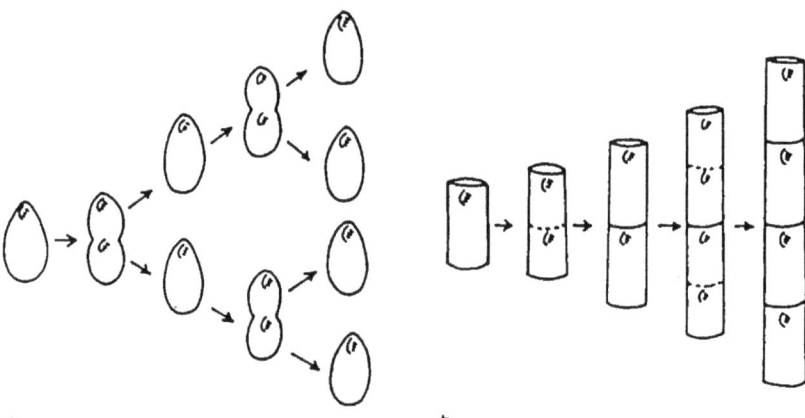

Fig. 5. A schematic of clonal growth in ciliates. (a) Two division cycles in *Tetrahymena*. (b) The same visualized as a longitudinally growing cylinder that undergoes periodic segmentation. Oral structures are shown. From Frankel & Nelsen (1981).

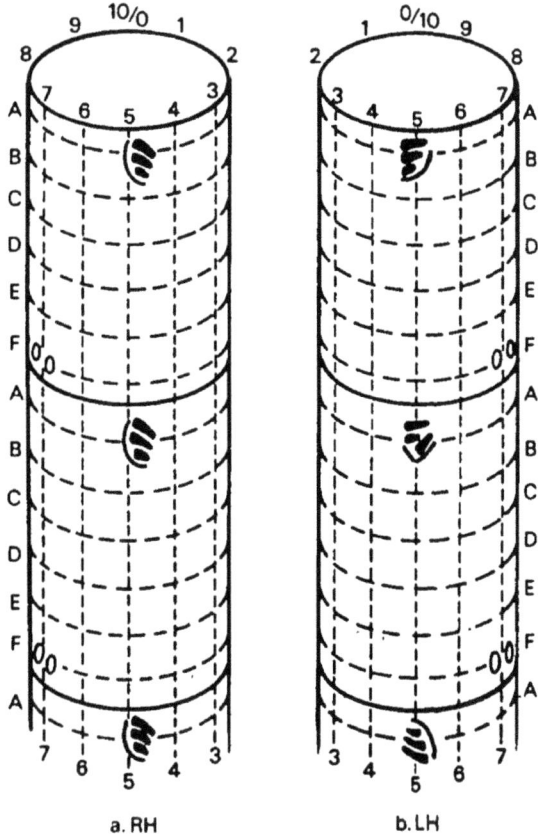

Fig. 6. Orthogonal coordinates superimposed on the clonal cylinder for (a) 'right-handed' (RH) and (b) 'left-handed' (LH) cells. Solid lines indicate cell boundaries, while dashed lines indicate hypothetical latitudes (letters) and longitudes (numbers). Oral apparatuses are shown at the coordinate 'A5', contractile vacuole pores at 'F7'. The oral apparatuses of RH cells are shown in their typical, invariant form, while those of LH cells are drawn to emphasize the large variations in internal organization that are encountered (Nelsen & Frankel, 1989). For further explanation, see the text.

configuration. This direction would control the arrangement of surface structures and influence the internal organization of complex structures such as the oral apparatus (Nelsen *et al* 1989). The instability of the LH configuration is a consequence of the topological transitions that are already known to occur in regulating doublets (Frankel & Nelsen, 1986a, 1987; Nelsen & Frankel, 1986), amplified by the tremendous growth advantage of RH cells.

The concept of propagated intracellular positional values is based on a transposition of the idea of positional information (Wolpert, 1969, 1971) and its corollary, nonequivalence (Lewis & Wolpert, 1976), to the intracellular realm. It also is specifically related to the 'polar coordinate' model of French *et al.* (1976). Thus, we conceptualize the transitions between RH and LH forms in terms of reverse intercalation within the contracting perimeter of regulating doublets (Nelsen

& Frankel, 1989), an intracellular and morphallactic application of the shortest-distance intercalation rule as initially formulated in the polar coordinate model of French et al. (1976). This permits us to explain the genically and non-genically inherited differences in large-scale asymmetry within a single model: the non-genically inherited cases involve reverse intercalation provoked by abnormal distributions or confrontations of positional values, the genic cases arise from reverse intercalation provoked by an incapacity of the genetic system to maintain certain subsets of these values (Frankel & Nelsen, 1986b). The difference is not in the underlying mechanisms but in the proximate causes for bringing the mechanisms into play.

This concept can be extended to the speculative idea that the underlying substratum of positional information might be similar in ciliates and eggs (Frankel, 1989, chapter 11). However, even if this is true, there is one essential difference: while the geometry of ciliate growth allows a positional system to extend across division products and thereby perpetuate itself, eggs are formed anew in each sexual generation, with no direct topological continuity to the organization of the mother. For this reason, differences of large-scale asymmetry are directly inheritable in ciliates, while in multicellular organisms they must be rebuilt in every generation, with the choice between alternatives depending on information supplied by the maternal genome. The importance of the ciliate example is that it demonstrates that a system of positional information has the potential for direct inheritance; the usual absence of such inheritance in animals might then be an accidental consequence of the newly evolved geometry of sexual reproduction. In the rare cases of direct asexual propagation, as in the fission of flatworms, the direct perpetuation of positional values might well be retained (Sonneborn, 1930).

The authors would like to thank Drs Robert E. Malone and Gary N. Gussin for provocative discussions, and Dr Malone as well as Drs Anne W. K. Frankel, Norman E. Williams, and Mr Eric Cole for critical comments on various drafts of this manuscript. The research was supported by grant HD-08485 from the US National Institutes of Health.

References

ALLEN, S. L. (1967). Genomic exclusion: a rapid means for inducing homozygous diploid lines in *Tetrahymena Science* **155**, 575–577.

ALLEN, S. L. & GIBSON, I. (1973). Genetics of *Tetrahymena*. In *Biology of Tetrahymena* (ed. A. M. Elliott), pp. 307–373. Stroudsburg, PA: Dowden, Hutchinson, and Ross.

AUFDERHEIDE, K. J. (1982). An improvement of the protargol technique of Ng and Nelsen. *Trans. Am. microsc. Soc.* **101**, 100–104.

BEISSON, J. & SONNEBORN, T. M. (1965). Cytoplasmic inheritance of the organization of the cell cortex of *Paramecium aurelia. Proc. natn. Acad Sci. U.S.A.* **53**, 275–282.

BRUNS, P. J. (1986). Genetic Organization of *Tetrahymena*. In *The Molecular Biology of Ciliated Protozoa* (ed. J. G. Gall). pp. 27–44. New York: Academic Press.

BRUNS, P. J. & BRUSSARD, T. B. (1974). Positive selection for mating with functional heterokaryons in *Tetrahymena pyriformis*. *Genetics* **78**, 831–841.

BRUNS, P. J., BRUSSARD, T. B. & KAVKA, A. B. (1976). Isolation of homozygous mutants after induced self-fertilization in *Tetrahymena*. *Proc. natn. Acad. Sci. U.S.A.* **73**, 3243–3247.

BRUNS, P. J. & PALESTINE, R. F. (1975). Costimulation in *Tetrahymena pyriformis*: a developmental interaction between specially prepared cells. *Devl Biol.* **42**, 75–83.

DOERDER, F. P. & BERKOWITZ, M. S. (1987). Nucleo-cytoplasmic interaction during macronuclear differentiation in ciliate protists: genetic basis for cytoplasmic control of *serH* expression during macronuclear development in *Tetrahymena thermophila*. *Genetics* **117**, 13–23.

DOERDER, F. P., LIEF, J. H. & DEBAULT, L. E. (1977). Macronuclear subunits of *Tetrahymena* are functionally haploid. *Science* **198**, 946–948.

EPSTEIN, L. M. & FORNEY, J. D. (1984). Mendelian and non-mendelian mutations affecting surface antigen expression in *Paramecium tetraurelia*. *Molec. Cell Biol.* **4**, 1583–1590.

FRANKEL, J. (1984). Pattern formation in ciliated protozoa. In *Pattern Formation: A Primer in Developmental Biology* (ed. G. M. Malacinski & S. V. Bryant), pp. 163–196. New York: Macmillan.

FRANKEL, J. (1989). *Pattern Formation: Ciliate Observations and Models*. New York: Oxford University Press (in press).

FRANKEL, J. & HECKMANN, K. (1968). A simplified Chatton-Lwoff silver impregnation procedure for use in experimental studies with ciliates. *Trans. Am. microsc. Soc.* **87**, 317–321.

FRANKEL, J. & JENKINS, L. M. (1979). A mutant of *Tetrahymena thermophila* with a partial mirror-image duplication of cell surface pattern. II. Nature of genic control. *J. Embryol. exp. Morph.* **49**, 203–227.

FRANKEL, J., JENKINS, L. M. & BAKOWSKA, J. (1984). Selective mirror-image reversal of ciliary patterns in *Tetrahymena thermophila* homozygous for a *janus* mutation. *Wilhelm Roux's Arch, devl Biol.* **194**, 107–120.

FRANKEL, J., JENKINS, L. M., DOERDER, F. P. & NELSEN, E. M. (1976). Mutations affecting cell division in *Tetrahymena pyriformis*. I. Selection and genetic analysis. *Genetics* **83**, 489–506.

FRANKEL, J. & NELSEN, E. M. (1981). Discontinuities and overlaps in patterning within single cells. *Phil. Trans. Roy. Soc. Lond.* B **295**, 525–538.

FRANKEL, J. & NELSEN, E. M. (1986a). Intracellular pattern reversal in *Tetrahymena thermophila*. II. Transient expression of a *janus* phenocopy in balanced doublets. *Devl Biol.* **114**, 72–86.

FRANKEL, J. & NELSEN, E. M. (1986b). How the mirror-image pattern specified by a *janus* mutation of *Tetrahymena* comes to expression. *Devl Genet.* **6**, 213–238.

FRANKEL, J. & NELSEN, E. M. (1987). Positional reorganization in compound *janus* cells of *Tetrahymena thermophila*. *Development* **99**, 51–68.

FRANKEL, J., NELSEN, E. M. & JENKINS, L. M. (1987). Intracellular pattern reversal in *Tetrahymena thermophila*: *janus* mutants and their geometrical phenocopies. In

Genetic Regulation of Development, Society for Developmental Biology Symposium, vol. 45 (ed. W. F. Loomis), pp. 219–244. New York: Alan R. Liss.

FREEMAN, G. H. & LUNDELIUS, J. W. (1982). The developmental genetics of dextrality and sinistrality in the gastropod *Lymnaea peregra. Wilhelm Roux' Arch, devl Biol.* **191,** 69–83.

FRENCH, V., BRYANT, P. J. & BRYANT, S. V. (1976). Pattern regulation in epimorphic fields. *Science* **193,** 969–981.

GRIMES, G. W., MCKENNA, M. E., GOLDSMITH-SPOEGLER, C. M. & KNAUPP, E. A. (1980). Patterning and assembly of ciliature are independent processes of hypotrich ciliates. *Science* **209,** 281–283.

HARUMOTO, T. (1986). Induced change in a non-mendelian determinant by transplantation of macronucleoplasm in *Paramecium tetraurelia. Molec. Cell Biol.* **6,** 3498–3501.

JERKA-DZIADOSZ, M. (1981). Patterning of ciliary structures in *janus* mutant of *Tetrahymena* with mirror-image cortical duplications. An ultrastructural study. *Acta Protozool.* **20,** 337–356.

JERKA-DZIADOSZ, M. (1983). The origin of mirror-image symmetry doublet cells in the hypotrich ciliate *Paraurostyla weissei. Wilhelm Roux' Arch, devl Biol.* **192,** 179–188.

JERKA-DZIADOSZ, M. & FRANKEL, J. (1979). A mutant of *Tetrahymena thermophila* with a partial mirror-image duplication of cell surface pattern. I. Analysis of the phenotype. *J. Embryol. exp. Morph.* **49,** 167–202.

LEWIS, J. H. & WOLPERT, L. (1976). The principle of non- equivalence in development. *J. theoret. Biol.* **62,** 479–490.

MAYO, K. A. & ORIAS, E. (1985). Lack of expression of micronuclear genes determining two different enzymatic activities in *Tetrahymena thermophila. Differentiation* **28,** 217–224.

MOHLER, J. & WIESCHAUS, E. F. (1986). Dominant maternal-effect mutations of *Drosophila melanogaster* causing the production of double abdomen embryos. *Genetics* **112,** 803–822.

NANNEY, D. L. (1964). Macronuclear differentiation and subnuclear assortment in ciliates. In *The Role of Chromosomes in Development, Society for Developmental Biology Symposium,* vol. 23 (ed. M. Locke), pp. 253–273, New York: Academic Press.

NANNEY, D. L. (1983). The ciliates and the cytoplasm. *J. Hered.* **74,** 163–170.

NANNEY, D. L. & DUBERT, J. M. (1960). The genetics of the H serotype system in variety 1 of *Tetrahymena pyriformis. Genetics* **45,** 1335–1358.

NELSEN, E. M. & DEBAULT, L. E. (1978). Transformation in *Tetrahymena pyriformis*: Description of an inducible phenotype. *J. Protozool.* **25,** 113–119.

NELSEN, E. M. & FRANKEL, J. (1986). Intracellular pattern reversal in *Tetrahymena thermophila*. I. Evidence for reverse intercalation in unbalanced doublets. *Devl Biol.* **114,** 53–71.

NELSEN, E. M. & FRANKEL, J. (1989). Maintenance and regulation of cellular handedness in *Tetrahymena. Development* **105,** 457–471.

NELSEN, E. M., FRANKEL, J. & MARTEL, E. (1981). Development of the ciliature of *Tetrahymena thermophila*. I. Temporal coordination with oral development. *Devl Biol.* **88**, 27–38.

NELSEN, E. M., FRANKEL, J. & WILLIAMS, N. E. (1989). Effects of cellular handedness on oral assembly in *Tetrahymena*. *J. Protozool.* (submitted).

NG, S. F. & FRANKEL, J. (1977). 180° rotation of ciliary rows and its morphogenetic implications in *Tetrahymena pyriformis*. *Proc. natn. Acad. Sci. U.S.A.* **74**, 1115–1119.

NG, S. F. & NELSEN, E. M. (1977). The protargol staining technique: an improved version for *Tetrahymena pyriformis*. *Trans. Am. microsc. Soc.* **96**, 369–376.

NÜSSLEIN-VOLHARD, C. (1977). Genetic analysis of pattern formation in the embryo of *Drosophila melanogaster*. Characterization of the maternal effect mutation *bicaudal*. *Wilhelm Roux' Arch, devl Biol.* **183**, 249–268.

ORIAS, E. (1986). Ciliate conjugation. In *The Molecular Biology of Ciliated Protozoa* (ed. J. G. Gall), pp. 45–84. New York: Academic Press.

ORIAS, E. & BRUNS, P. J. (1976). Induction and isolation of mutants in *Tetrahymena*. In *Methods in Cell Biology* (ed. D. M. Prescott), vol. 13, pp. 247–282. New York: Academic Press.

ORIAS, E. & FLACKS, M. (1975). Macronuclear genetics of *Tetrahymena*. I. Random distribution of macronuclear copies in *T. pyriformis*, syngen 1. *Genetics* **79**, 187–206.

ORIAS, E., HAMILTON, E. P. & FLACKS, M. (1979). Osmotic shock prevents nuclear exchange and produces whole-genome homozygotes in conjugating *Tetrahymena*. *Science* **203**, 660–663.

PREER, J. R. (1988). Foreword. In *Paramecium* (ed. H.-D Görtz). pp. v–xvi. Berlin: Springer Verlag.

SAPP, J. (1987). *Beyond the Gene. Cytoplasmic Inheritance and the Struggle for Authority in Genetics.* New York: Oxford University Press.

SCHOLNICK, S. B. & BRUNS, P. J. (1982a). A genetic analysis of *Tetrahymena* that have aborted normal development. *Genetics* **102**, 29–38.

SCHOLNICK, S. B. & BRUNS, P. J. (1982b). Conditional lethality associated with macronuclear development in *Tetrahymena thermophila*. *Devl Biol.* **93**, 216–225.

SIMON, E. M. & FLACKS, M. (1975). Preparation, storage, and recovery of free-living non-encysting protozoa. In *Cryogenic Preservation of Cell Cultures* (ed. A. P. Rinfret & B. LaSalle), pp. 37–49. Washington D.C.: National Academy of Sciences.

SONNEBORN, T. M. (1930). Genetic studies in *Stenostomum incaudatum*. II. The effects of lead acetate on the hereditary constitution. *J. exp. Zool.* **57**, 409–439.

SONNEBORN, T. M. (1947). Recent advances in the genetics of *Paramecium* and *Euplotes Adv. Genet.* **1**, 263–358.

SONNEBORN, T. M. (1963). Does preformed cell structure play an essential role in cell heredity? In *The Nature of Biological Diversity* (ed. J. M. Allen), pp. 165–221. New York: McGraw-Hill.

SONNEBORN, T. M. (1977). Genetics of cellular differentiation: stable nuclear differentiation in eukaryotic cells. *A. Rev. Genet.* **11**, 349–367.

STURTEVANT, A. H. (1923). Inheritance of the direction of coiling in *Limnaea*. *Science* **58**, 269–270.

SUGANUMA, Y., SHIMODE, C. & YAMOMOTO, H. (1984). Conjugation in *Tetrahymena*: formation of a special junction area for conjugation during the co-stimulation period. *J. Electron Microsc.* **33**, 10–18.

SUHAMA, M. (1982). Homopolar doublets of the ciliate *Glaucoma scintillans* with a reversed oral apparatus. I. Development of the oral primordium. *J. Sci. Hiroshima Univ., Ser. B, Div. 1* **30**, 51–65.

SUHAMA, M. (1985). Reproducing singlets with an inverted oral apparatus in *Glaucoma scintillans* (Ciliophora, Hymenostomatida). *J. Protozool.* **32**, 454–459.

TAM, L.-W. & NG, S. F. (1987). Genetic analysis of heterokaryons in search of active micronuclear genes in stomatogenesis of *Paramecium aurelia*. *Eur. J. Protistol.* **23**, 43–50.

TARTAR, V. (1962). Morphogenesis in *Stentor*. *Adv. Morphogen.* **2**, 1–26.

TCHANG, T.-R., SHI, X.-B. & PANG, Y.-B. (1964). An induced monster ciliate transmitted through three hundred and more generations. *Scientia Sinica* **13**, 850–853.

WOLPERT, L. (1969). Positional information and the spatial pattern of cellular differentiation. *J. theoret. Biol.* **25**, 1–47.

WOLPERT, L. (1971). Positional information and pattern formation. *Curr. Top. devl Biol.* **6**, 183–224.

ZUBENKO, G. S., PARK, F. J. & JONES, E. W. (1982). Genetic properties of mutations at the PEP4 locus in *Saccharomyces cerevisiae*. *Genetics* **102**, 679–690.

(*Accepted 1 December 1988*)

Case Study 9: Don't Blame Mom

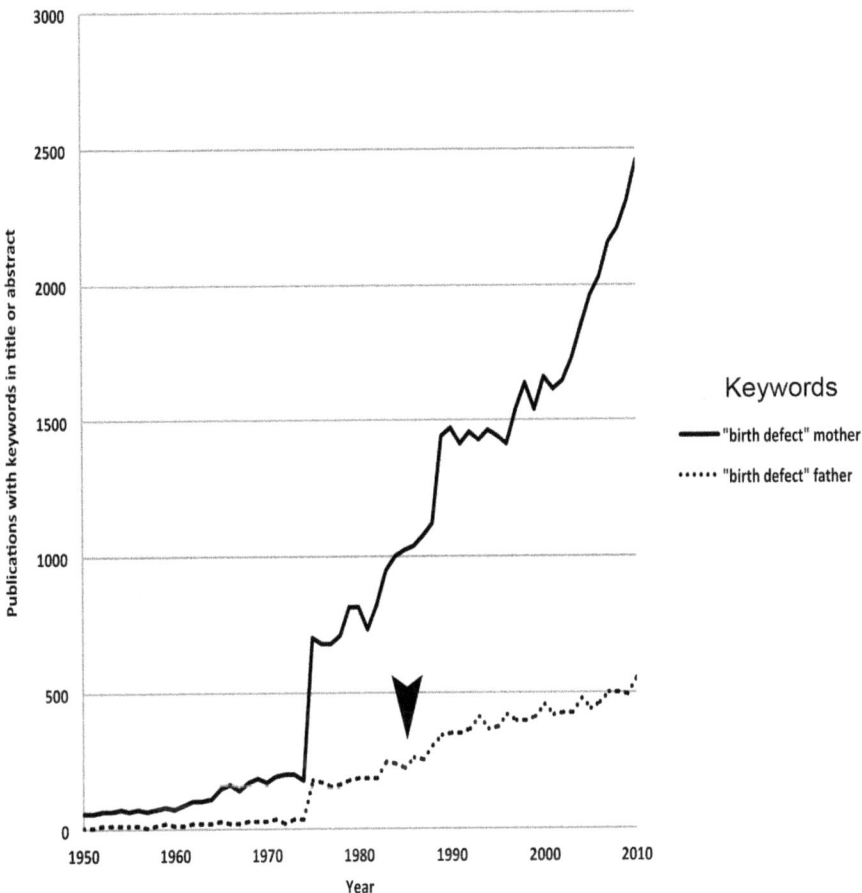

Friedler (1985) Effects of limited paternal exposure to xenobiotic agents on the development of progeny.

Perspective on *Effects of limited paternal exposure to xenobiotic agents on the development of progeny*

Scott F. Gilbert
Professor Emeritus
Swarthmore College
Author of *Developmental Biology*

In the late 1970s, there was no context in which to put Gladys Freidler's work. Epigenetics had not framed a theory where a chemical could affect DNA by causing changes in its expression rather than in its sequence. Teratogens worked by disrupting normal embryonic development, and in mammals, this took place within the embryo. The egg was also susceptible to alterations, since eggs were the survivors of a perishing population, accumulating mutations and potential non-disjunctions as the mothers aged. Sperm, however, were seen as motile nuclei, newly made, and impervious to the environment. If a sperm had a serious mutation, it would probably be at a disadvantage for fertilizing the egg, since such was the prerogative of only the quickest, most virile, of sperm. According to Wyer, biologists who tried to pursue a father-fetal connection had difficulty getting funding, recognition, or publication. Friedler persisted, showing that male mice exposed to morphine and alcohol fathered pups with birth defects. Moreover, these carefully controlled experiments found that males of the affected generation could pass the defect to the next. When an epigenetic context was formulated, such sperm epimutations were not only found, but so were the mechanisms of their intergenerational inheritance.

Cited:

Wyer M, Barbercheck M, Cookmeyer D, Ozturk H and Wayne M (ed) 2013 *Women, Science, and Technology: A Reader in Feminist Science Studies* 3rd Ed (New York: Routledge)

Effects of Limited Paternal Exposure to Xenobiotic Agents on the Development of Progeny

GLADYS FRIEDLER

Boston University School of Medicine, 80 East Concord Street, Boston MA 02118

FRIEDLER. G. *Effects of limited paternal exposure to xenobiotic agents on the development of progeny.* NEUROBEHAV TOXICOL TERATOL 7(6) 739–743, 1985.—This report represents a series of investigations of the postnatal functional sequelae of paternal exposure to opioids and other xenobiotic agents. In separate studies, young adult male mice were exposed to (I) morphine; (2) levorphanol or its nonanalgesic isomer dextrorphan; or (3) 80% nitrous oxide/20% oxygen. Males (5–8/group) were injected twice daily for 5½ or 8½ days with opioids or saline, or received a single 4 hour inhalation exposure to nitrous oxide/oxygen or compressed air. At 6–8½ days post-treatment, each male was housed with 3 drug-naive nulliparous females. With the exception of birth weights of litters, no alterations in reproductive indices were observed. Changes in reproductive endocrine parameters included marked attenuation of serum luteinizing hormone response to castration in both young adult (8 week) and older (18 week) F_1 morphine offspring, when compared with saline-derived groups. The morphine progeny showed decrements in body weight and delayed onset of maturational indices through four generations of selective inbreeding. Similar developmental delays occurred in F_1, offspring originating from paternal levorphanol, dextrorphan and nitrous oxide groups, when compared with their respective controls. Behavioral study of F_1, offspring revealed alterations in: water maze performance and learning, if preceded by unavoidable footshock, in morphine offspring at 8 and 18 weeks; aberrant swim patterns in both levorphanol and dextrorphan progeny at 6½ and 8½ weeks. Nitrous oxide progeny showed a blunting of hypothermic response to pharmacologic challenge at 16 weeks, in comparison to compressed air controls. These results suggest that limited paternal exposure of mice to xenobiotic agents can induce permanent changes in the functional maturation of resultant progeny, including alterations in growth and in behavioral and neuroendocrine integrity.

Maturation   Inhalation   anesthetics   Nitrous   oxide   Xenobiotic   Opioids   Dextrorphan   Paternal   Male-mediated   Growth retardation   Intergenerational effects   Neuroendocrine   Behavioral decrements   Progeny

SUCCESSFUL reproduction in all mammals depends upon the genetic, anatomic, and endocrine integrity of the reproductive tract and gametes. Because of the cryptic and periodic nature of human reproduction, a variety of influences, including drugs and other xenobiotic agents or environmental stressors, may impair reproduction or progeny outcome, yet remain undetected. The major focus of reproductive investigations has been on the consequences for the fetus and neonate of maternal exposure during pregnancy. Interest in and investigation of the sequelae of paternal exposure upon progeny is of recent origin. Prior to the 1977 report of sterility and infertility among male workers exposed to the pesticide dibromochloropropane, occupational studies of male reproductive effects were virtually nonexistent in the United States [9, 27]. Epidemiologic studies now have identified a number of drugs and other xenobiotic agents which may adversely affect the reproductive process. Exposure of the male to numerous agents has been associated with adverse pregnancy outcome in their unexposed wives [10, 31, 36]. However, unless the effects are of overwhelming proportions, exogenous factors that damage the male germ cell are difficult to detect by epidemiologic study. As a consequence, the information on the sequelae of paternal exposure to a variety of potential toxins is limited.

It has been suggested that low sperm count and an increase in morphologic abnormalities may be related to an increased incidence of spontaneous abortions [25] although the relationship of sperm defects to heritable changes in the human is not well understood [38]. In addition to the prevention of fetal deaths, of equal concern is paternal exposure to exogenous factors which result in structural or functional alterations in live births and may also influence future generations. Postnatal functional deficits, often delayed and expressed as subtle behavior anomalies, are frequently not recognized as prenatal in origin. It is therefore not surprising that there is little information on the functional correlates in progeny of human males exposed to xenobiotic agents. The recent reports of neurophysiologic deficits among sons of alcoholic fathers and abstinent mothers appear to be an important breakthrough in this regard [4, 12, 32].

In the rodent, paternal exposure to gametotoxic agents has been reported to adversely affect reproductive indices and behavior of offspring [1,7]. More surprising

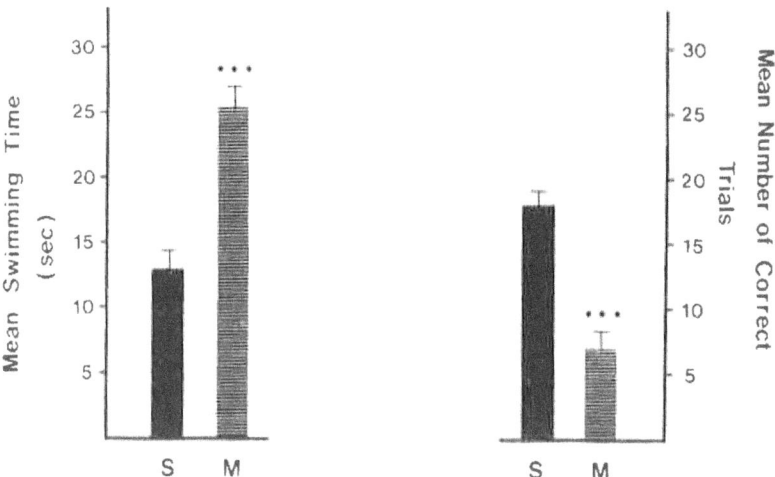

Fig. 1. Water maze behavior in F_1 offspring at 18 weeks after prior exposure to unavoidable shock (0.16 mA. 2 sec) at 12 weeks. Sires received progestational injections of morphine (M) or saline (S). Each point represents mean (± S.E.) for 6–8 mice (4 litters). Significant differences (*t*-tests) indicated by *** ($p < 0.001$). Reprinted with permission from *Pharmacol Biochem Behav Suppl* **11**: 23–28, 1979.

are the findings of subtle yet long-term alterations in the normal maturational pattern of progeny derived from sires which have been exposed to minimal endocrine [3, 24, 34] or drug [13, 19–23] manipulation. Our laboratory has been investigating the protracted effects of limited exposure of the male mouse to opioids and other xenobiotic agents upon the functional integrity of his progeny. The studies evolved from our findings that persisting alterations in growth and behavior of progeny followed the progestational exposure of the female rodent to morphine [16, 18]. Although an extensive literature now documents functional as well as morphologic alterations following prenatal exposure to a variety of agents, the observations were unexpected since the premating experimental design precluded any direct drug exposure of mother or fetus during prenatal development.

The presence of alterations in a second (F_2) generation of progeny suggested that the premating exposure exerted a profound and probably permanent influence upon both the female rodent and her descendants [13] and provided the impetus for study of any male-mediated effects. Paternal exposure minimizes the likelihood of effects due to direct drug exposure or to the multiple maternal and placental factors which could contribute to any observed deficits. It thus permits closer examination of the comparative role of genetic and environmental influences.

In separate studies, adult male mice (CD-1, Charles River Laboratories), 10–12 weeks of age, were injected SC twice daily for 5½ or 8½ days with morphine, levorphanol or dextrorphan, or saline vehicle. Drug doses were increased in daily increments to a maximum of 240 mg/kg (morphine) or 42 mg/kg (levorphanol, dextrorphan). Other groups received a single 4 hour inhalation exposure of 80% nitrous oxide/20% oxygen or compressed air. Within each study, control and experimental males (5–8/group) were mated at identical post-exposure intervals (6–8½ days) depending upon the individual study. Each male was mated with 3 treatment-naive nulliparous females (8–10 weeks) for a single estrous cycle and females were isolated at one week prior to parturition. Births were recorded within 12 hours and litters culled to 8 pups (4M, 4F) using procedures detailed elsewhere [16]. All studies used a blind procedure throughout, including injection of males and the analysis of data, in order to eliminate any source of handling or observer bias.

GROWTH EFFECTS; INBREEDING STUDIES

In the first group of studies, male mice were injected with morphine (5 ½ days) and mated to drug-naive females at 6 days following the final injection. Retardation in growth of F_1 generation offspring of the morphine-treated males and delays in pinna unfolding and eye opening were apparent when compared with offspring of saline-derived sires. Extensive karyotyping of F_1 offspring did not detect any consistent aberrations in chromosome morphology among offspring of paternal morphine derivation. The modal number of chromosomes corresponded to that of the normal diploid mouse and the frequency of other aberrations was well within the range of a comparable control series [23].

Progeny derived from paternal drug and control groups (8–10 litters/group) were then selectively inbred for four generations by choosing single brother-sister pairs from each litter [26]. To minimize selection bias unrelated to paternal treatment,

TABLE 1. SERUM LH LEVELS IN F_1 MALE OFFSPRING*

Age (weeks)	Group	Morphine	Serum LH† Saline
8	sham	13.9 ± 3.05‡	12.7 ± 2.00
	castrate	362 ± 46.4§	545 ± 565$$
18	sham	23.8 ± 2.42	30.1 ± 4.02
	castrate	113 ± 16.9§	226 ± 42.3

* Derived from paternal morphine and saline lineage. Sacrificed at 48 hr post-surgery.
† ng/ml.
‡ Mean ± S.E. of 8–10 males from separate litters.
§ $p < 0.05$ compared with saline controls.

each paternal drug pair was weight-matched with a litter pair of control lineage. Intergenerational delays in growth and maturational landmarks were evident in progeny of the paternal morphine line [13], with body weight differences between control and opiate groups increasing with the continued inbreeding, Maturational indices were severely retarded in F_1 litters of morphine lineage. At three weeks of age, the physical appearance and weight of these progeny were comparable to saline-derived pups at 10 day s of age. This highly visible developmental delay stands in marked contrast to prior pregestational maternal or paternal studies in which differences between parental treatment groups were apparent only after data analysis. These selective inbreeding findings suggest a role of genetic components in the male-mediated effect on growth and early maturation.

BEHAVIORAL MEASURES IN F_1 OFFSPRING

Behavioral alterations of F_1 offspring of paternal opioid lineage are also apparent over a wide age span and include effects on both learned and unlearned behaviors [14, 17, 22]. Mice of paternal morphine and saline parentage were tested in a single alley water maze (18.5 °C) at 18 weeks after prior exposure to an unavoidable footshock (12 weeks) using procedures previously detailed [22]. Mice were individually tested for 20 trials (30 sec cutoff, 20 sec intertrial interval). Swim speed, the time from the starting point to the exit ramp, was used as a measure of performance. The number of error-free trials was used as a measure of maze learning.

Female offspring of paternal morphine groups showed an increase in swim time and decrease in errorless trials, as compared with control mice (Fig. 1). In a second study, these alterations were also evident in young adults (8 weeks) of both sexes following footshock (0.16 mA, 2 sec) two weeks earlier. Differences in swim behavior between paternal treatment groups were not apparent unless the task was preceded by an earlier experience with unavoidable shock [22]. These results, together with prior findings of decrements in exploratory and locomotor behavior [15] and attenuation of response on two aversive procedures [13,14], suggest possible alterations in responsiveness to stress-producing tasks in progeny of drug parentage.

ENDOCRINE PARAMETERS

Overt endocrine abnormalities are not evident in parent or progeny of paternal opioid groups. With the exception of lower birth weights, no alterations in reproductive indices (including fertility, gestation period, sex ratio, litter size, incidence of stillbirths) have been noted. These latter observations are surprising since the disruptive effects of opioids on reproductive function have been well documented for several species. It was therefore decided to examine reproductive endocrine parameters in male mice of paternal opioid and control parentage.

In the normal animal, hypothalamic LHRH stimulates the pituitary to release luteinizing hormone (LH) which in turn simulates testosterone production. Testosterone then exerts negative feedback control on the release of LH. In the castrated mouse, the absence of this control results in both production and release of LH at near maximum and thus provides a model system which facilitates the detection of any subtle deficits in the normal functioning of the hypothalamic-pituitary-gonadal (HPG) axis.

Sexually mature sires were injected (8½ days) with saline or morphine and were mated after a similar drug-free interval with drug-naive females. Resultant offspring of the two paternal groups were castrated or sham-operated (18 weeks). At 48 hours post-surgery, they were sacrificed in preparation for neuroendocrine evaluation of their HPG axis, as detailed elsewhere [8].

Serum LH levels were determined by radioimmunoassay [29]. A slight decrease in basal levels of LH was apparent in sham-operated mice of morphine lineage, as compared with saline controls. However, the response of the two paternal groups to castration differed markedly [20]. The anticipated surge of serum LH which normally accompanies castration was present in control mice and was in marked contrast to the blunted LH response to castration in those of morphine parentage. These findings were replicated in a second study in 8 week offspring (Table 1). This attenuated response to castration in both young adult and older mice of paternal morphine groups suggests a tenacious imprint of the paternal exposure on male reproductive endocrinology, despite the absence of obvious reproductive deficits.

Depression of both seminal vesicle weight and serum testosterone levels and a slight decrease in pituitary LH were evident in morphine pups. Pituitary response to exogenous LHRH, hypothalamic LHRH content and opioid receptor profiles in hypothalamus and whole brain did not differ significantly for the two paternal groups.

SPECIFICITY OF PATERNAL IMPRINT

More recent investigation indicates that the tenacious imprint of paternal exposure is not unique to morphine and its congeners [17]. In a comparison of paternal effects of a potent opioid (levorphanol) with its analgesically inactive enantiomer (dextrorphan), both agents resulted in significant depression of birth weights and delayed onset of developmental indices (pinna unfolding, eye opening). Alterations in the normal pattern of swim behavior were also evident in offspring of both groups. Using criteria described elsewhere [29], several types of swim movement (including front and hind limb positioning and motion, freezing, splaying) were rated independently by two "blind" observers. A higher incidence of movement aberrations occurred in both stereoisomer groups when compared with saline-derived pups at 45 and 60 days of age, and in both sexes. These observations indicate that male-mediated alterations are not restricted to opioid analgesics.

The latter findings, together with epidemiologic and experimental studies which suggest that paternal exposure to inhalation anesthetics may adversely affect reproduction, led to investigation of the sequelae of paternal exposure to nitrous oxide (N_2O). Male mice received a single 4 hour inhalation exposure of either N_2O or compressed air (flow rate 0.1 liter/min) using a modification of described procedures [28]. Mice were mated at 7 days post-exposure.

Paternal exposure to N_2O resulted in a significant depression of litter birth weights ($p < 0.01$) and delay in appearance of maturational indices. As in prior studies, these changes occurred in the absence of alterations in other reproductive parameters. Differences in preweaning or early postweaning behaviors between paternal groups were not observed.

Pharmacologic challenge can often unmask functional deficits which may otherwise remain undetected. Since morphine is frequently employed as a probe to explore thermoregulatory function, hypothermic response of mice of the two paternal groups to morphine (10 mg/kg) or vehicle challenge was assessed at 16 weeks, at 25°C ambient temperature. Rectal temperature was recorded by thermistor probes at 30 min intervals for 60 min prior to injection and for 180 min post-challenge, and changes from preinjection baseline determined. Drug-induced hypothermia was less evident in N_2O offspring, when compared with offspring of compressed air groups. Paternal treatment differences in response to drug challenge were significant, $F(1,48) = 5.13$, $p < 0.028$, with the effect more pronounced in males [21].

This attenuated response to hypothermic challenge in N_2O offspring suggests that paternal exposure may affect central control of thermoregulation. Further evidence of a profound effect of paternal exposure is the blunted response of the HPG axis in paternally exposed progeny, findings which indicate alterations in the functional integrity of this neuroendocrine system. As suggested previously [22], the decrements in behavioral response to a stressor such as unavoidable shock could also reflect alterations in neuroendocrine function.

It is difficult to propose a mechanism by which exposure of the male can result in alterations in the functional maturation of his progeny, and in particular in the absence of evidence of any reduction in fertility or the induction of dominant lethal effects. In the mouse, a complete cycle of spermatogenesis is 34½ days [30] and mature spermatozoa may be stored for an additional 10–14 days prior to ejaculation. Thus, in all experiments presented in this paper, the time interval encompassing the exposure and drug-free periods prior to mating would restrict effects to either epididymal spermatozoa or to gametes retained within the vas deferens. However, additional observations in our laboratory have demonstrated that functional changes in offspring occur following matings at prolonged periods (55–60 days) after paternal exposure to both opioids [19] and their analgesically inactive enantiomers (unpublished studies). This latter time frame involves exposure at the earliest stages of gamete development and indicates that a paternal influence is not restricted to postmeiotic cells.

Although the etiology of the paternal effect is not presently understood, some process of gamete selection may be involved. The altered integrity of the HPG axis in progeny suggests that xenobiotic-induced changes in paternal endocrine function could participate in such a process. It is conceivable that xenobiotic-induced changes in receptors on spermatozoa affect sperm motility or metabolism and thereby select a certain population of gametes for participation in the fertilization process, although such receptors have yet to be identified. Alternately, if the normal process of sperm maturation which occurs during transport through the epididymis is altered in some way by the paternal exposure, this might influence the specific pool involved in fertilization. Should semen composition itself be altered, either through effects on absorption or secretion within the epididymis or in other sex accessory organs, this might also have a selective influence on the pool of gametes available for fertilization. It is even more difficult to speculate on the role that such selective processes might play following paternal exposure during premeiotic stages of germ cell maturation.

There is precedence for a selective effect on sperm populations through alterations in the expected transmission ratio of alleles at the t-locus in the mouse [5,11]. Distortion of this ratio results in the preferential selection of one allele over others. It is a major departure from conventional genetics and occurs only in the male. The locus has striking and unorthodox features which include distortion of embryonic development and alterations in sperm differentiation and function [33]. This phenomenon has intrigued geneticists since it was first described by the eminent geneticist L. C. Dunn in the 1930's. Professor Dunn devoted a considerable portion of his professional career to characterizing this locus and unravelling mechanisms responsible for the selective transmission [6,11]. The latter remain elusive.

Spermatogenesis, an intricate and complex process, and the evidence that some selective process may also occur within the female reproductive tract are beyond the purview of this discussion. The possible influence of these and other reproductive endocrinologic factors in the paternal effect, is further complicated by the rapidly developing information on the complexity of the genome. It is now recognized that a variety of environmental influences (including other cells, and small molecules such as hormones and drugs) impact on genomic expression. Such observations emphasize the importance of determining whether or how receptors on the cell surface identify these environmental components and produce signals which affect other cells. For example, antisera against testicular cells have been developed which can recognize unique antigens on male germ cells [2].

SUMMARY AND CONCLUSIONS

Xenobiotic agents, including those with pharmacologically or structurally distinct properties, can induce protracted and apparently permanent alterations in progeny after limited paternal exposure in the mouse. Despite the vulnerability of both premeiotic and postmeiotic stages of spermatogenesis to this effect, sequelae are frequently subtle in character. However, the fact that a number of functional parameters are affected illustrates the pervasive nature of the influence. A single paternal injection of morphine can also interfere with growth and developmental maturation of mouse offspring (unpublished observations) and further documents the susceptibility of the male organism to insult. Although the etiology of the paternal influence is unknown, there is no reason to assume that a male-mediated effect is restricted to the agents studied to date. In fact, the broad spectrum of changes recorded in these studies, in consort with reports of long-term and intergenerational alterations in the phenotypic expression of rodent progeny which follow manipulations of the paternal endocrine milieu, suggest that male-mediated effects may be far more ubiquitous than recognized heretofore. Mechanism(s) for the paternal imprint have yet to be elucidated, but the findings appear to reflect participation of both genetic and environmental components.

REFERENCES
1. Adams, P. M., J. D. Fabricant and M. S. Legator. Cyclophosphamide-induced spermatogenic effects detected in the F_1 generation by behavioral testing. *Science* **211** 80–82, 1981.
2. Artzt. K. and D. Bennett. Serological analysis of sperm of antigenically cross-reacting T/t haplotypes and their recombinants. *Immunogenetics* **5**: 97–107, 1977.
3. Bakke, J. L., N. L. Lawrence, S. Robinson and J. Bennett. Observations on the untreated progeny of hypothyroid male rats. *Metabolsim* **25**: 437–444, 1976.
4. Begleiter, H., B. Porjesz, B. Bihari and B. Kissin. Event-related brain potentials in boys at risk for alcoholism. *Science* **225**: 1493–1496, 1984.
5. Bennett, D. L. C. Dunn and his contribution to t-locus genetics. *Annu Rev Genet* **11**: 1–12, 1977.
6. Bennett. D. and L. C. Dunn. Studies of effects of l-alleles in the house mouse on spermatozoa. *J Reprod Fertil* **13**: 421–428, 1967.
7. Brady. K., Y. Herrera and H. Zenick. Influence of parental lend exposure on subsequent learning ability of offspring. *Pharmacol Biochem Behav* **3**: 561–565, 1975.
8. Cicero, T. J. Opiate and opioid modulation of reproductive endocrinology in the male and female: development and pregestational aspects. Proceedings of 46th Annual Meeting of Committee on Problems of Drug Dependence. NIDA Research Monograph **55**: 14–23. 1984.
9. Cohen, E. N., B. W. Brown, D. L. Bruce, H. F. Cascorbi, T. H. Corbett, T. W. Jones and C. E. Whitcher. Occupational disease among operating room personnel. *Anesthesiology* **41**: 341–344, 1974.
10. Cohen. E. N., B. W. Brown. M. L. Wu, C. E. Whitcher, J. B. Brodsky. H. C. Gift, W. Greenfield, T. W. Jones and E. J. Driscoll. Occupational disease in dentistry and exposure to anesthetic gases. *J Am Dent Assoc* **101**: 21–31, 1980.
11. Dunn, L. C. Abnormalities associated with a chromosome region in the mouse. I. Transmission and population genetics of the t-region. *Science* **144**: 260–263, 1964.

12. Elmasian. R., H. Neville. D. Woods. M. Schuckit and F. Bloom. Event-related brain potentials are different in individuals at high and low risk for developing alcoholism. *Proc Natl Acad Sci USA* **79:** 7900–7903, 1982.
13. Friedler, G. Long-term effects of opiates. In: *Perinatal Pharmacology: Problems and Priorities*, edited by J. Dancis and J. C. Huang. New York: Raven Press, 1974, pp. 207–216.
14. Friedler, G. Effects of progestational administration of morphine to mice on behavior of their offspring. *Pharmacologist* **16:** 203, 1974.
15. Friedler, G. Effect of progestational morphine and methadone administration to mice on the development of their offspring. *Fed Proc* **36:** 1001, 1977.
16. Friedler, G. Progestational administration of morphine sulfate to female mice: long-term effects on the development of progeny. *J Pharmacol Exp Ther* **205:** 33–39, 1978.
17. Friedler. G. The effect of preconception administration of opioid stereoisomers in the male mouse on progeny. *8th Int Pharmacol Congr* **857:**, 1981.
18. Friedler. G. and J. Cochin. Growth retardation in offspring of female rats treated with morphine prior lo mating. *Science* **175:** 654–656, 1972.
19. Friedler, G. and C. A. Crescenzi. Developmental alterations in offspring of male mice mated after a prolonged opioid-free interval, *Fed Proc* **40:** 264, 1981.
20. Friedler, G., M. Gerrity and T. J. Cicero. Alterations in the hypothalamic-pituitary-gonadal axis of offspring of male mice exposed to morphine before mating. *Fed Proc* **41:** 1301, 1982.
21. Friedler, G. and M.-E. Meadows. Paternal exposure to nitrous oxide: effects on the development of offspring. *Neurotoxicology* **7:** in press, 1986.
22. Friedler, G. and H. S. Wheeling. Behavioral effects in offspring of male mice injected with opioids prior to mating. *Pharmacol Biochem Behav Suppl* **II:** 23–28, 1979.
23. Friedler, G. and D. Wurster-Hill. Influence of morphine administration to male mice on their progeny. Proceedings of 36th Annual Meeting of Committee on Problems of Drug Dependence. National Academy of Sciences, National Research Council. 869–875, 1974.
24. Fujii, T. and M. Sakamoto. Parathyroid transplantation in male rats affects the functional differentiation of the testis in their first generation offspring. *Biomed Res* **1:** 556–559, 1980.
25. Furuhjelm, M., B. Jonson and G. G. Lagergren. The quality of human semen in spontaneous abortion. *Int J Fertil* **7:** 17–21, 1962.
26. Green, E. L. Breeding systems. In: *Biology of the Laboratory Mouse*, edited by E. L. Green. New York: Dover Publications, 1975, pp. 11–32.
27. Knill-Jones, R. P., B. J. Newmann and A. A. Spence. Anesthetic practice and pregnancy. Controlled survey of male anesthetists in the United Kingdom. *Lancet* **2:** 807–809, 1975.
28. Land, P. C., E. L. Owen and H. W. Linde. Morphologic changes in mouse spermatozoa after exposure to inhalation anesthetics during early spermatogenesis. *Anesthesiology* **54:** 47–50, 1981.

29. Niswender, G. P., L. E. Reichert, A. R. Midgley and A. V. Nalbandov. Radioimmunoassay for bovine and ovine luteinizing hormone. *Endocrinology* **84**: 1166–1173, 1969.
30. Oakberg, E. F. Duration of spermatogenesis in the mouse and timing of stages of the cycle of the seminiferous epithelium. *Am J Anal* **99**: 507–516, 1956.
31. Peters, J. M., S. Preston-Martin and M. C. Yu. Brain tumors in children and occupational exposure of parents. Science **213**: 235–237, 1981.
32. Schmidt, A. L. and H. Neville. Language processing in men at risk for alcoholism: an event-related potential study. I. *Alcohol* **2**: 529–533, 1985.
33. Silver. L. M. Genetic organization of the mouse t complex. *Cell* **27**: 239–240, 1981.
34. Spergel. G., L. J. Levy and M. G. Goldner. Glucose intolerance in the progeny of rats treated with a single subdiabetogenic dose of alloxan. *Metabolism* **20**: 401–413, 1971.
35. Spyker, J. M., S. B. Sparber and A. M. Goldberg. Subtle consequences of methylmercury exposure: Behavioral deviations in offspring from treated mothers. *Science* **177**: 621–623, 1972.
36. Strobino. B. R., J. Kline and Z. Stein. Chemical and physical exposures of parents: effects on human reproduction and offspring. *Early Hum Dev* **1**: 371–399, 1978.
37. Whorton, D., K. M. Krauss, S. Marshall and T. H. Milby. Infertility in male pesticide workers. *Lancet* **2**: 1259–1261, 1977.
38. Wyrobek, A. J., L. A. Gordon, J. G. Burkhart, M. W. Francis, R. W. Kapp, G. Letz, H. V. Malling, J. C. Topham and M. D. Whorton. An evaluation of human sperm as indicators of chemically induced alterations of spermatogenic function. *Mutat Res* **115**: 148–149, 1983.

Case Study 10: Memory Survives Decapitation

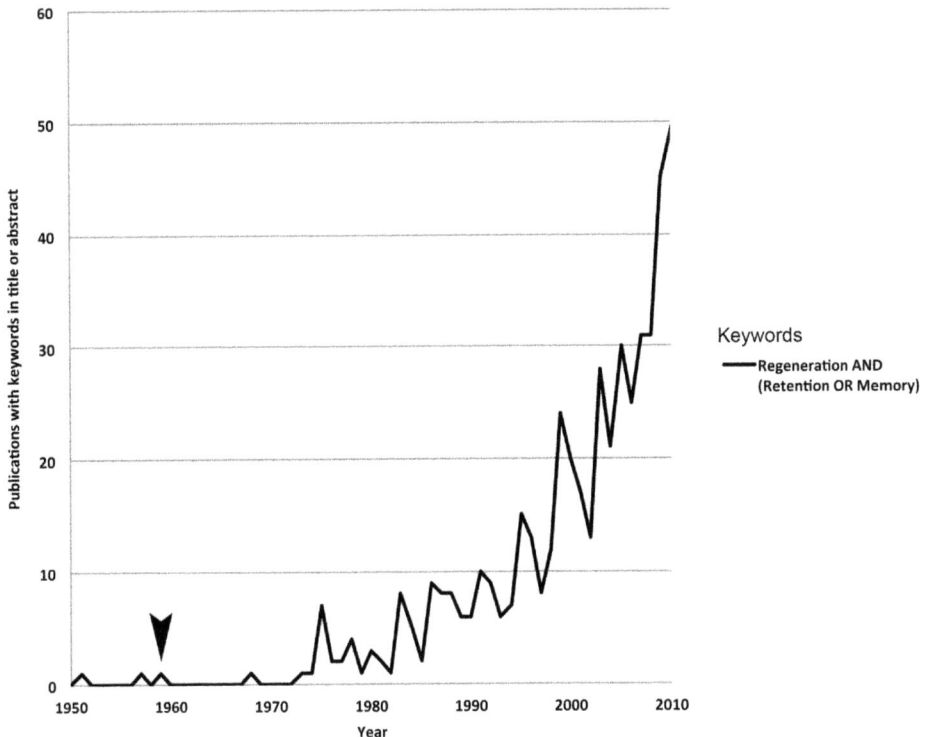

McConnell *et al* (1959) The effects of regeneration upon retention of a conditioned response in the planarian.

Perspective on *The effects of regeneration upon retention of a conditioned response in the planarian*

Lawrence Stern
College Collin

The Reception of Extraordinary Scientific Claims: The Search for the Elusive Engram

Scientific claims that depart in significant ways from prevailing cognitive frameworks pose formidable problems in science. If borne out to the satisfaction of the relevant practitioners working in the problem-area, claims such as these will force major, often revolutionary, revisions of those frameworks—and a plethora of honorific awards, perhaps even a Nobel prize, might be in the offering to those nailing the claim down. It is likewise true, however, that claims such as these, initially appearing to have strong evidential support, have led scientists far astray and down false paths. Since there is no sure way of knowing whether such claims will turn out to be a 'boon' or a 'bust', scientists in their research and gate-keeping roles—and the scientific community at large—are faced with a serious dilemma. How are such claims to be handled? Should these lines of research be funded and the claims granted access to the reputable journals in the field? If so, for how long? Correlatively, should researchers invest their time, energy, scarce resources, and perhaps their reputation in the pursuit of such claims?

These problems are not only formidable, they are constantly occurring. They are part and parcel of the inner workings of the scientific community. Nearly all analysts of science stress that it is essential that science reach a balance between its reliance on the traditions of the day and the vigorous pursuit and accommodation of new novel findings. If a proposed novelty were judged solely on its relation to existing theories, very few major transformations would occur. For such transformations to take place the novel finding must gain some first supporters who will develop the idea to the point of general persuasiveness and attract other workers And, as sociological fine-grained analysis of case studies have indicated, this is where much of the action occurs in science.

Rather than take this as a matter of methodological precept—how science 'ought' to operate—sociologists argue that this is the inevitable outcome of specifiable social processes. The reward system in science provides institutionalized incentives that motivate scientists to challenge prevailing frameworks. A high premium is placed on originality, and the rewards scientists receive—recognition by qualified peers and the promotions, honorific awards, and material resources for future research that derive from this recognition—are usually somewhat proportionate to the perceived 'impact' of their research. Although scientists who propose novel ideas or report unusual findings have the burden of persuading their peers—and in keeping with the maxim that 'extraordinary claims require extraordinary evidence' these claims will come under more intense scrutiny than those of lesser significance—the introduction of such claims, so long as they conform to the methodological canons of the time, is not only appropriate but approved.

My research focuses on controversy in science, and I study how both social and cognitive factors, in varying degrees, affect the development and reception of what I refer to as 'extraordinary scientific claims'—claims that are initially perceived to depart in significant ways from prevailing cognitive frameworks while at the same time being widely recognized as having an enormous impact—in both basic and applied senses—on the development of science and its benefits to society. James V. McConnell's claim (McConnell *et al* 1959), reprinted here, that memory survives regeneration—more so—that memory is encoded in molecular form and resides not solely in the brain but, rather, throughout the body of an organism, is a prime example of this type of claim. It is important to note that McConnell's planarian regeneration studies were but one phase of an overarching research program that used other test animals (newts, mynah birds, and, of course, rats) and procedures (cannibalism, shuttle boxes, injections of brain extracts) in attempts to demonstrate that memories are encoded in specific forms of RNA—that RNA is an 'archival' molecule.

How, then, was this claim developed and how was it received?

The 'search for the *engram*'—the physical representation of memory in an organism—has a rather long and checkered past. Some of the sharpest minds in science have been drawn to the problem and research has been—and currently is—decidedly pluralistic. During the 1950s, the period during which McConnell conducted his research on regeneration in planarians, the field was in a turbulent and agitated state.

At this time, the conventional wisdom held that memory was primarily a matter of electrical impulses traveling along neural pathways. For those who took the notion of a neural code seriously, attention was directed to deciphering electrical wave patterns recorded from neural structures, such as the hippocampus. The underlying assumption was that these patterns, like the electromagnetic waves in the telecommunication systems, were the main carriers of information. In other words, it was the frequency and amplitude of electrical waves traveling over *specific neural pathways* that most likely accounted and coded for memories. These were dubbed 'connectionist' or 'switchboard' theories.

What was less certain was precisely *where* in the brain memories were located. Which were the relevant neural pathways? The issue that divided researchers was whether or not the so-called memory trace was localized or more widely distributed throughout the cortex. Karl Lashley took the latter position, denying the importance of specific neural connections. The memory trace, he argued, was a purely dynamic pattern of electrical activity and was reflected in unique spatiotemporal patterns of discharges of electrical activity rippling throughout the brain without any reference to specific neural pathways. Memories, then, were coded in the 'bioelectric field' that was generated by the aggregate activity of countless neurons. Indeed, Lashley argued that the same set of neurons might be involved in quite different learning tasks and/or the storage of different memories as electrical waves of differing frequencies or amplitudes.

The publication of Donald Hebb's *The Organization of Behavior* (1949) was clearly a watershed event. Here, Hebb (a former student of Lashley) presented a dual trace theory that can be seen as an attempt to reconcile the so-called connectionist or switchboard theories with Lashley's field theory. The trace for short term or transient memories, Hebb argued, was to be located in the reverberatory electrical activity of cell assemblies that were thought to be widely distributed in the brain. To account for long-term memory, however, Hebb argued that some structural change in the brain's circuitry must occur. In his view, this entailed either the growth of new synapses or the alteration of those that already existed. Moreover, although he was somewhat vague about this, Hebb also suggested that this change or alteration in synaptic connections was brought about by the preceding reverberatory activity. This view gained currency when it was found that long-term memory was not abolished by complete cessation of electrical activity—as observed in hypothermia or by its violent perturbation during electroconvulsive shock. The synapse was now where the action was thought to be.

Such was the state of affairs when James V. McConnell, then a graduate student at University of Texas in Austin, was approached by fellow graduate student Robert Thompson some time in 1953. Having just read Hebb's book, Thompson was eager to test the notion that synaptic connections were implicated in learning. Since the planarian—the common flatworm—is the simplest animal on the phyletic tree with true synapses, it should, Thompson reasoned, be capable of learning.

Convinced that they had demonstrated classical conditioning—they paired the appearance of a bright light with an electric shock—they submitted their results to the *Journal of Comparative and Physiological Psychology,* a reputable and mainstream journal, where it appeared, without a hitch, in 1955. To be sure, there were some who thought that the idea that a lowly worm could learn sounded far-fetched, but the full brunt of that criticism remained to be seen.

When McConnell arrival at the University of Michigan in the fall of 1956, his chairman promptly informed him that there were certain expectations—that the old bugaboo that one must 'publish or perish' was indeed true. Looking for a publishable experiment, and intrigued by his worm studies, McConnell set up a worm lab and recruited two students, Dan Kimble and Allan Jacobson, to work with him. Knowing that planarian are capable of regeneration, McConnell wondered which half of a bisected beast would retain a conditioned response.

As will be seen in the full article following this Commentary, McConnell and his students classically conditioned planarians of the species *Dugesia dorotocephala* using light as a conditioned stimulus and shock from a Harvard inductorium as an unconditioned stimulus. After the animals had reached the criterion of 23 responses out of 25 consecutive trials, they were cut in half and allowed to regenerate for a month. McConnell wrote,

> 'In all honesty I must admit that we did not obtain the results we had expected. We had assumed that the regenerated heads would show fairly complete retention of the response for, after all, the head section retained the primitive brain and 'everybody knows' that the brain is where memories are located. And the heads did show just as much retention as did the uncut control animals. We had also hoped, in our hearts, that perhaps the tails would show a

slight but perhaps significant retention of some kind, merely because we thought this would be an interesting finding. We were astounded, then, to discover that the tails not only showed as much retention as did the heads, but in many cases did much better than the heads and showed absolutely no forgetting whatsoever' (McConnell 1962:44).

As before, McConnell submitted the experimental report to the *Journal of Comparative and Physiological Psychology*. But this time, there was a hitch—in fact, several hitches. After sending the report out for review, Harry Harlow, the editor, informed McConnell that one reviewer referred to the work as 'patent nonsense', while another questioned the work on methodological and procedural matters. McConnell revised the paper, adding material to satisfy the referees and despite their grumblings the paper appeared in early 1959.

These results led McConnell to think more seriously about the chemical nature of memory. But at this point, McConnell still didn't know quite what to make of his results. Yes, they suggested that memory wasn't necessarily stored in the beast's primitive brain—that it had apparently migrated throughout the body down to the tail end . . . but in what form? He was, in fact, still thinking in Hebbian terms and didn't think that his data did too much damage to Hebb's theory of cell assemblies. After all, the cell assemblies could easily be built up throughout the worm's body, and when the tail section was retrained, it would learn faster since parts of it would contain cell assemblies already set up in the original animal.

The results of the *next* experiment, however, convinced McConnell that chemicals were more important that neural structures (McConnell *et al* 1959). Along with Reeva Jacobson he showed—to his satisfaction—that a trained worm could be cut in *several pieces*, and each piece would retain the training after regeneration. More importantly, they trained a worm, cut off (and discarded) the original head and let the original tail grow a new head and then cut off the original tail and let the new head grow a new tail. Memory survived total regeneration. That is to say, *a planarian that contained none of the structure of the originally trained animal retained the memory*. In McConnell's view, Hebb's theory—in which memory was coded in neural structures—had fallen and the only way he could imagine the memory's being retained in the completely regenerated animal was if the engram was contained in a chemical, not in a neural structure.

Once more, McConnell submitted the experimental report to the *Journal of Comparative and Physiological Psychology*. Once again, the reviewers' comments were quite harsh, leading Harlow to write that in view of these comments it would be impossible to publish the article until McConnell could in some way or another establish the conditioned response beyond any reasonable doubt of his fellow psychologists. Rather than deal with what McConnell believed to be recalcitrant —and biased—reviewers, he decided to withdraw the paper.

Undeterred, McConnell became more firm in his belief that after conditioning, the 'engram' was be stored throughout the planarian's body, not just in the animal's brain, and that the storage mechanism was chemical in nature. Another regeneration study, this one conducted by E. Roy John and his student William Corning at the

University of Rochester (Corning and John 1961), reinforced McConnell's notions and led him to focus on RNA as the putative memory molecule. In their experiment, which appeared in the journal *Science*, Corning and John trained flatworms using the light/shock conditioning technique McConnell had developed and, after cutting them in half, allowed some of the pieces to regenerate in the usual pond water, while others did so in a weak solution of ribonuclease, an enzyme that breaks up RNA. Although both heads and tails that regenerated in pond water showed evidence of retention, the tails regenerating in ribonuclease showed no evidence of the original training. To McConnell, it seemed as if the enzyme had somehow erased the blackboard on which the memory was written.

Taken with the notion that RNA was somehow involved, McConnell began to think about ways to get the RNA from a trained flatworm into the body of one that was untrained.

First he tried to graft the head of a trained worm onto the tail of a naïve worm—but the head kept falling off. Next he tried grinding up trained worms and injecting them into naïve recipients and that didn't work either. The hypodermic needles were too big—getting one inside a flatworm was like trying to impale a prune with a javelin—and if, by chance, the needle was positioned well enough to inject the planarian-puree, it either oozed out or, if not, caused the worm to explode.

When fellow worm runner Jay Boyd Best wrote McConnell about the cannibalistic tendencies of a particular species of planarians, McConnell designed his infamous transfer experiments. Both trained and untrained worms were cannibalized and recipients eating their educated brethren performed better than those ingesting untrained donors. Soon thereafter, McConnell reported that RNA taken from trained worms and injected into naïve recipients produced the trained behavior. Taken together, these results convinced McConnell that specific memories are independent of neural structures and, instead, were encoded in the structure of unique variants of RNA.

It is important to note that McConnell was by no means alone in speculating that RNA played a pivotal role in memory storage. In fact, expectations ran high and work proceeded along a number of collateral paths. The spectacular success of Watson and Crick led some to ask: If genetic information is stored in nucleic acids and proteins, why not acquired information as well? Although many neurophysiologists thought this equivalence nothing more than a bad pun, a number of molecular biologists, thinking that the time was ripe to apply their tools and analytic approach to the study of memory processes, began to seriously discuss these issues at various conferences, meetings and workshops.

Holger Hyden (1962), for example, comparing the brain chemistry of animals trained to perform a specific task to those of controls, reported both quantitative and qualitative differences in neuronal RNA. Others, such as Bernard Agranoff (1965), Louis Flexner (1964), and Samuel Barondes (1966) claimed that the administration of various inhibitors of RNA and protein synthesis either preceding or just after training had significant adverse effects on retention. Still others, like James McGaugh (1966) administered drugs to stimulate RNA and protein synthesis either preceding or just after training and claimed an enhancement effect.

But these investigators still adhered to the conventional view that memory was primarily a matter of electrical impulses traveling along specific neural pathways—and that if biochemical agents played any role at all, it was merely to grease the wheels of neural processing.

McConnell, of course, thought otherwise and at the Midwestern Psychological Meetings in Chicago, Illinois, on 2 May 1963, he unveiled his 'tape recorder theory of memory' which, he announced, was '99% wild speculation and had only what might euphemistically be called 'heuristic value" (McConnell [1963] 1965:3).

Memories, McConnell suggested, were recorded on RNA molecules much the same as voices were registered on magnetic tape by a tape recorder. Although he stated that he was unable to even imagine—much less explain—how the memory was actually encoded on the RNA, how the RNA code stored in a neuron was 'read out' or 'decoded', and how it eventually translated into behavior, he nevertheless 'theorized' that new variants of RNA were permanent, archival molecules, capable of being synthesized continuously and of migrating throughout the body of an organism. Moreover, and in direct opposition to the developing core of canonical knowledge in neurophysiology, the molecule *was* the memory and existed independently of any neural circuitry.

Reactions to McConnell's work during the fifteen plus years that the episode unfolded ranged from bemused and polite dismissal to outright ridicule and scorn. Some referred to it as idiotic, pure rubbish, or a waste of time. Nobel laureate James Watson called it, simply, 'sheer unmitigated rot'. Some who published positive findings were referred to—in print—as 'charlatans' and 'crackpots', having their competency, and on more than one occasion, their veracity called into question. Most working in the area were more than a bit leery when McConnell talked about the *inheritance of learning* or acquired characteristics—it smacked too much of Lamarckianism. And yet, as one commented, 'There is no doubt that the theory is crazy; the only question is whether it is crazy enough to be true'. Another simply stated that the transfer claims were 'just too damn important to ignore'.

It was not just the impact this work would have on prevailing notions of memory processes. The applied implications of this research stretched far beyond the halls of academia and were staggering. At the same time that McConnell's work was being debated, Cameron (Cameron *et al* 1963) reported that injections of yeast-RNA into elderly patients improved their memory. If drugs could be found that would either enhance learning or if specific information could be introduced into an organism chemically, new therapies dealing with the treatment of senility, dementia, learning disabilities, and various forms of mental retardation could be developed. To no one's surprise, pharmaceutical companies quickly entered the field. Two were testing patentable memory drugs in the early 1960s: Abbott Laboratories of Chicago was conducting tests of the efficacy of a drug it named 'cylert', a stimulant commonly known as pemoline combined with a magnesium compound, while International Chemical and Nuclear Corporation of Los Angeles was testing 'ribanimol', a drug derived from yeast cells. Smith, Kline and French pharmaceutical laboratory, as well as Squibb and Wyeth, also conducted research in this vein. Talk of new wonder drugs and memory pills filled the airwaves and the popular press.

It should come as no surprise, then, that, all told, roughly 175 independent researchers or research teams invested differing amounts of their time, energy, and resources conducting what came to be known as memory-transfer experiments—more than two dozen were in hot pursuit, devoting considerable resources in serious and concerted efforts.

To circumvent the criticism that worms were simply not capable of learning—or that they could not be trained reliably, the argument of Nobel laureate Mac Calvin—most researchers turned to rats. No one could argue that rats could not learn! But, in actuality, twenty-three different types of experimental subjects, including various squishies (octopus), scrunchies (praying mantis and cockroaches), fluffies (baby chicks) and furries (rats) were used in a variety of different learning paradigms.

And it is important to note that those conducting transfer experiments were, to an appreciable extent, fairly reputable scientists. At least one-quarter came from the upper reaches of the stratification system in science. Using the number of times a scientist's work is cited by others as a crude approximation of their standing in their field, we find that transfer workers received, on average, nearly 18 citations the year they conducted their experiments. This is three times higher than the average yearly citation rate for all published authors at the time and actually higher than the number of citations an average author received over a five-year period! Moreover, while only 2.3% of cited authors received more than 30 citations in any given year, 19.5% of those conducting these experiments fell into this category.

This, too, should come as no surprise. Risk-taking in science is not merely a matter of psychological proclivity. One's structural position and access to resources matters, too. Scientists in the upper tiers do not work on *a* research problem. They work simultaneously on an entire set of different problems and typically have access to graduate students, post-docs, and equipment. Given the rewards that would accrue, it makes sense to include at least one if not more 'long-shots' or 'high-flyers' in their repertoire, allocating some small portion of their resources to such matters.

This is not to say, of course, that these researchers believed that the phenomenon was valid—that so-called 'memory molecules' existed—but, rather, that it was something definitely worth pursuing. Indeed, nearly all of these researchers that obtained positive results thought that McConnell's notion of RNA as an archival memory molecule was far off the mark, arguing, instead, that these molecules acted more like 'sign-posts' that directed the flow of neural impulses along specific neural circuits.

Perhaps more important, even though the transfer work was generally perceived to radically contradict the fundamental tenets of orthodoxy, it was granted considerable access to the government research funds and the reputable journals of the field. Although, to be sure, not all grant applications were funded, government agencies (National Institutes of Health, National Science Foundation, National Institute of Education, Atomic Energy Commission, Office of Naval Research) awarded over $1 million (more than $8 million in 2016 dollars) specifically to support memory transfer work over the course of the episode.

Early on, McConnell's regeneration studies were handsomely supported. Although he was turned down by the National Science Foundation, between the years 1958 and 1965 the Atomic Energy Commission provided McConnell $113,759 to study 'The Effects of Radiation on Learning and Regeneration in Planarian'. At the same time, between June 1959 and September 1964 the National Institutes of Mental Health provided $41,876 to study 'Learning and Regeneration in the Planarian'. These two grants, totaling $155,635, amounted to $1,252,094 in 2016 dollars!

Publication space, too, was available (though, again, to be sure, not all submissions were accepted). Two-hundred-forty-seven research reports appeared in print and 75% of these papers—the overwhelming majority of these papers—appeared in reputable journals such as *Science* and *Nature*.

As the editors of this volume attest, the popular conception of science as a continuous, steady, upward climb of progress is inadequate. The history of science shows us that the development of knowledge is marked by all sorts of discontinuities. Some claims are 'premature' (Stent 1972). Others claims are 'post-mature' (Zuckerman and Lederberg 1986).

An analysis of the reception accorded to McConnell's initial claims that memory survives regeneration, that it is encoded in a molecular form and may be stored throughout the body of the organism—even with its Lamarckian overtones—may shed light on a recurring though poorly understood pattern in the history of scientific ideas that Robert K. Merton (1984) has dubbed the 'Phoenix Phenomenon': 'the continuing resiliency of theories or theoretically derived hypotheses even though they have been periodically subjected to much and allegedly conclusive demolition'. As Merton notes, it is important to draw a clear distinction between alleged refutations of a scientific claim, no matter how decisive it might seem to logicians, and the actual rejection of that claim, which is a purely behavioral phenomenon.

Recent work by Levin and his colleagues at the Tufts Center for Regenerative and Developmental Biology is certainly relevant in this regard as it resurrects—and significantly improves upon and extends—McConnell's research conducted some fifty years ago.

But why follow this mostly discredited path? First and foremost, let me be clear in stating that this is only one strand of a much larger research program on the mechanisms underlying regeneration and development—and perhaps the most controversial of all of Levin's work. But he argues that significant advances, particularly modern discoveries indicating that epigenetic modifications occur in many cell types, not just in the central nervous system, makes McConnell's ideas more plausible than they once were. Our understanding of the dynamics and mechanisms of neural and developmental processing have dramatically improved over the past half-century. So, too, have the research tools at ones' disposal.

Moreover, planarians, given their regenerative abilities, can still serve as an ideal system to investigate how specific memories are encoded and stored in biological tissues and, more so, to probe brain remodeling during regeneration. And the applied medical benefits are as important—if not more so—than ever. If true, these findings will have important clinical implications for stem cell treatments of degenerative and other neurological brain disorders.

In one recent report (Shomrat and Levin 2013), Levin and his colleague provide evidence that planarian learned to recognize a familiar environment and that the memory survives long enough to allow for regeneration after amputation, such that that memory traces survive entire brain regeneration in a 'savings' paradigm. More recently, in a paper titled 'Vertically- and horizontally-transmitted memories—the fading boundaries between regeneration and inheritance in planaria' (Neuhof *et al* 2016), the authors conclude,

> 'In planaria, and other organisms that reproduce by fission, producing and maintaining variation between fragments after asymmetric division may be adaptive (much like the beneficial increase in variation following sexual reproduction and recombination). Therefore, the theoretical ability of asymmetric division to create variability in an otherwise isogenic population could be considered as a tool for producing evolutionary progress. Thus, asymmetric fission is a mechanism that challenges our current view of what defines the temporal axis of evolution, since epigenetic processes, environmental cues, biochemical gradients and generation of a complete individual from a community of cells can generate natural variation, without requiring so called 'distinct' generations. It is likely that we have only begun to glimpse the prevalence and variety of long-term memory in somatic tissues during lifespan and across reproduction throughout phyla. The continued future analysis of such instructive interactions is likely to have profound implications for understanding evolution. Moreover, a mature understanding of these fascinating processes will drive numerous applications in regenerative medicine and bioengineering that exploit the rich informational plasticity of tissues for the rational control of form and function.

How this work will be evaluated and received by researchers in relevant problem-area remains to be seen. That work in this program has been adequately funded and published shows, once more, that extraordinary scientific claims are typically given room to breathe.

But I should like to end on a related, and to my mind, equally important note about the role of error in science. I take the position that even those scientific claims that do not stand up to the exacting scrutiny of practitioners in the problem-area nevertheless have important positive consequences for the development of science.

It is widely recognized that false assumptions and outright errors often lead to important new discoveries. Concrete examples abound. To cite but two, drawn from the history of the medical sciences, the discovery of insulin originated in what, by all accounts, was a wrongly conceived, wrongly conducted, and wrongly interpreted series of experiments, while, as noted by Fleck ([1935] 1979), Wasserman was led to the discovery of the test to detect syphilis that bears his name by proceeding quite systematically on the basis of current knowledge, assumptions, and reasoning which was subsequently proven wrong. Some errors, then, are quite 'fruitful'.

Extraordinary scientific claims have a *focusing effect*. They often draw attention to certain limitations and difficulties in the prevailing cognitive framework or identify new problems that were not generally perceived by practitioners. In a word, they may be considered as extremely 'provocative'. They thus function as *catalysts* that force those who embrace the conventional view to further clarify, strengthen, or otherwise improve their arguments as they grapple and then come to grips with the challenging claim. This, of course, leads to a deepened understanding of the phenomenon under scrutiny.

Such was the case when the physicist George Gamow proposed, soon after the publication of Watson and Crick's famous paper, a scheme to explain how the sequential structure of DNA could directly and physically order the sequential structure of proteins. Crick saw at once that Gamow had made errors, but as Crick has since stated,

> '... it drew attention to what it was you had to discover - namely what you would now call the genetic code. Which wasn't a thing which most people realized must exist' (in Judson 1979:278).

But perhaps the most important potential benefit derived from the accommodation of extraordinary scientific claims, even if they turn out to be false, is that they often contribute to an important—and too little studied—mode of cognitive change: the socially patterned selective and partial incorporation of some of the constitutive claims of one research program by another through the give-and-take of cognitive disagreements. Even as some controversies become quite heated, it is often the case that the interchanges between scholars holding quite different views, rhetoric and posturing aside, lead to the further development of knowledge as competing groups take over ideas or procedures from one another. Even as one rival research program vanquishes another, that which is considered to be of cognitive value is retained and pressed into service by the victors. The accommodation of extraordinary scientific claims, then, may trigger a process of 'creative disagreement' whereby selected aspects of these claims may be embraced and/or interpreted differently than intended and lead to new insights that will considerably strengthen or improve the prevailing framework.

The plain truth, then, is that truth can be—and often is—generated from error. I close with a comment by the 19th century economist-sociologist Vilfredo Pareto (commenting on Kepler):

> 'Give me a fruitful error any time, full of seeds, bursting with its own corrections. You can keep your sterile truth for yourself'.

Bibliography

Agranoff Bernard, Davis Roger E and Brink John J 1965 Memory Fixation in the Goldfish *Proc. Nat. Acad. Sci.* **54** 789–93

Barondes Samuel and Cohen Harry D 1966 Puromycin Effect on Successive Phases of Memory Storage *Science* **151** 594–5

Cameron D Ewen, Sved S, Solyom B, Wainrib B and Barik H 1963 Effects of ribonucleic acid on memory defect in the aged *American Journal of Psychiatry* **120** 320–5

Corning William and John E R 1961 Effect of ribonuclease on retention of conditioned response in regenerated planarians *Science* **134** 1363–4

Feyerabend Paul 1975 *Against Method* (Verso Books)

Fleck Ludwik 1935 *Genesis and Development of a Scientific Fact* University of Chicago Press) 1979

Flexner L B, Flexner J B, Roberts R B and De La Haba G 1964 Loss of Recent Memory in Mice as Related to Regional Inhibition of Cerebral Protein Synthesis *Proceedings of the National Academy of Sciences* **52** 1165–9

Hebb Donald 1949 *The Organization of Behavior: A Neuropsychological Theory* John Wiley & Sons)

Hyden Holger 1962 Nuclear RNA Changes of Nerve Cells During a Learning Experiment in Rats *Proceedings of the National Academy of Sciences* **48** 1366–73

Judson Horace Freeland 1979 *The Eighth Day of Creation: Makers of the Revolution in Biology* (Simon and Schuster)

Kuhn Thomas S 1962 *The Structure of Scientific Revolutions* (University of Chicago Press) 1970

Lakatos Imre 1970 Falsification and the Methodology of Scientific Research Programmes pp 91–195 ed Imre Lakatos *Criticism and the Growth of Knowledge* (Cambridge University Press)

McConnell James V 1963 1965:3 A tape recorder theory of memory *Worm Runner's Digest* **7** 3–10

McConnell James V, Jacobson Allan J and Kimble Daniel P 1959 The effects of regeneration upon retention of a conditioned response in the planarian *The Journal of Comparative and Physiological Psychology* **52** 1–5

McConnell James V, Jacobson Reeva and Maynard D M 1959 Apparent retention of a conditioned response following total regeneration in the planarian *American Psychologist* **14** 410

McGaugh James L 1966 Time Dependent Processes in Memory Storage *Science* **153** 1351–8

Merton Robert K 1984 The Fallacy of the Latest Word: The Case of 'Pietism and Science *American Journal of Sociology* **85** 1091–121

Neuhof Moran, Levin Michael and Rechavi Oded 2016 Vertically- and horizontally-transmitted memories – the fading boundaries between regeneration and inheritance in planaria *Biology Open* **5** 1177–88

Popper Karl 1935 *The Logic of Scientific Discovery* (Routledge) 1959

Shomrat T and Levin M 2013 An automated training paradigm reveals long-term memory in planarians and its persistence through head regeneration *J. Exp. Biol.* **216** 3799–810

Stent Gunther 1972 Prematurity and Uniqueness in Scientific Discovery *Scientific American* **227** 84–93

Zuckerman Harriet and Lederberg Joshua 1986 Postmature Scientific Discovery? *Nature* **324** 629–31

THE EFFECTS OF REGENERATION UPON RETENTION OF A CONDITIONED RESPONSE IN THE PLANARIAN[3]

JAMES V. McCONNELL, ALLAN L. JACOBSON,[4] AND DANIEL P. KIMBLE

University of Michigan

In a recent study, Thompson and McConnell (1955) demonstrated that classical conditioning could be established in the planarian or common marine flatworm, *D. dorotocephala*. From the standpoint of evolution, the planarian is an especially significant animal. Such evolutionary advancements as true synaptic nervous transmission, definite encephalization, and bilateral symmetry, to mention only a few, appear for the first time in planaria. These characteristics alone make the flatworm an interesting animal to study. For those psychologists interested in learning theory, however, perhaps an even more striking characteristic is the planarian's great regenerative ability, A single organism can be cut into as many as six transverse sections and, under optimal conditions, each section will regenerate a complete organism. If a planarian were conditioned, then cut in half and each half were allowed to regenerate, would either section show any retention of the conditioned response (CR)? If the tail section showed any retention of the CR at all, once it had regenerated into a complete organism, would it have retained as much as the head section?

The present study is an attempt to answer the above questions. Planaria were conditioned to a set criterion of responding, then immediately cut in half and allowed to regenerate. When regeneration was complete, each half was retested by the method of savings to determine the degree of retention of the original training.

METHOD

Subjects

Fifteen planaria (*D. dorotocephala*), varying in length from 10 to 24 mm., were studied. They were obtained locally and placed in individual glass aquaria, the dimensions of which were 8 in. by 5 in. by 5 in. The water in the aquaria was continually aerated and was maintained at room temperature, approximately 22°C.

Apparatus

The apparatus in which the Ss were conditioned and later retested consisted of a semicircular plastic trough, 12 in. long and $\frac{1}{2}$ in. in diameter. The trough was suspended between two vertical supports which were anchored to a wooden base 1 in. thick. The trough was filled to the top with aquarium water. Mounted at both ends of the trough were electrodes, which transmitted the US (weak electric shock)

---

[3] This study was assisted by a grant from the Faculty Research Fund of the Horace H. Rackham School of Graduate Studies of the University of Michigan.
[4] Now at Harvard University.

through the water in the trough. Current for the US was provided by storage batteries and was passed through an inductorium, which provided the means for controlling the intensity of the stimulation.

A double lamp, housing two 100-w. frosted bulbs, was placed about 6 in. above the trough, with one bulb located near each end of the trough. Light from this lamp constituted the CS. The only other source of illumination was the overhead lighting of the laboratory, which was of a much weaker intensity than the CS.

Procedure

Only one S was studied at a time. The S was transported from the aquarium to the trough in a pipette. The S was allowed to "explore" the trough for 5 min. before training was begun. After an initial period of disturbed activity, during which the animal's locomotion was ditaxic or ameboid in character, it generally began to move in a normal fashion on the bottom of the trough. This movement may be described as "gliding," accomplished by ciliary action. When S reached the end of the trough, it would turn and retrace the trough. Occasionally S would come to rest and would remain motionless. If this "sleep" state continued for more than 1 min., S was prompted to locomotion by water currents set up by a pipette. Trials were given only when S was gliding in a straight line and was oriented toward either of the electrodes.

Each trial was of 3-sec. duration. For the first 2 sec., the light (CS) alone was presented. During the third second both light and shock (US) were presented. The US consistently evoked a longitudinal contraction of the animal's entire body. All responses occurring during the initial 2-sec. period were observed and "called" by one member of the experimental team, while another recorded the responses and handled the switches for presentation of the CS and the US. A minimum of 1 min. was allowed to pass between trials. A maximum of 50 trials per day was given, with training continuing until S reached the criterion of 23 responses in 25 consecutive trials.

The Ss were divided roughly in terms of length into three groups of five Ss each: the experimental group (Group E), the regeneration control group (Group RC), and the time control group (Group TC).

When Ss in Group E had satisfied the criterion of conditioning, they were immediately removed from the trough and cut transversely in half as closely as possible with a razor blade. The two halves were then placed in separate aquaria and allowed to regenerate. The external appearance and the behavior of the animals suggested that regeneration was usually complete within 10 to 14 days. In order to make sure that internal structures were completely regenerated, however, an additional two weeks was allowed for recovery. Retesting for savings thus began approximately four weeks after separation. Both halves were retested, and so the experimental group served partially as its own control.

Since we feared that the processes of cutting and regeneration might in some way "sensitize" the animals, and thus make them subject to a more rapid conditioning, Ss in Group RC were given no experimental training of any kind before being cut. Naive animals were cut as soon as obtained, and each half was placed in a separate aquarium and allowed approximately four weeks to regenerate. Following regeneration, each half was conditioned separately to the same criterion as used for Group E.

Since we had no notion of how much extinction, or forgetting, might take place spontaneously during the four-week regeneration period, Ss in Group TC were conditioned to the criterion, then placed *uncut* in their individual aquaria and allowed to "rest" for a period of approximately four weeks. Following this "rest" period, they were retrained to the same criterion.

We did not think it necessary to control for the effects of sensitization to either light or shock, since Thompson and McConnell (1955) had already demonstrated that the CR they established was not the result of repeated exposure to either of these stimuli. It did seem possible, however, that we were establishing CRs in our Os rather than in our experimental Ss. In order to determine the reliability of our observations, three Es[5] observed the conditioning of one animal, making independent observations. When the planarian had reached the criterion, and only then, the observations were compared.

RESULTS

Two types of responses were observed in the Ss during the initial stages of conditioning, with a third type appearing generally during the latter stages. The first two types were as follows; (*a*) a sharp turning of the cephalic region to one side or the other and (*b*) a longitudinal contraction of the entire body. The third type was essentially a combination of the first two, in which the cephalic region turned sharply to one side or the other while the caudal region contracted sharply. In general, as the S neared criterion, the third type of response predominated—so much so, in fact, that its first appearance almost always heralded an immediate reaching of criterion. For the purposes of this study, however, any of the three types was scored as a "response."

Table 1 presents the raw data for the experimental group in terms of the number of trials necessary to reach criterion, both for the initial conditioning and for retest of the regenerated animals. A t test for comparison of means shows that the difference between the number of trials for original training and for later retraining of both the head and the tail halves is significant at the .01 level of confidence. There was no significant difference between the head and the tail sections, in terms of the mean number of trials required for each to retrain to criterion following regeneration.

Table 2 presents the raw data for the regeneration control group in terms of the number of trials necessary to reach criterion for both the head and the tail sections. As might be guessed from a cursory inspection of the data, the head and tail sections do not differ significantly from each other, but both groups took significantly longer to reach criterion (.01 level of confidence) than did the regenerated sections of the experimental animals. What is perhaps more surprising is the fact that the animals in Group RC also took more trials to reach criterion than did the uncut animals in Group E. While this difference is not statistically significant (.09 level of confidence),

---

[5] We wish to thank John Burns and Paul Cornwell for their assistance in running the reliability study.

TABLE 1 Number of Trials to Criterion for Group E

S	Original Training	Retest Head	Retest Tail
E-1	99	50	51
E-2	191	37	24
E-3	97	48	72
E-4	83	35	44
E-5	200	30	25
M	134	40	43.2

TABLE 2 Number of Trials to Criterion for Group RC

S	Head	Tail
RC-1	134	150
RC-2	188	179
RC-3	276	85
RC-4	395	300
RC-5	250	325
M	248.6	207.8

TABLE 3 Number of Trials to Criterion for Groan TC

S	Original Training	Retest
TC-1	123	24
TC-2	153	25
TC-3	195	62
TC-4	131	43
TC-5	325	45
M	185.4	39.8

it is probably large enough for us to conclude that the processes of cutting and regeneration in no way "sensitize" the animals.

The raw data for the time control group are presented in Table 3, Again, even a rough inspection of the data indicates that the TC animals showed a significant savings of the CR (.01 level of confidence) following the four-week "rest" period. There is no significant difference between the mean trials to criterion of the TC group following the "rest" period and the mean of either the regenerated head or tail sections of Group E.

The average intertrial interval was approximately 84 sec.

TABLE 4 Per Cent Agreement of Three Os Making Independent Observations During Conditioning of One Animal

Os	Agreement for 238 Trials (%)	Agreement for Final 25 Trials (%)
A-B	85	100
A-C	86	96
B-C	89	96

The experimental animal used in the reliability study took 238 trials to reach criterion. The three sets of independent observations were then compared for per cent agreement among the Os for the entire series of 238 trials. The results are shown in the first part of Table 4. While the agreement thus obtained is fairly high for this kind of observational datum, it was apparent from the data that reliability increased as conditioning progressed. At the beginning of training, planaria often make what might be called "minimal responses," so that it is sometimes difficult to know whether the animal has actually responded or not. Toward the end of training, however, when the vigorous and distinctive "third type" of response (mentioned earlier) begins to appear, the responses become much easier to judge. Hence, we also calculated per cent agreements for the final 25 (criterional) trials. As the data in Table 4 indicate, two of the three Os agreed perfectly on the final 25 trials, while the third O differed from the other two on only one response in the 25 trials.

DISCUSSION

Perhaps the one most startling result of this study is the fact that the tail sections in the experimental group showed such a great savings of the initial conditioning. The results of Group RC indicate that this phenomenon is not due to any "sensitization" due to cutting and regeneration. Had the tails showed only some slight savings, it might have passed with little notice. That the tails showed at least as much savings as did the heads, and that both heads and tails showed as much retention as did uncut animals, calls for considerable cogitation and perhaps even some reorganization of our thoughts concerning the mechanisms of learning and retention in such organisms as the planarian.

A brief consideration of the nervous system of the planarian might well be of service here. As can be seen in Figure 1, "The nerve cell bodies are contained in two masses of nervous tissue, the cerebral ganglia, commonly referred to as the brain. From this concentrated point two longitudinal nerve cords pass posteriorly and two short nerves extended anteriorly to connect with the eyes. Along the two longitudinal cords are many transverse nerves, which are distributed to the internal structures of the body...." Thus we find in the planarian "the beginnings of a definite central nervous system" (Elliott, 1952, pp. 164–165). This has important implications for explanations of the behavior of the animal. The cephalic ganglion receives impulses from the sensitive elements of the snout and head, i.e., the eyespots, which are sensitive to changes of illumination but have no lenses and consequently form no

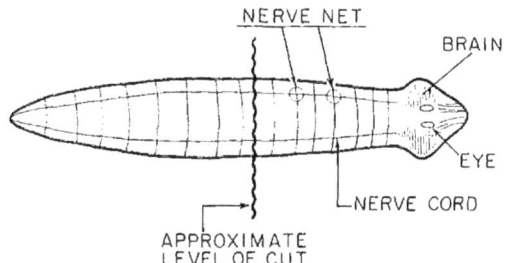

Fig. 1. The nervous system of the planarian (magnified many times). Adapted from Maier and Schneirla (1935, p. 80).

image, and the auricles on the sides of the head region, which "may be chemoreceptors of 'taste' or 'smell' " (Storer, 1943, p. 328). This means, then, that "the head end leads in behavior, and through its nerve-ganglion impulses originating in the head receptors are conducted in two principal nerve strands and in their branches to other parts of the body" (Maier & Schneirla, 1935, p. 79).

This short description makes apparent the importance of the cephalic ganglia for the organization and direction of behavior in the planarian. This would lead one to believe that the ganglia should be crucial for establishment of learning in this animal. Hovey, who established a type of "negative adaptation" in these animals, reports that "following extirpation of the cerebral ganglia, worms did not learn to move less in the light, and repeated hindrance to such movement had no effect. Evidently the brain is involved in associative learning in polyclads" (Hovey, 1929, p. 332).

Although it may be dangerous to generalize too freely from Hovey's data, we are presently of the opinion that the cephalic ganglia may be necessary for *acquisition* but not for *retention* of a conditioned response in the planarian. This conclusion leaves us several interesting problems to face. To begin with, how does the tail section retain anything? We must conclude that if structural changes are to account for the learning, these changes must occur throughout the nervous system, not solely in the cerebral ganglia. An even more intriguing question is this: Since the tail section must grow a new anterior half, including eyespots and cephalic ganglia, how does the relearning become established so quickly? It seems apparent that these cephalic centers must be involved in the rather complicated response that eventually becomes the CR. And yet two of the tail sections in Group E gave this highly complex response in their very first trials! Did the tail sections then regenerate new ganglia with the CR already "built in"?

A concerted program of experimentation is now underway in an effort to answer these and other questions.

SUMMARY

The purpose of the present experiment was to study the effects of regeneration upon retention of a conditioned response in the planarian (*D. dorotocephala*). A classical conditioning situation was employed in which light was the CS and shock the US. Fifteen animals were divided into three equal groups, an experimental group and two control groups. The experimental animals were first conditioned to a criterion of

23 responses in 25 consecutive trials. They were then cut in half transversely and allowed to regenerate. Following regeneration, both head and tail sections were retrained to the original criterion.

It was found that both head and tail sections showed significant retention of the CR and to the same degree. That these findings were not the result of sensitization due to the processes of cutting and regeneration was indicated by the performance of one of the control groups. That the small amount of "forgetting" of the CR which occurred in the experimental animals was probably due merely to the passage of time was indicated by the performance of the other control group.

It was concluded that in planaria the rudimentary brain is necessary for learning to take place but not for retention of the learned response. Two intriguing questions posed by these findings are: How does the tail retain anything? and When the tail section regenerates a new brain, is the CR "built into" the cerebral ganglia?

REFERENCES

ELLIOTT, A. M. *Zoology.* New York: Appleton-Century-Crofts, 1952.

HOVEY, H. B. Associative hysteresis in flatworms. *Physiol. Zool.,* 1929, **2**, 322–333.

MAIER, N. R. F., & SCHNEIRLA, T. C. *Principles of animal psychology.* New York: McGraw-Hill, 1935.

STORER, T. I. *General zoology.* New York: McGraw-Hill, 1943.

THOMPSON, R., & MCCONNELL, J. V. Classical conditioning in the planarian, *Dugesia dorotocephala. J. comp. physiol. psychol.,* 1955, **48**, 65–68.

Received January 31, 1958.

IOP Publishing

Ahead of the Curve
Hidden breakthroughs in the biosciences
Michael Levin and Dany Spencer Adams

Chapter 4

Physiology and cancer

Case Study 11: Random Expression Profiles Predict Breast Cancer

Perspective on *Most Random Gene Expression Signatures Are Significantly Associated with Breast Cancer Outcome*

Dany Spencer Adams
Research Professor
Tufts University

This is a relatively recent paper, thus we have no evidence for or against its acceptance by those using gene expression profiles to understand the etiology and progression of cancer. As of this writing, it has been cited by 171 articles in journals monitored by PLOS. The electronic version has been viewed 37,164 times, saved 442 times, and shared 42 times. For comparison, a fast and unscientific PubMed search of the citation counts for similar papers published around the same time indicates that this paper is getting plenty of attention. Whether or not it is favorable we cannot say.

This is an important paper not just for studies of breast cancer, or even cancers in general, but for any protocol exploiting advanced numerical and/or statistical methods. Looking for correlations in big data sets is that kind of protocol. It is all too easy to get distracted, enamored, or befuddled by the mathematics and miss the basics such as positive and negative controls.

This paper describes a negative control for expression profiling, one that should have been standard from the very beginning, even though, or more accurately because, it is an example of the most humble statistical analysis: compare your candidate cause to chance. In the case of profiling, the hypothesis is that the suite of mutations and/or the change in profile causes the cancer or predicts its progression. The negative control is, therefore, to look for correlations between cancer incidence or progression and random mutations or expression changes of random genes. This control should become part of every S.O.P. for profiling.

Statistical packages will do what you tell them to do; it remains up to the scientist to tell them to do the correct thing, the whole correct thing, and nothing but the correct thing.

Most Random Gene Expression Signatures Are Significantly Associated with Breast Cancer Outcome

David Venet[1], Jacques E. Dumont[2], Vincent Detours[2,3]*

[1]IRIDIA-CoDE, Université Libre de Bruxelles (U.L.B.), Brussels, Belgium, [2]IRIBHM, Université Libre de Bruxelles (U.L.B.), Campus Erasme, Brussels, Belgium, [3]WELBIO, Université Libre de Bruxelles (U.L.B.), Campus Erasme, Brussels, Belgium

Abstract

Bridging the gap between animal or *in vitro* models and human disease is essential in medical research. Researchers often suggest that a biological mechanism is relevant to human cancer from the statistical association of a gene expression marker (a signature) of this mechanism, that was discovered in an experimental system, with disease outcome in humans. We examined this argument for breast cancer. Surprisingly, we found that gene expression signatures—unrelated to cancer—of the effect of postprandial laughter, of mice social defeat and of skin fibroblast localization were all significantly associated with breast cancer outcome. We next compared 47 published breast cancer outcome signatures to signatures made of random genes. Twenty-eight of them (60%) were not significantly better outcome predictors than random signatures of identical size and 11 (23%) were worst predictors than the median random signature. More than 90% of random signatures >100 genes were significant outcome predictors. We next derived a metagene, called meta-PCNA, by selecting the 1% genes most positively correlated with proliferation marker PCNA in a compendium of normal tissues expression. Adjusting breast cancer expression data for meta-PCNA abrogated almost entirely the outcome association of published and random signatures. We also found that, in the absence of adjustment, the hazard ratio of outcome association of a signature strongly correlated with meta-PCNA ($R^2 = 0.9$). This relation also applied to single-gene expression markers. Moreover, >50% of the breast cancer transcriptome was correlated with meta-PCNA. A corollary was that purging cell cycle genes out of a signature failed to rule out the confounding effect of proliferation. Hence, it is questionable to suggest that a mechanism is relevant to human breast cancer from the finding that a gene expression marker for this mechanism predicts human breast cancer outcome, because most markers do. The methods we present help to overcome this problem.

Citation: Venet D, Dumont JE, Detours V (2011) Most Random Gene Expression Signatures Are Significantly Associated with Breast Cancer Outcome. PLoS Comput Biol 7(10): e1002240. doi:10.1371/journal.pcbi.1002240

Editor: Isidore Rigoutsos, Jefferson Medical College/Thomas Jefferson University, United States of America

Received April 27, 2011; **Accepted** September 7, 2011; **Published** October 20, 2011

Copyright: © 2011 Venet et al. This is an open-access article distributed under the terms of the Creative Commons Attribution License, which permits unrestricted use, distribution, and reproduction in any medium, provided the original author and source are credited.

Funding: DV was funded by the IRSIB Brussels Region-Capitale ICT-Impulse 2006 program 'InSilico wet lab'. The funders had no role in study design, data collection and analysis, decision to publish, or preparation of the manuscript.

Competing Interests: The authors have declared that no competing interests exist.
* E-mail: vdetours@ulb.ac.be

Introduction

Ethics limits experimental investigation on human subjects. Hence, most experimental biomedical research is performed on animal and/or *in vitro* models. Proving that findings from model systems are relevant to human health is a major bottleneck.

Hundreds of studies in oncology have suggested the biological relevance to human of putative cancer-driving mechanisms with the following three steps: 1) characterize the mechanism in a model system, 2) derive from the model system a marker whose expression changes when the mechanism is altered, and 3) show that marker expression correlates with disease outcome in patients—the last figure of such paper is typically a Kaplan-Meier plot illustrating this correlation.

Breast cancer has been a test bed in oncogenomics. Several landmark studies (reviewed in ref. [1]) uncovered multi-gene mRNA markers of disease recurrence, which are independent of classical clinical markers and may provide useful information to guide treatment. These clinically motivated multi-genes markers, also called signatures, were derived from compendia of genome-wide mRNA tumoral profiles by selecting genes whose expression correlated with outcome [2–5], or with known aggressiveness markers such as proliferation [6–9] or grade [10–12].

Beyond clinical utility, many signatures were derived as markers of specific mechanisms and/or biological states and their association with outcome was evaluated in the context of studies structured along the 3-steps outlined above. These include signatures of stem cells [13–15], aneuploidy [16], wound healing [17,18], hypoxia [19,20], stromal component [21], epithelial-mesenchymal transition [22–24]; of mutations in TP53 [25], ALK5 [26]; of loss of PTEN [27]; of perturbations of E2F1 [28], bromodomain 4 [29], mir31 targets [30], p18^{ink4c} [31], retinoic acid receptor [32]; of anchorage-independent growth [33], activation of modules related to the proteasome and mitochondrions [34], etc. Contrasting with this diversity, meta-analyses of several outcome signatures have shown that they have essentially equivalent prognostic performances [35,36], and are highly correlated with proliferation [7–8,37], a predictor of breast cancer outcome that has been used for decades [38–40].

This raises a question: are all these mechanisms major independent drivers of breast cancer progression, or is step #3 inconclusive because of a basic confounding variable problem? To take an example of complex system outside oncology, let us suppose we are trying to discover which socio-economical variables drive people's health. We may find that the number of TV sets per household is positively correlated with longer life expectancy. This, of course, does not imply that TV sets improve health. Life expectancy and TV sets per household are both correlated with the gross national product per capita of nations, as are many other causes or byproducts of wealth such as energy consumption or education. So, is the significant association of say, a stem cell signature, with human breast cancer outcome informative about the relevance of stem cells to human breast cancer?

Author Summary
Proving that research findings from *in vitro* or animal models are relevant to human diseases is a major bottleneck in medical science. Hundreds of researchers have suggested the human relevance of oncogenic mechanisms from the statistical association between gene expression markers of these mechanisms and disease outcome. Such evidence has become easier to obtain recently with the advent of microarray screens and of large public-domain genome-wide expression datasets with patient follow-up. We demonstrated that in breast cancer any set of 100 genes or more selected at random has a 90% chance to be significantly associated with outcome. Thus, investigators are bound to find an association however whimsical their marker is. For example, we could establish outcome associations for a signature of post-prandial laughter and a signature of social defeat in mice. Association was not stronger than expected at random for 28 (60%) of 47 published breast cancer signatures. The odds of association are 5–17% with random single gene markers—a finding relevant to older breast cancer studies. We explained these results by showing that much of the breast cancer transcriptome is correlated with proliferation, which integrates most prognostic information in this disease.

Resolving this issue has become more pressing recently. Several large cohorts with genome-wide tumoral expression profiles and patient follow-ups are available in the public domain. Servers resting on these data [41,42] make step #3 accessible to anyone with an Internet connection. Genome-wide expression profiling has also considerably lowered the barrier to step #2. The search for markers is reduced to a nearly automated screen by comparing microarray profiles in situations where the putative cancer-driving mechanism is active or inactive. The end result is an increasing number of signatures.

Few studies using the outcome-association argument present negative controls to check whether their signature of interest is indeed more strongly related to outcome than signatures with no underlying oncological rationale. In statistical terms, these studies typically rest on H_0 assuming a background of no association with outcome. The negative controls we present here prove this assumption wrong: a random signature is more likely to be correlated with breast cancer outcome than not. The statistical explanation for this phenomenon lies in the correlation of a large fraction of the breast transcriptome with one variable, we call it meta-PCNA, which integrates most of the prognostic information available in current breast cancer gene expression data.

Results

Most signatures not biologically related to cancer are statistically associated with breast cancer outcome

In order to assess whether association with outcome was specific, we tested the association with breast cancer outcome of three signatures whose rationale does not suggest any connection with cancer: a signature of the effect of postprandial laughter on peripheral blood mononuclear cells [43], a signature of skin fibroblast localization [44] and a signature of social defeat obtained from mice brains [45]. For the sake of simplicity, and because this is the most commonly used setup in the field, we focused on the 295 patients of the Netherlands Cancer Institute (a.k.a. NKI) cohort [2] and the overall survival end-point. Details on the procedure used to estimate association with outcome are provided in Supporting Information (Text S1). Surprisingly, the three control signatures were significantly associated with outcome (Figure 1, panels A–C).

To check that these were not anecdotal observations, we downloaded all signatures from MSigDB database [46] belonging to the c2 category and assessed their association with outcome. MSigDB c2 signatures are manually curated from the literature on gene expression and also include gene sets from curated pathways databases such as KEGG. Trivial single-gene signatures were removed. The 1890 signatures examined in MSigDB c2 encompass all the fields of biomedical sciences, nevertheless we discovered that 67% of them were associated with breast cancer outcome at $p < 0.05$, 23% at $p < 10^{-5}$ (Figure 1D).

Cancer is a major subject matter of biomedical research, thus MSigDB c2 may be enriched for cancer-related signatures. To rule out the potential effect of a cancer bias, we generated for each signature in MSigDB c2 a signature of identical size but selected its genes randomly in the human genome. Although they are completely devoid of any biological rationale, 77% of these signatures were associated with outcome at $p < 0.05$, and 30% at $p < 10^{-5}$ (Figure 1D).

Thus, nominal p-values should not be used directly because a signature associated with outcome with a significance of 10^{-5} and even more so, 0.05, is not more related to outcome than a random set of genes.

Most published breast cancer signatures are not more strongly associated with breast cancer outcome than sets of random genes

Although most random signatures are significantly associated with breast cancer outcome, the association could be much stronger for published breast cancer signatures and provide valid statistical support for their relevance.

We compiled 47 signatures from the literature. Association with outcome has been reported for most of them (Supporting Information, Text S1), either for the purpose of finding better prognostic tools, or, in most cases, to suggest biological relevance. We compared the outcome association of each signature to that of 1000 random signatures of identical size (Figure 2). We confirmed the outcome association of 42 in these 47 signatures. Yet, 11 of them (23%) showed a weaker association than the median of random signatures. Abiding to statistical standard, one may consider a signature biologically relevant if its association with outcome is stronger than the association of the best 5% random signatures. Only 18 signatures in 47 (40%) met this criterion.

Figure 2 reveals that larger signatures are more significant. More than 90% of the signatures >100 genes we generated were significant at $p < 0.05$. For the two largest ones, 714 and 1345 genes respectively, all 1000 random signatures tested were significant.

At the other end of the size spectrum, we found that 26% of individual genes printed on the NKI arrays were associated with outcome at $p < 0.05$. Thus, a single gene study has 26 chances in 100 to yield a significant association. When we applied a q-value correction [47]—relevant to genome-wide studies—17% of all genes were associated with outcome at $q < 0.05$. A comparable calculation was presented by Ein-Dor et al. [48]: 1234 genes among 5852 that passed their initial filter were associated with outcome with a false discovery rate <10%.

Meta-PCNA integrates most of the outcome-related signal contained in the breast cancer transcriptome

Proliferation is a well-known breast cancer prognostic marker [38–40]. Cycling cells express thousands of specific genes [49], thus genome-wide expression profiles are likely to capture the fraction of cycling cells within a tissue. A proliferation cluster was noticed in early breast cancer microarray studies [50–52], and proliferation is the major variable behind gene expression-based breast cancer prognosis [7–9]. We devised a new metagene, meta-PCNA, in order to investigate further the role of proliferation.

The proliferating cell nuclear antigen, PCNA, is a ring-shaped protein that encircles DNA and regulates several processes leading to DNA replication [53]. As suggested by its name, this is one of the most widely used antigen target for immunohistochemical measures of the fraction of proliferating cells in tissues. Ge et al. [54] profiled with microarrays 36 tissues from normal, healthy, individuals encompassing 27 organs. We call 'meta-PCNA' the signature composed of the 1%

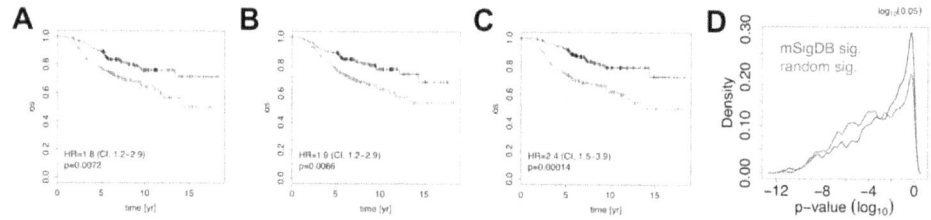

Figure 1. Association of negative control signatures with overall survival. In plots A–C the NKI cohort was split into two groups using a signature of post-prandial laughter (panel A), localization of skin fibroblasts (panel B), social defeat in mice (panel C). In panels A–C, the fraction of patients alive (overall survival, OS) is shown as a function of time for both groups. Hazard ratios (HR) between groups and their associated p-values are given in bottom-left corners. Panel D depicts p-values for association with outcome for all MSigDB c2 signatures and random signatures of identical size as MSigDB c2 signatures. doi:10.1371/journal.pcbi.1002240.g001

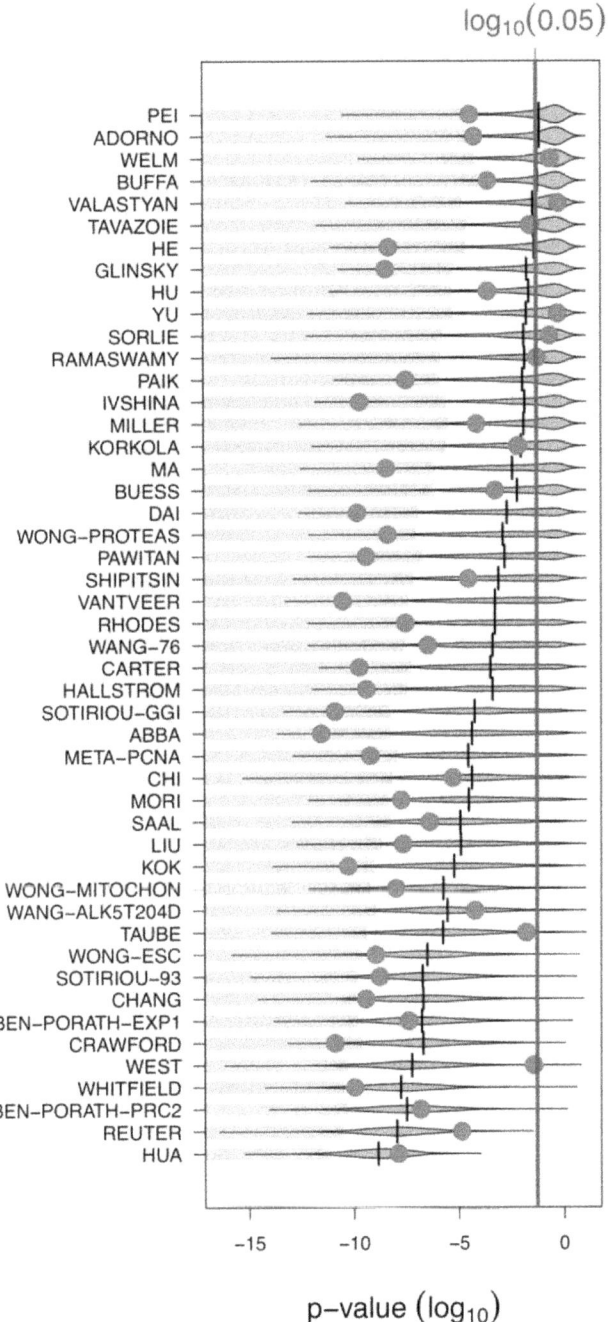

Figure 2. **Most published signatures are not significantly better outcome predictors than random signatures of identical size.** The x-axis denotes the p-value of association with overall survival. Red dots stand for published signatures, yellow shapes depict the distribution of p-values for 1000 random signatures of identical size, with the lower 5% quantiles shaded in green and the median shown as black line. Signatures are ordered by increasing sizes. doi:10.1371/journal.pcbi.1002240.g002

genes the most positively correlated with PCNA expression across these 36 tissues (Table S1). In plain language, meta-PCNA genes are consistently expressed when PCNA is expressed in normal tissues and consistently repressed when PCNA is repressed. We define the meta-PCNA index as the median expression of meta-PCNA genes. Beside PCNA itself, meta-PCNA includes other canonical proliferation markers such as MKI67, TOP2A, MCM2, etc.

We next compared for each one of the 47 published signatures its association with outcome in the original NKI data set and after adjustment of expression levels for the meta-PCNA index (Figure 3, Kaplan-Meier plots in Supporting Information, Text S1). Their association with outcome dropped dramatically after adjustment, although a few signatures remained strongly outcome associated. Any transformation damaging expression data will trivially decrease the association between outcome and expression. To control that was not the case with our adjustment procedure we reran the same analysis, except that meta-PCNA values were permuted randomly among patients prior to adjustment. In contrast with the adjustment of the actual non-permuted index, outcome association was not affected (Supporting Information, Text S1).

We plotted the hazard ratios of the 47 signatures against the absolute correlation of their first principal component with the meta-PCNA index. The more a signature was correlated with meta-PCNA, the higher its hazard ratio ($R^2 = 0.9$, Figure 4A, details for each data point in Supporting Information, Text S1).

Since only a limited set of genes is included in the 47 signatures, we plotted the distribution of correlations with the meta-PCNA index of all genes significantly associated with outcome and, as a negative control, of all genes printed on the microarrays (Figure 4B). Among the 17% of genes associated with outcome at $q < 0.05$, 91% were significantly correlated with meta-PCNA. Thus, any predictor resting on a linear combination of genes associated with outcome has a high probability to be confounded by proliferation.

More than 50% of the breast cancer transcriptome is correlated with meta-PCNA, hence removing cell cycle genes from a signature cannot rule out proliferation as a confounder

The potential confounding effect of proliferation has been recognized by a number of authors who attempted to rule it out by removing known proliferation genes from expression data [17,14,15]. These genes have been defined in various ways, including the Gene Ontology 'cell cycle' category, the genes periodically regulated in a cell-cycle time course [49], or genes of the breast cancer 'proliferation cluster' [55].

Following Ben Porath et al. [14], we defined as cell-cycle genes any gene present in at least one of these three categories. We calculated the distributions of correlations between the meta-PCNA index and genes of the Embryonic Stem Cell Module (ESCM) of Wong et al. [15], with and without the cell cycle genes (Figure 5). Purging these genes out of the ESCM did eliminate signals in the highest correlation range, but the ESCM remained vastly more correlated with meta-PCNA than the bulk of genes printed on the arrays ($p = 10^{-25}$).

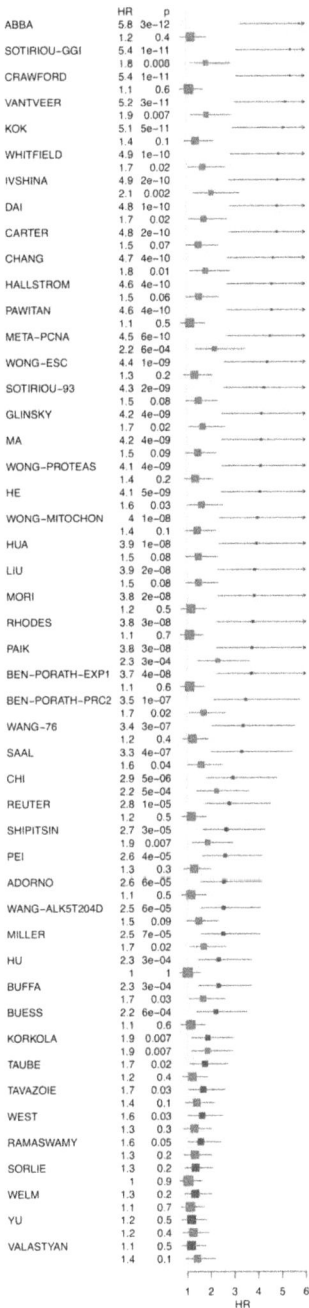

Figure 3. Meta-PCNA adjustment decreases the prognostic abilities of published signatures. Hazard ratios for overall survival association of 48 signatures in the original dataset (blue) and the meta-PCNA-adjusted dataset (red). Box sizes are inversely related to the size of the confidence intervals. Related Kaplan-Meier plots are available in the Supporting Information (Text S1). doi:10.1371/journal.pcbi.1002240.g003

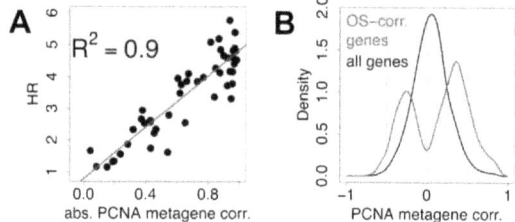

Figure 4. Most prognostic transcriptional signals are correlated with meta-PCNA. A) Each point denotes a signature. The x-axis depicts the absolute value of the correlation of the first principal component of the signatures with meta-PCNA, the y-axis depicts the hazard ratio for outcome association. Details of the analysis for each data point are available in the Supporting Information (Text S1). B) Distribution of the correlations of individual genes with meta-PCNA, for genes significantly associated with overall survival (red) and for all the genes spotted on the microarrays (black). doi:10.1371/journal.pcbi.1002240.g004

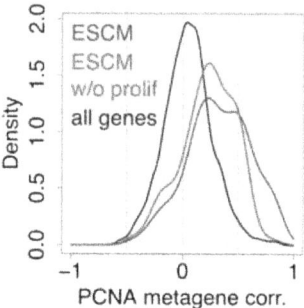

Figure 5. Purging cell cycle genes from a signature does not rule out proliferation signals. Distribution of the correlations with meta-PCNA of genes in the Embryonic Stem Cell Module (blue, ref. [15]), of the correlations of the same module with its cell cycle genes removed (red) and of all of the genes spotted on the microarray (black). doi:10.1371/journal.pcbi.1002240.g005

Moreover, 58% of the genes printed on the array were significantly correlated with the meta-PCNA index in the NKI cohort. Thus, the correlations with meta-PCNA extend far beyond cell cycle genes. Removing these genes fails to rule out the confounding effect of proliferation. Similarly, a signature does not have to be enriched with known cell cycle genes to convey a strong cell proliferation signal.

Results are reproducible across cohorts and end-points

Previous sections rested on the NKI data set and the overall survival end-point. Are our observations specific of this popular, but not universal, setup? We reran the analyses using recurrence-free survival, and on another cohort [56] using both overall survival and relapse-free survival.

We calculated hazard ratios for the 47 published signatures using all combinations of end-points and cohorts. Correlation between hazard ratios among the different cohorts/end-points was ≥ 0.97 (Figure 6). Thus, the ranking of the

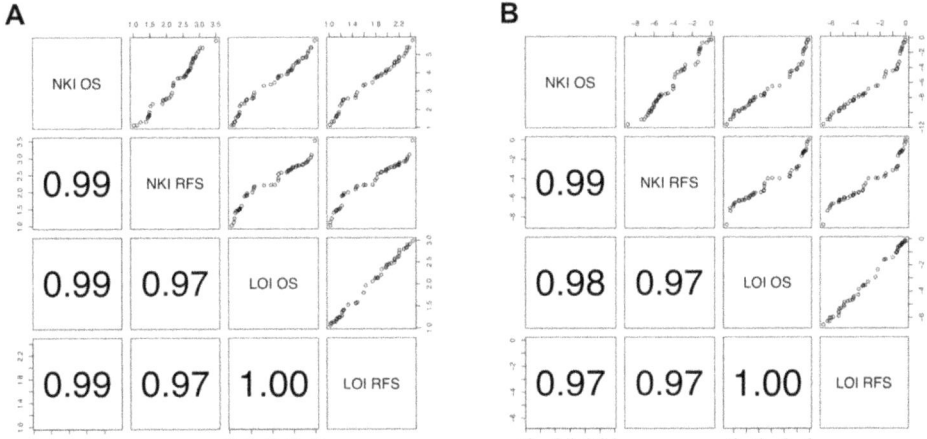

Figure 6. **Reproducible outcome predictions across end-points and cohorts.** Each dot represents a published signature. A) Hazard ratios. B) Log rank p-values. Lower panels give correlation coefficients for corresponding scatter plots in the symmetric upper panels. OS, overall survival; RFS, recurrence-free survival. NKI, data from ref. [2]; LOI, data from ref. [56].

signatures with respect to association with outcome was highly reproducible. However, the combination of NKI data and overall survival gave hazard ratios ~1.3 units higher (median HR = 3.8 in NKI and OS vs. <2.5 in other setups). Accordingly, p-values were ~4 orders of magnitude smaller than when association with outcome was estimated from the overall survival in the cohort of Loi et al. [56], although it included ~30% more patients. This difference between the 2 cohorts is less marked with relapse-free survival. Nevertheless, our analysis (summarized Table 1) reveals that, irrespective of the specific setup, at least 40% of MSigDB c2 signatures and 5% of all genes are associated with outcome, and at most 40% of the 47 published signatures are better than the 5% best same-size random signatures.

Discussion

There are many ways to estimate association between the expression of a multi-gene marker and disease outcome, and different studies have taken different routes. Our goal to compare signatures and assess them against negative controls, however, implied a uniform statistical framework. We present a comparison of a number of such methods in the Supporting Information (Text S1). A popular approach used in the studies we reviewed consists in stratifying the patients by hierarchical clustering in the signature subspace [57,21,29,24,28,15,58]. In most cases, our method of choice (using the first component of a Principal Components Analysis of a signature as a prognostic score) reveals stronger outcome associations than this approach. Our method is validated by the fact that we could reproduce the outcome association of most published signatures, which, conversely, validates the prognostic value of those signatures. The choice of association method is of course important, as there is a possibility that it misses some signals captured by specific combinations of signatures and models. However, most papers use similarly simple methods as ours.

Furthermore, the strength of such association might be doubted if it depended on an elaborate algorithm, as it is likely to be caused by spurious signals arising from model selection biases.

The main message of this paper is that, if the purpose of a study is to assert the biological relevance to human cancer of a signature, the association between this signature and outcome cannot rest on the nominal p-values, as obtained on breast cancer by the Cox analysis. This follows from elevated likelihood that random sets of genes are related to the outcome. Thus, an investigator finding that her/his signature is associated with outcome with a significance of 10^{-5}, and even more so, 0.05, gains no specific information because sets of random genes would likely yield similar, or better, results. Nominal p-values do not answer the appropriate statistical question: the question is not whether a given set of genes is related to survival, but whether it is more related to survival than random sets of genes.

This problem extends to single-gene markers and therefore to many studies published in the pre-genomic era. Claims similar to those concerning signatures have been made, that single genes, important in a model system, are relevant for human cancer progression based on differential expression between short- and long-survival groups. As 26% of the genes are related to survival at $p < 0.05$ (17% at $q<0.05$), much tighter p-values than commonly used should be imposed to demonstrate such a relation.

Several studies in the panel of 47 we investigated developed arguments independent of outcome association. For example, Hu et al. [59] used outcome association not as a validation argument, but as an exploratory tool to discover driver DNA copy number aberrations, which were then directly investigated. However, most of these studies, and many more not reviewed here, extrapolated the results from animal or highly artificial *in vitro* models to human *in vivo* cancer on the basis of questionable association statistics alone.

The present study addresses purely correlative association between gene-expression and disease outcome. We have shown that proliferation integrates most of the prognostic information contained in the breast cancer transcriptome. Yet—we cannot stress this enough—we have *not* shown that proliferation is a core driving force behind breast cancer progression. Disentangling the role of a biological process in cancer progression *in vivo* from the role of proliferation and from the role of the other processes associated with it is a crucial issue. The adjustment methodology we propose may be useful in assessing whether markers of biological processes do or do not rest on association with proliferation. Our results also imply that such markers should be evaluated against the outcome association of comparable negative control markers.

Our study questions the biological interpretation of the prognostic value of published breast cancer signatures, but has no bearing on their usefulness in the clinic: a marker may be accurate without yielding interesting biological insight regarding the mechanism of disease progression. Nevertheless, the prominence of proliferation should be taken into account in future clinical research. Are there transcriptional signals in breast cancer that are prognostic, but independent of proliferation? Is there any hope to perform better than the 70 genes NKI signature

[2]? The studies we reviewed assessed outcome prediction from gene expression measured in bulk tumors sampled from a relatively wide spectrum of patients, thus prognostic transcriptional signals detectable in specific tumor cells and/or specific patient groups were out of scope. Yet, proliferation-related signals are prognostic mostly in ER+ tumors [1]. Immunological genes convey prognostic information in ER- tumors and in tumors with HER2 amplification [8,60–64]. This information is unquestionably independent of proliferation since it improves prognostic accuracy beyond the abilities of proliferation-driven signatures and classical clinical markers [65]. Larger cohorts allowing the analysis of patients sub-groups and expression profiling of specific tumor cells/tumor areas may lead to better prognostic tools in the future.

In conclusion, we have shown that 1) random single- and multiple-genes expression markers have a high probability to be associated with breast cancer outcome; 2) most published signatures are not significantly more associated with outcome than random predictors; 3) the meta-PCNA metagene integrates most of the outcome-related information contained in the breast cancer transcriptome; 4) this information is present in over 50% of the transcriptome and cannot be removed by purging known cell- cycle genes from a signature.

Methods
Software setup
All analyses were run with R 2.9.0 [66] with packages specified in the following sections. Functions were run with default parameters unless specified otherwise.

Code and data availability

The code and data underlying the results and figures of this study are available as a Bzip2-compressed tar bundle from the *PLoS Computational Biology* web site (Dataset S1, size is 87 MB). The scripts assume a UNIX/LINUX environment.

Expression data

All the data were available from public sources:
- Ge *et al.* [54] data were downloaded from NCBI's Gene expression Omnibus (www.ncbi.nlm.nih.gov/geo; accession, GDS1096). We renormalized the raw data (CEL files) using Bioconductor [67] package gcrma [68].
- Loi *et al.* [56] data were downloaded from NCBI's Gene expression Omnibus (accession GSE6532). We used the Rdata file.
- The NKI, a.k.a. van de Vijver *et al.* [2], data set was downloaded from the Rosetta Inpharmatics web site on April 26th 2007 (www.rii.com, this site is now defunct, the dataset is available in the supplementary code and data tar bundle). Probe annotations were reconstructed using Bioconductor [67] package annotate. Contigs not mapped to genes in the original data set were recovered as much as possible using the table ArrayNomenclature_contig_accession.xls, also

on Rosetta web site. We used the original authors normalization, but ignored the flags.

Probes mapping to the same genes were averaged in each one of the three datasets.

Literature signatures

Whenever possible, the signatures were compiled from the publications online supplementary tables. When not available, the gene symbols were automatically read with an optical character recognition system from the papers tables and figures. In rare instances, signatures were encoded manually and double-checked. Because gene names and symbols are changing over time, the gene symbols of all signature genes were updated to match the HUGO nomenclature and therefore maximize the match with microarray gene annotations. HUGO gene symbols and their older aliases were obtained from the file gene_info as available on May 9th 2007 from the NCBI ftp server.

MSigBD 2.0 c2 signatures were downloaded as a *.gmt file from the Broad Institute page www.broadinstitute.org/gsea/msigdb/index.jsp.

Meta-PCNA index

We computed the Pearson correlation between PCNA and all the genes in the Ge et al. [54] dataset and selected the 1% most positively correlated, i.e., 131 genes out of 13,077, to form the meta-PCNA signature (Table S1). The meta-PCNA index of a tissue was computed from its expression profile by taking the median expression of these genes.

Adjusting data for the meta-PCNA index

The expression of each gene was fitted with R's 'lm' function and each expression measurement was substituted by the sum of its residual and its mean expression across the cohort.

Association of signatures with outcome

In order to systematically compare the published signatures to random signatures and evaluate the relation between outcome association and meta-PCNA, we needed an outcome association estimation procedure that is robust and fully automated. We systematically compared three procedures and selected among them the most sensitive and stable one. This is described in Supporting Information (Text S1), only the selected method is described here. It consists in computing the first principal component (PC1) of the signature (with R's prcomp) and then split the cohort according to the median of PC1. Probes mapping to the same gene were averaged and, following Ramaswamy et al. [57], data were median polished (R's medpolish) before the dimension reduction step.

Table 1. Summary of analysis with different cohorts and end points.

	NKI OS	NKI RFS	LOI OS	LOI RFS
Fraction of patient experiencing an event	79/295	101/295	96/380	139/393
% MSigDB c2 with p<0.05	67%	56%	52%	45%
% of all genes with p<0.05	17%	9%	8%	5%
% BC signatures better than 5% best random signatures of same size	40%	35%	29%	31%
Correlation of BC signatures HR with their association with meta-PCNA	0.9	0.9	0.9	0.9

Supporting Information has not been reprinted here, but is available online: doi:10.1371/journal.pcbi.1002240

Acknowledgments

This work rests almost entirely on open source software and data. Contributors are gratefully acknowledged.

Author Contributions

Conceived and designed the experiments: VD. Performed the experiments: DV VD. Analyzed the data: DV VD. Wrote the paper: DV JED VD.

References

1. Sotiriou C, Pusztai L (2009) Gene-expression signatures in breast cancer. N Engl J Med 360: 790–800. doi:10.1056/NEJMra0801289.
2. van de Vijver MJ, He YD, van't Veer LJ, Dai H, Hart AAM, et al. (2002) A gene-expression signature as a predictor of survival in breast cancer. N Engl J Med 347: 1999–2009. doi:10.1056/NEJMoa021967.
3. Paik S, Shak S, Tang G, Kim C, Baker J, et al. (2004) A multigene assay to predict recurrence of tamoxifen-treated, node-negative breast cancer. N Engl J Med 351: 2817–2826. doi:10.1056/NEJMoa041588.
4. Pawitan Y, Bjöhle J, Amler L, Borg A-L, Egyhazi S, et al. (2005) Gene expression profiling spares early breast cancer patients from adjuvant therapy: derived and validated in two population-based cohorts. Breast Cancer Res 7: R953–964. doi:10.1186/bcr1325.
5. Korkola JE, Blaveri E, DeVries S, Moore DH, Hwang ES, et al. (2007) Identification of a robust gene signature that predicts breast cancer outcome in independent data sets. BMC Cancer 7: 61. doi:10.1186/1471-2407-7-61.
6. Dai H, van't Veer L, Lamb J, He YD, Mao M, et al. (2005) A cell proliferation signature is a marker of extremely poor outcome in a subpopulation of breast cancer patients. Cancer Res 65: 4059–4066. doi:10.1158/0008-5472.CAN-04-3953.

7. Wirapati P, Sotiriou C, Kunkel S, Farmer P, Pradervand S, et al. (2008) Meta-analysis of gene expression profiles in breast cancer: toward a unified understanding of breast cancer subtyping and prognosis signatures. Breast Cancer Res 10: R65. doi:10.1186/bcr2124.
8. Desmedt C, Haibe-Kains B, Wirapati P, Buyse M, Larsimont D, et al. (2008) Biological processes associated with breast cancer clinical outcome depend on the molecular subtypes. Clin Cancer Res 14: 5158-5165. doi:10.1158/1078- 0432.CCR-07-4756.
9. Haibe-Kains B, Desmedt C, Sotiriou C, Bontempi G (2008) A comparative study of survival models for breast cancer prognostication based on microarray data: does a single gene beat them all? Bioinformatics 24: 2200–2208. doi:10.1093/bioinformatics/btn374.
10. Ma X-J, Salunga R, Tuggle JT, Gaudet J, Enright E, et al. (2003) Gene expression profiles of human breast cancer progression. Proc Natl Acad Sci USA 100: 5974–5979. doi:10.1073/pnas.0931261100.
11. Ivshina AV, George J, Senko O, Mow B, Putti TC, et al. (2006) Genetic reclassification of histologic grade delineates new clinical subtypes of breast cancer. Cancer Res 66: 10292–10301. doi:10.1158/0008-5472.CAN-05-4414.
12. Sotiriou C, Wirapati P, Loi S, Harris A, Fox S, et al. (2006) Gene expression profiling in breast cancer: understanding the molecular basis of histologic grade to improve prognosis. J Natl Cancer Inst 98: 262–272. doi:10.1093/jnci/djj052.
13. Glinsky GV, Berezovska O, Glinskii AB (2005) Microarray analysis identifies a death-from-cancer signature predicting therapy failure in patients with multiple types of cancer. J Clin Invest 115: 1503–1521. doi:10.1172/JCI23412.
14. Ben-Porath I, Thomson MW, Carey VJ, Ge R, Bell GW, et al. (2008) An embryonic stem cell-like gene expression signature in poorly differentiated aggressive human tumors. Nat Genet 40: 499–507. doi:10.1038/ng.127.
15. Wong DJ, Liu H, Ridky TW, Cassarino D, Segal E, et al. (2008) Module map of stem cell genes guides creation of epithelial cancer stem cells. Cell Stem Cell 2: 333–344. doi:10.1016/j.stem.2008.02.009.
16. Carter SL, Eklund AC, Kohane IS, Harris LN, Szallasi Z (2006) A signature of chromosomal instability inferred from gene expression profiles predicts clinical outcome in multiple human cancers. Nat Genet 38: 1043–1048. doi:10.1038/ng1861.
17. Chang HY, Sneddon JB, Alizadeh AA, Sood R, West RB, et al. (2004) Gene expression signature of fibroblast serum response predicts human cancer progression: similarities between tumors and wounds. PLoS Biol 2: E7. doi:10.1371/journal.pbio.0020007.
18. Chang HY, Nuyten DSA, Sneddon JB, Hastie T, Tibshirani R, et al. (2005) Robustness, scalability, and integration of a wound-response gene expression signature in predicting breast cancer survival. Proc Natl Acad Sc U S A 102: 3738–3743. doi:10.1073/pnas.0409462102.
19. Chi J-T, Wang Z, Nuyten DSA, Rodriguez EH, Schaner ME, et al. (2006) Gene expression programs in response to hypoxia: cell type specificity and prognostic

significance in human cancers. PLoS Med 3: e47. doi:10.1371/jour- nal. pmed.0030047.
20. Buffa FM, Harris AL, West CM, Miller CJ (2010) Large meta-analysis of multiple cancers reveals a common, compact and highly prognostic hypoxia metagene. Br J Cancer 102: 428–435. doi:10.1038/sj.bjc.6605450.
21. West RB, Nuyten DSA, Subramanian S, Nielsen TO, Corless CL, et al. (2005) Determination of stromal signatures in breast carcinoma. PLoS Biol 3: e187. doi:10.1371/journal.pbio.0030187.
22. Taube JH, Herschkowitz JI, Komurov K, Zhou AY, Gupta S, et al. (2010) Core epithelial-to-mesenchymal transition interactome gene-expression signature is associated with claudin-low and metaplastic breast cancer subtypes. Proc Natl Acad Sci U S A 107: 15449–15454. doi:10.1073/pnas.1004900107.
23. Welm AL, Sneddon JB, Taylor C, Nuyten DSA, van de Vijver MJ, et al. (2007) The macrophage-stimulating protein pathway promotes metastasis in a mouse model for breast cancer and predicts poor prognosis in humans. Proc Natl Acad Sci U S A 104: 7570–7575. doi:10.1073/pnas.0702095104.
24. Buess M, Nuyten DSA, Hastie T, Nielsen T, Pesich R, et al. (2007) Characterization of heterotypic interaction effects in vitro to deconvolute global gene expression profiles in cancer. Genome Biol 8: R191. doi:10.1186/gb-2007-8-9-r191.
25. Miller LD, Smeds J, George J, Vega VB, Vergara L, et al. (2005) An expression signature for p53 status in human breast cancer predicts mutation status, transcriptional effects, and patient survival. Proc Natl Acad Sci U S A 102: 13550–13555. doi:10.1073/pnas.0506230102.
26. Wang SE, Xiang B, Guix M, Olivares MG, Parker J, et al. (2008) Transforming growth factor beta engages TACE and ErbB3 to activate phosphatidylinositol-3 kinase/Akt in ErbB2-overexpressing breast cancer and desensitizes cells to trastuzumab. Mol Cell Biol 28: 5605–5620. doi:10.1128/MCB.00787-08.
27. Saal LH, Johansson P, Holm K, Gruvberger-Saal SK, She Q-B, et al. (2007) Poor prognosis in carcinoma is associated with a gene expression signature of aberrant PTEN tumor suppressor pathway activity. Proc Natl Acad Sci USA 104: 7564–7569. doi:10.1073/pnas.0702507104.
28. Hallstrom TC, Mori S, Nevins JR (2008) An E2F1-dependent gene expression program that determines the balance between proliferation and cell death. Cancer Cell 13: 11–22. doi:10.1016/j.ccr.2007.11.031.
29. Crawford NPS, Alsarraj J, Lukes L, Walker RC, Officewala JS, et al. (2008) Bromodomain 4 activation predicts breast cancer survival. Proc Natl Acad Sci U S A 105: 6380–6385. doi:10.1073/pnas.0710331105.
30. Valastyan S, Reinhardt F, Benaich N, Calogrias D, Szasz AM, et al. (2009) A pleiotropically acting microRNA, miR-31, inhibits breast cancer metastasis. Cell 137: 1032–1046. doi:10.1016/j.cell.2009.03.047.
31. Pei X-H, Bai F, Smith MD, Usary J, Fan C, et al. (2009) CDK inhibitor p18 (INK4c) is a downstream target of GATA3 and restrains mammary luminal progenitor cell proliferation and tumorigenesis. Cancer Cell 15: 389–401. doi:10.1016/j.ccr.2009.03.004.

32. Hua S, Kittler R, White KP (2009) Genomic antagonism between retinoic acid and estrogen signaling in breast cancer. Cell 137: 1259–1271. doi:10.1016/j.cell.2009.04.043.
33. Mori S, Chang JT, Andrechek ER, Matsumura N, Baba T, et al. (2009) Anchorage-independent cell growth signature identifies tumors with metastatic potential. Oncogene 28: 2796–2805. doi:10.1038/onc.2009.139.
34. Wong DJ, Nuyten DSA, Regev A, Lin M, Adler AS, et al. (2008) Revealing targeted therapy for human cancer by gene module maps. Cancer Res 68: 369–378. doi:10.1158/0008-5472.CAN-07-0382.
35. Fan C, Oh DS, Wessels L, Weigelt B, Nuyten DSA, et al. (2006) Concordance among gene-expression-based predictors for breast cancer. N Engl J Med 355: 560–569. doi:10.1056/NEJMoa052933.
36. Haibe-Kains B, Desmedt C, Piette F, Buyse M, Cardoso F, et al. (2008) Comparison of prognostic gene expression signatures for breast cancer. BMC Genomics 9: 394. doi:10.1186/1471-2164-9-394.
37. Mosley JD, Keri RA (2008) Cell cycle correlated genes dictate the prognostic power of breast cancer gene lists. BMC Med Genomics 1: 11. doi:10.1186/1755-8794-1-11.
38. Bloom HJ, Richardson WW (1957) Histological grading and prognosis in breast cancer; a study of 1409 cases of which 359 have been followed for 15 years. Br J Cancer 11: 359–377.
39. Tubiana M, Pejovic MJ, Renaud A, Contesso G, Chavaudra N, et al. (1981) Kinetic parameters and the course of the disease in breast cancer. Cancer 47: 937–943. doi:10.1002/1097-0142(19810301)47:5<937::AID-CNCR2820470520>3.0.CO;2-6.
40. Elston CW, Ellis IO (1991) Pathological prognostic factors in breast cancer. I. The value of histological grade in breast cancer: experience from a large study with long-term follow-up. Histopathology 19: 403–410.
41. Györffy B, Lanczky A, Eklund AC, Denkert C, Budczies J, et al. (2010) An online survival analysis tool to rapidly assess the effect of 22,277 genes on breast cancer prognosis using microarray data of 1,809 patients. Breast Cancer Res Treat 123: 725–731. doi:10.1007/s10549-009-0674-9.
42. Ringnér M, Fredlund E, Hakkinen J, Borg A, Staaf J (2011) GOBO: Gene Expression-Based Outcome for Breast Cancer Online. PLoS ONE 6: e17911. doi:10.1371/journal.pone.0017911.
43. Hayashi T, Urayama O, Kawai K, Hayashi K, Iwanaga S, et al. (2006) Laughter regulates gene expression in patients with type 2 diabetes. Psychother Psychosom 75: 62–65. doi:10.1159/000089228.
44. Rinn JL, Bondre C, Gladstone HB, Brown PO, Chang HY (2006) Anatomic demarcation by positional variation in fibroblast gene expression programs. PLoS Genet 2: e119. doi:10.1371/journal.pgen.0020119.
45. Krishnan V, Han M-H, Graham DL, Berton O, Renthal W, et al. (2007) Molecular adaptations underlying susceptibility and resistance to social defeat in brain reward regions. Cell 131: 391–404. doi:10.1016/j.cell.2007.09.018.

46. Subramanian A, Tamayo P, Mootha VK, Mukherjee S, Ebert BL, et al. (2005) Gene set enrichment analysis: a knowledge-based approach for interpreting genome-wide expression profiles. Proc Natl Acad Sci U S A 102: 15545–15550. doi:10.1073/pnas.0506580102.
47. Storey JD, Tibshirani R (2003) Statistical significance for genomewide studies. Proc Natl Acad Sci U S A 100: 9440-9445. doi:10.1073/pnas.1530509100.
48. Ein-Dor L, Kela I, Getz G, Givol D, Domany E (2005) Outcome signature genes in breast cancer: is there a unique set? Bioinformatics 21: 171–178. doi:10.1093/bioinformatics/bth469.
49. Whitfield ML, Sherlock G, Saldanha AJ, Murray JI, Ball CA, et al. (2002) Identification of genes periodically expressed in the human cell cycle and their expression in tumors. Mol Biol Cell 13: 1977–2000. doi:10.1091/mbc.02-02-0030.
50. Perou CM, Jeffrey SS, van de Rijn M, Rees CA, Eisen MB, et al. (1999) Distinctive gene expression patterns in human mammary epithelial cells and breast cancers. Proc Natl Acad Sci U S A 96: 9212–9217.
51. Perou CM, Sorlie T, Eisen MB, van de Rijn M, Jeffrey SS, et al. (2000) Molecular portraits of human breast tumours. Nature 406: 747–752. doi:10.1038/35021093.
52. Whitfield ML, George LK, Grant GD, Perou CM (2006) Common markers of proliferation. Nat Rev Cancer 6: 99–106. doi:10.1038/nrc1802.
53. Moldovan G-L, Pfander B, Jentsch S (2007) PCNA, the maestro of the replication fork. Cell 129: 665–679. doi:10.1016/j.cell.2007.05.003.
54. Ge X, Yamamoto S, Tsutsumi S, Midorikawa Y, Ihara S, et al. (2005) Interpreting expression profiles of cancers by genome-wide survey of breadth of expression in normal tissues. Genomics 86: 127–141. doi:10.1016/j.ygeno.2005.04.008.
55. Hu Z, Fan C, Oh DS, MarronJS, He X, et al. (2006) The molecular portraits of breast tumors are conserved across microarray platforms. BMC Genomics 7: 96. doi:10.1186/1471-2164-7-96.
56. Loi S, Haibe-Kains B, Desmedt C, Lallemand F, Tutt AM, et al. (2007) Definition of clinically distinct molecular subtypes in estrogen receptor-positive breast carcinomas through genomic grade. J Clin Oncol 25: 1239–1246. doi:10.1200/JCO.2006.07.1522.
57. Ramaswamy S, Ross KN, Lander ES, Golub TR (2003) A molecular signature of metastasis in primary solid tumors. Nat Genet 33: 49–54. doi:10.1038/ng1060.
58. Reuter JA, Ortiz-Urda S, Kretz M, Garcia J, Scholl FA, et al. (2009) Modeling inducible human tissue neoplasia identifies an extracellular matrix interaction network involved in cancer progression. Cancer Cell 15: 477–488. doi:10.1016/j.ccr.2009.04.002.
59. Hu G, Chong RA, Yang Q, Wei Y, Blanco MA, et al. (2009) MTDH activation by 8q22 genomic gain promotes chemoresistance and metastasis of poor-prognosis breast cancer. Cancer Cell 15: 9–20. doi:10.1016/j.ccr.2008.11.013.

60. Alexe G, Dalgin GS, Scanfeld D, Tamayo P, Mesirov JP, et al. (2007) High expression of lymphocyte-associated genes in node-negative HER2+ breast cancers correlates with lower recurrence rates. Cancer Res 67: 10669–10676. doi:10.1158/0008-5472.CAN-07-0539.
61. Schmidt M, Böhm D, von Torne C, Steiner E, Puhl A, et al. (2008) The humoral immune system has a key prognostic impact in node-negative breast cancer. Cancer Res 68: 5405–5413. doi:10.1158/0008-5472.CAN-07-5206.
62. Reyal F, van Vliet MH, Armstrong NJ, Horlings HM, de Visser KE, et al. (2008) A comprehensive analysis of prognostic signatures reveals the high predictive capacity of the proliferation, immune response and RNA splicing modules in breast cancer. Breast Cancer Res 10: R93. doi:10.1186/bcr2192.
63. Teschendorff AE, Caldas C (2008) A robust classifier of high predictive value to identify good prognosis patients in ER-negative breast cancer. Breast Cancer Res 10: R73. doi:10.1186/bcr2138.
64. Teschendorff AE, Gomez S, Arenas A, El-Ashry D, Schmidt M, et al. (2010) Improved prognostic classification of breast cancer defined by antagonistic activation patterns of immune response pathway modules. BMC Cancer 10: 604. doi:10.1186/1471-2407-10-604.
65. Haibe-Kains B, Desmedt C, Rothé F, Piccart M, Sotiriou C, et al. (2010) A fuzzy gene expression-based computational approach improves breast cancer prognostication. Genome Biol 11: R18. doi:10.1186/gb-2010-11-2-r18.
66. R Development Core Team (n.d.) R: A Language and Environment for Statistical Computing. 1: ISBN 3-900051-07-0.
67. Gentleman RC, Carey VJ, Bates DM, Bolstad B, Dettling M, et al. (2004) Bioconductor: open software development for computational biology and bioinformatics. Genome Biol 5: R80. doi:10.1186/gb-2004-5-10-r80.
68. Wu Z, Irizarry RA, Gentleman R, Martinez-Murillo F, Spencer F (2004) A model-based background adjustment for oligonucleotide expression arrays. J Amer Statistical Assoc 99: 909–917.

Case Study 12: The Bright Side of Infection

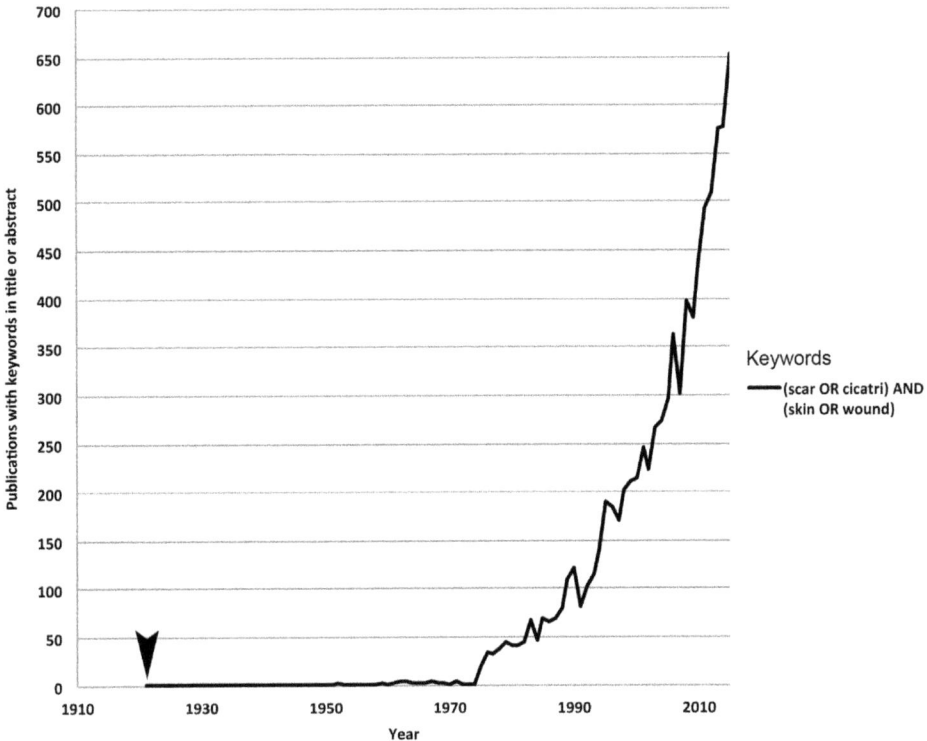

Carrel (1921) Cicatrization of wounds: XII. Factors initiating regeneration.

Perspective on *Cicatrization of Wounds: XII. Factors Initiating Regeneration*

Dany Spencer Adams
Research Associate Professor
Tufts University

The above graphic results from a search for *any* papers covering scarring ('cicatrix' is another name for a scar, and in this context, 'regeneration' refers to the growth of new tissue during scarring). The curve shows that to study the mechanisms by which scarring was initiated was by itself rare in 1921; however, this paper describes experiments to determine the roles of *external* factors, such as the presence of bacteria:

1. " ...[U]nder the conditions of the experiments, a wound, protected by a non-irritant dressing, shows no granulation tissue or beginning of contraction for 25 days at least.
2. *Local application of certain irritants, such as...staphylococci, reduces the duration of the latent period to less than 2 days.*
3. *Regeneration is apparently initiated, not by an internal, but by an external factor."*

These data are less startling in the context of the hygiene hypothesis and new information about the role of our microbiota in keeping us healthy. Still, the idea that the bacteria could be necessary and/or sufficient to initiate an important process involving gene expression changes has not been addressed in any systematic way. I have long thought that the presence of bacteria was as inevitable and ubiquitous as other non-genetic inducers of specific physiologic or genetic programs – think temperature dependent sex determination in turtles, cAMP signaling among *Disctyostelium discoideum* amoebae, and storing more fat when it is cold. Indeed, many mutualisms between bacteria and animals have been demonstrated, the role of *Wolbachia* in mosquito sex determination, for example. It seems obvious that natural selection would favor an organism that could exploit omnipresent, bacterial lemons to regulate normal, required lemonade. It would not surprise me in the least if we discover that bacteria have necessary functions not just in processes like scarring that protect us from bacteria, but in the normal processes of development and homeostasis.

CICATRIZATION OF WOUNDS.
XII. Factors Initiating Regeneration.
By ALEXIS CARREL, M.D.

(From the Laboratories of The Rockefeller Institute for Medical Research.)
(Received for publication, May 24, 1921.)

The nature of the factors which initiate cicatrization after an injury is not as yet exactly known. The resumption of cell proliferation in the wounded tissues of an adult animal may be attributed to the removal of resistance to growth, in consequence of the defect resulting from loss of tissue.[1] In other words, the removal of the products of growth, that is of a portion of the tissues, immediately reinaugurates the growth process, just as the removal of the products of a balanced chemical reaction at equilibrium immediately reinitiates the forward reaction.[2] This means that regeneration, being a direct consequence of the injury, is started by forces within the organism. But the same phenomenon may also be logically attributed to the action of an external factor. According to this hypothesis, the cells would be directly stimulated to growth and multiplication by forces without the organism, acting on tissues deprived of their natural protection by the injury.

I. Effect of Protection against Irritation.

If regeneration is a direct consequence of the loss of tissue and initiated by an internal factor, the cicatrization of a wound protected against all external irritation must take place normally. But if this hypothesis be not true, the wound should not begin to cicatrize. It was observed in 1908 that the latent period of cicatrization of a wound dressed with dead connective tissue or plasma clot was abnormally prolonged. This fact suggested that regeneration was not initiated directly by the loss of tissue and that, if the surface of the wound were effectively protected against mechanical, chemical, and bacterial irritations, the setting in motion of the process of cicatrization would be indefinitely postponed. In order to ascertain in what measure the onset of regeneration could be delayed by adequate protection of the surface of the wound, five experiments were performed. Two circular wounds of equal size were made on the dorsal region of dogs, according to a technique previously described.[3,4] The control wound was covered with a paste containing chloramine-T in a concentration which had been shown to be non- irritating for the tissues, and to keep them in a sterile condition.[5] The experimental wound was dressed with subcutaneous connective tissue, excised from the lumbar region of a dog and kept in cold storage. Circular flaps, slightly larger than the wound and

---
[1] Welch, W. H., *Science,* 1897, v, 813.
[2] Robertson, T. B., Principles of biochemistry for students of medicine, agriculture and related sciences, Philadelphia, 1920, 482.
[3] Carrel, Alexis, and Hartmann, A., *J. Exp. Med,* 1916, xxiv, 429.
[4] All operations were performed under ether anesthesia.
[5] Carrel, Alexis, du Noüy, P. L., and Carrel, Anne, *J. Exp. Med.*, **1917**, xxvi, 279.

Table I. Action of Connective Tissue Dressing on the Latent Period.

Experiment No.	Animal No.	Date.	Control wound.			Experimental wound.			Remarks.
			Area	Bacteria per field.	Dressing.	Area.	Bacteria per field.	Dressing.	
		1921	sq. cm.			sq. cm.			
1	1	Apr. 5	4.0		Chloramine-T.	4.6		Connective tissue.	Connective tissue was heated at 56°C. for 3 hrs.
		" 19	1.9	16	"	1.4	1		Inner dressing slipped from both wounds.
2	2	Mar. 29	3.6		"	4.1		Connective tissue.	Text-fig. 1.
		Apr. 18	0.5	0	"	2.8	0	Chloramine-T.	Inner dressing slipped from both wounds. Connective tissue dressing partially disappeared
3	3	" 22	0.1	0	"	1.6	0	"	Inner dressing slipped.
		Mar. 31	4.0		"	3.9		Connective tissue.	Infection of both wounds.
4	4	Apr. 19	0.7		Chloramine-T.	2.8	∞	Connective tissue.	Text-fig. 2. No displacement of dressing.
		Mar. 31	3.5	∞	"	4.7			
		Apr. 7	3.0	0	"	4.7	0	Dry gauze.	Appearance of granulation issue.
		" 19	2.0	0	"	5.1		" "	
		" 25	1.7	0	"				
5	5	"	5.2		"	4.2		Connective tissue.	Text-fig. 3. Connective tissue heated at 56°C.
		" 20	3.3	0	"	4.0	0	Chloramine-T.	No displacement of dressing. Appearance of granulation tissue.
		" 25	3.0	0	0	4.5	0		

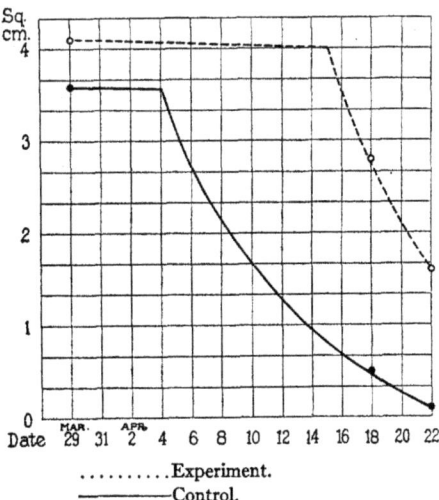

TEXT-FIG. 1. Experiment 2, Table I.

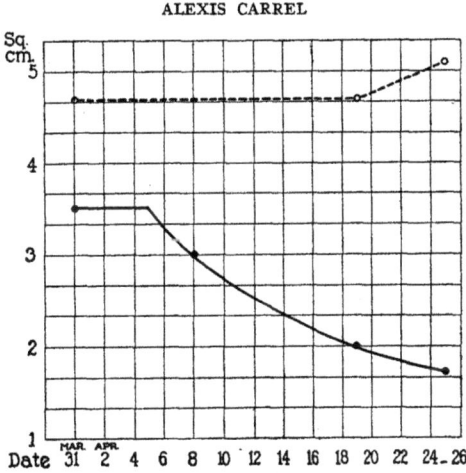

TEXT-FIG. 2. Experiment 4, Table I.

about 0.5 cm. thick, were prepared and fixed to the surface of the experimental wound by a few stitches. Both wounds were protected by a pad of dry gauze, sutured to the skin. Then the dressing was completed by a few other gauze pads, a large amount of cotton, a bandage, and a shirt. The animals were examined after a period of time varying from 13 to 25 days. As the examination involved the removal of the stitches holding the inner dressing to the skin, and also of the stitches fixed to the connective tissue placed on the wound, necessitating a considerable disturbance of the wound, the experiment was stopped after the second or third dressing.

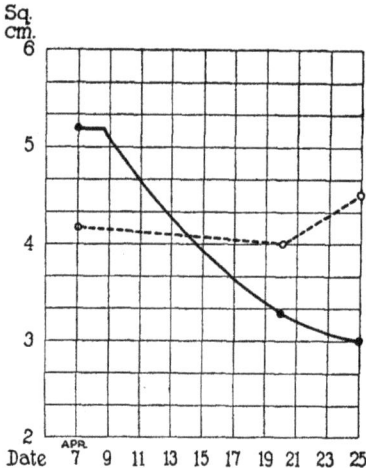

TEXT-FIG. 3. Experiment 5, Table I.

The results of the five experiments are summarized in Table I. Experiment 1 was unsuccessful because the gauze pads slipped from the surface of the wounds and infection occurred. 14 days after the operation, no difference was found in the condition of both control and experimental wounds. Experiments 2 and 3 succeeded partly. The protection given to the wounds by the connective tissue dressing was incomplete. When the wounds were inspected for the first time, 19 and 20 days respectively after the operation, the experimental wound was no longer covered by the connective tissue dressing, and cicatrization had started. In Experiment 2, the curve expressing the progress of regeneration showed that the latent period had lasted very much longer in the wound protected by connective tissue than in the control wound (Text-fig. 1). The duration of the latent period was probably 17 days, while in the control it was 6.

Although the connective tissue dressing did not remain at the surface of the wound for a long time, its effect was, however, manifest.

More significant results were obtained in Experiments 4 and 5. The connective tissue dressing remained exactly where it was applied and the surface of the wound was really protected against all outside irritation. The examination of the wounds was made 25 days after the operation in Experiment 4, and 18 days after the operation in Experiment 5. The period of contraction had not yet started and the area at that time was as large as at the time of the operation (Text-figs. 2 and 3). It was a striking fact that a wound, effectively protected by a non-irritant dressing, did not show any evidence of cicatrization 25 days after the operation.

II. Effect of Irritants.

In a second series of experiments, it was investigated whether the application of mild irritants on the surface of the wound would shorten the latent period. The experimental wound was covered by a gauze pad, soaked in turpentine and fixed

Table II. Action of Turpentine on the Latent Period.

Experiment No.	Animal No.	Control wound.				Experimental wound.			Remarks.
		Date.	Area.	Bacteria per field.	Dressing.	Area.	Bacteria per field.	Dressing.	
		1921	*sq. cm.*			*sq. cm.*			
6	6	Apr. 12	4.3		Chloramine-T.	4.6		Turpentine.	Text-fig. 4.
		" 15	4.2	0	Dry gauze.	3.6	0	Dry gauze.	No displacement of dressing.
		" 16	4.0	0	" "	3.1	0	" "	
		" 18	4.0	0	" "	3.1	0	Turpentine.	
		" 25	2.8	0		2.9	1		

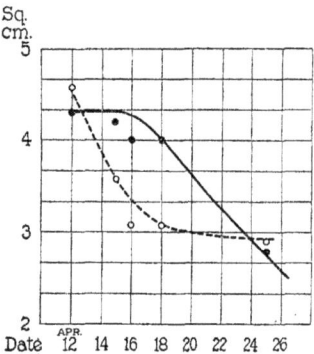

TEXT-FIG. 4. Experiment 6, Table II.

to the edges of the skin by a few stitches, while the control wound was dressed with chloramine paste. Both wounds were protected by gauze pads, sutured to the skin. The latent period of the experimental wound was very much shortened, lasting less than 2 days, while the latent period of the control wound lasted for about 5 or 6 days (Table II and Text- fig. 4).

In Experiment 7, chick embryo pulp was used instead of turpentine (Table III). 6 days after the operation, the contraction of the control wound had not started. On the contrary, in the experimental wound contraction began after a very short time, less than 2 days (Text-fig. 5). 6 days after the operation, the area of the experimental wound was about 50 per cent smaller than that of the control wound.

In four experiments (Table IV), the wounds were infected with staphylococci. Varied dilutions of a 24 hour culture of staphylococci in bouillon were used for inoculation. The control wound was dressed with chloramine paste, while the experimental wound was inoculated with 0.05 cc. of the dilution of staphylococcic culture, and dressed with dry gauze. The wounds remained in a condition of slight infection without swelling of the edges or abundant suppuration. The duration of the latent period was decreased, and often reduced to less than 2 days, as shown in Experiment 10 (Text-fig. 6).

III. Summary.

As long as the wounds were protected by a connective tissue dressing against mechanical, chemical, and bacterial irritations, no evidence of cicatrization was found. The complete or partial failure of four experiments was due to the slipping of the inner dressing from the wound, mechanical irritation by the gauze, and infection. In the two experiments in which the connective tissue was maintained at the surface of the wound, there was no beginning of cicatrization, although 25 and 18 days respectively had elapsed since the operation, while in the control wound the duration of the latent period did not exceed 5 or 6 days. The experiments were interrupted after the second or third inspection, on account of the teclyrical impossibility of again applying to the wounds a non-irritant dressing. It is probable that the wounds

Table III. Action of Chick Embryo Pulp on the Latent Period.

Experiment No.	Animal No.	Date.	Control wound.			Experimental wound.			Remarks.
			Area	Bacteria per field.	Dressing.	Area	Bacteria per field.	Dressing.	
7	7	1921 Apr. 21	sq. cm. 3.8		Chloramine-T.	sq. cm. 4.0		Chick embryo pulp.	Text-fig. 5.
		" 23	3.4	0	"	3.1	5	Dry gauze.	
		" 27	4.1	0	"	1.7	0	" "	

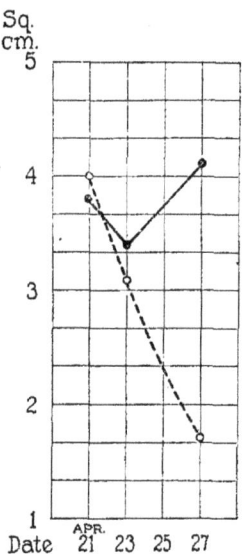

TEXT-FIG. 5. Experiment 7, Table III.

could have been kept for a much longer time in a condition of quiescence. While it is not known whether cicatrization could be prevented for an indefinite period, there is no doubt that the mechanism of regeneration is not set in motion at the usual time, when all external irritations are suppressed. It appears, therefore, that under ordinary conditions, cicatrization is not initiated by an internal factor.

On the contrary, the application of turpentine, chick embryo pulp, and staphylococci decreased markedly the length of the latent period, which was often reduced to less than 2 days. This fact demonstrated the importance of external factors in the initiation of cicatrization. It seems that the mechanism of regeneration has become adapted to the ordinary conditions of life of the animals. A small wound will begin to cicatrize sooner if slightly infected, as practically always happens, than if it were thoroughly protected by a non-irritant dressing.

IV. Conclusions.

1. It may be concluded that, tinder the conditions of the experiments, a wound, protected by a non-irritant dressing, shows no granulation tissue or beginning of contraction for 25 days at least.
2. Local application of certain irritants, such as turpentine, chick embryo pulp, and staphylococci, reduces the duration of the latent period to less than 2 days.
3. Regeneration is apparently initiated, not by an internal, but by an external factor.

Table IV. Action of Staphylococcic Infection on the Latent Period.

Experiment No.	Animal No.	Date.	Control wound.			Experimental wound.			Remarks.
			Area.	Bacteria per field.	Dressing.	Area.	Bacteria per field.	Dressing.	
		1921	*sq. cm.*			*sq. cm.*			
8	8	Apr. 19	5.0		Chloramine-T	4.0		1:100 dilution of staphylococcus suspension.	Mild infection; no edema; slight discharge.
		" 21	4.0	8	"	3.5	78	Dry gauze.	
		" 22	3.2	0	"	3.3	45	" "	
		" 25	2.8	0	"	1.8	24	" "	
		" 28	4.0	0	"	3.7	13	" "	
9	9	" 19						1:10 dilution of staphylococcus suspension.	
		" 21	3.7	4	"	3.4	8	Dry gauze.	
		" 22	3.4	0	"	3.9	85	" "	
		" 25	2.8	0	"	1.7	1	" "	
		" 28	3.7	0	"	4.0	20	" "	
10	10	" 19						1:50 dilution of staphylococcus suspension.	Text-fig. 6.
		" 22	3.7	0	"	2.5	50	Dry gauze.	
		" 25	3.2	0	"	2.3	0	" "	
		" 28	2.0	0	"	1.0	41	" "	
11	11	" 21	4.0		Chloramine-T.	3.5		Pure staphylococcus cultures.	
		" 23	2.6	0	"	2.4	6	Dry gauze.	
		" 25	2.5	0	"	2.2	0	"	
		" 28	2.3	0	"	0			

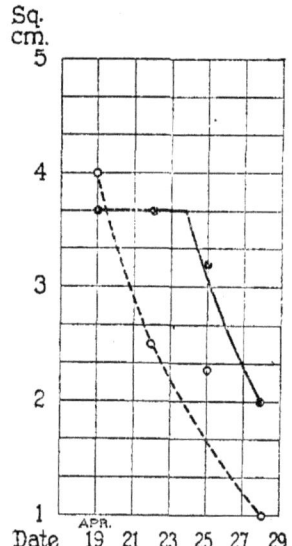

TEXT-FIG. 6. Experiment 10, Table IV.

Case Study 13: Innervation Suppresses Tumorigenesis

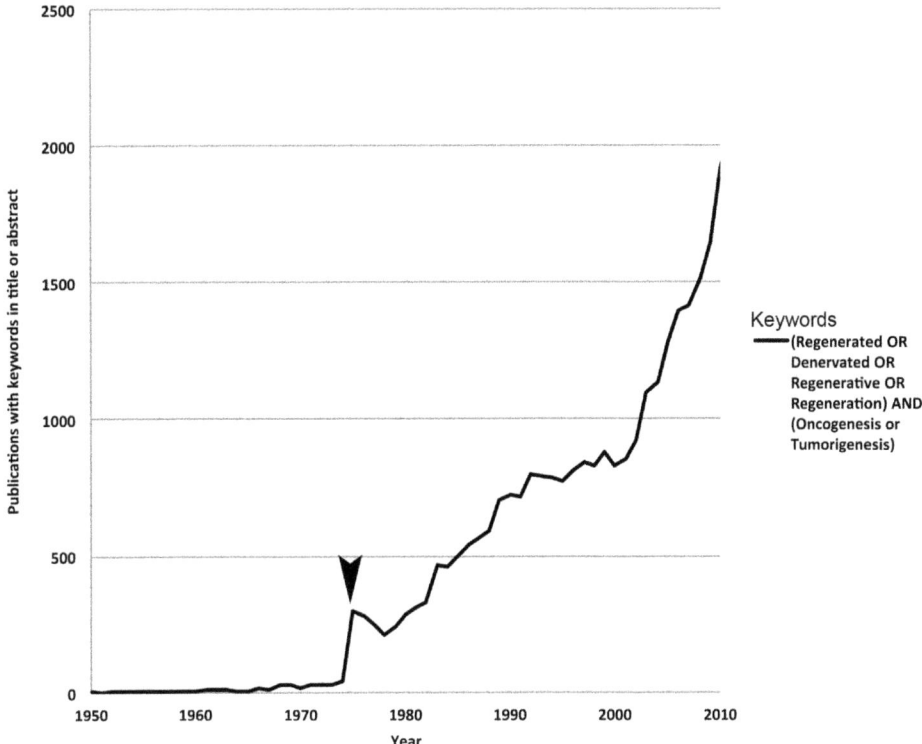

Outzen *et al* (1976) Influence of regenerative capacity and innervation on oncogenesis in the adult frog.

Perspective on: *Influence of regenerative capacity and innervation on oncogenesis in the adult frog*

Michael Levin, PhD
Allen Discovery Center,
Tufts University

Keeping Cells Harnessed Toward Correct Anatomy and Away from Tumorigenesis

Cancer biology comprises two fundamental paradigms. The mainstream view (the somatic mutation model) is that cancer cells are irreversibly damaged: they have accumulated genetic or epigenetic damage and have acquired cell-autonomous properties that underlie unlimited proliferation, tumorigenesis, and metastasis [1, 2]. Contrasting with this view is the idea that cells exhibit neoplastic behavior and form tumors due to a change of interactions with their environment. This view of cancer is focused on context – transforming physiological signals, or conversely lack of patterning signals that normally keep cells orchestrated towards normal anatomical upkeep and away from tumorigenesis [3–8]. The latter class of models ranges from simple growth suppressive molecules (morphostats) secreted by healthy tissue [9, 10] and averaging effects of cell neighbors that stabilize stochastic gene expression [11], to suppression of tumorigenesis by tissue-level organization [12, 13], to global models of whole-body morphogenetic information fields that maintain pattern throughout the lifespan [14–19].

While the mutation-centered paradigm has dominated work in this field for decades, increased attention is now focused on cancer as a progressive loss of the organization capacity of the environment over the heterogeneous behavior of isolated cells [18–24]. Interestingly, 'no cancer exhibits any trait which cannot be found in some normal tissue as the expression of normal genomic activity; no cancer grows faster than an embryo nor is any cancer cell more invasive than a macrophage nor are cancer cell lines more immortal than are germ lines. The only distinction is that, in the cancer, the expression or lack of expression of many traits may be inappropriate for the tissue in which the cancer occurs' [25]. This is a view of cancer as fundamentally a developmental disorder of cell regulation.

Large-scale shape, or the correct geometric arrangement of organs and tissues in an organism, is a key concept in biological growth and development. To achieve optimal health, organisms strive to maintain shape at all levels, from the single cell to the whole organism. Cancer can be seen as an error of geometry, because tumor cells grow, migrate, and function without regard for the orderly structure within which they occur [26]. This is seen most acutely in teratomas - embryonic tumors that display extensive differentiation of a number of tissues combined with a complete absence of orderly organization of the whole. The idea that cancer is a developmental disease is an old one [9, 10, 26–29]; Needham and Waddington speculated that cancers represented an escape from the control of the morphogenetic field [30–32]. On this view, tumors form when cells stop obeying the normal patterning cues of the body: 'cancer as part of an inexorable process in which the organism falls behind in its ceaseless effort to maintain order' [28].

Understanding cancer as a reversible physiological state of a multi-cellular dynamical system (as opposed to damage within single cancer stem cells) has significant medical implications because it suggests specific prevention and detection strategies focused on modulating the physiological interrelationships among many cells instead of looking for DNA markers in single cancer stem cells. A mechanistic dissection of these pathways may give rise to strategies that reboot [23] or normalize cancer, in contrast to current approaches that all seek to kill tumors and thus risk a compensatory proliferation response by rogue cells that still remain [33]. Thus, biologists are beginning to explore the idea that cancer is not a genetic disease of specific loci but rather a kind of attractor in a multi-dimensional transcriptional space describing cell states [17]: 'The topology of the attractor is the 'invisible hand' driving the system functions into coherent behavioral states: they are self-organizing structures and can capture the gene expression profiles associated with cell fates' [34].

A complete picture of cancer no doubt involves both an understanding of DNA damage and cell signaling dynamics, although considerable controversy exists as to the most appropriate level of organization at which to search for the origin and cure of cancer (ranging from genes, to stem cells, to tissues, to entire bodyplan organizing fields). The study by Outzen et al. reveals two key elements that suggest that the DNA damage view of cancer is not the whole story. They show that innervation is a key input into cells' decisions of whether to form tumors or not, and that regenerative state is able to counteract tumorigenesis.

References

[1] Vaux D L 2011 In defense of the somatic mutation theory of cancer *BioEssays* **33** 341–3
[2] Hanahan D and Weinberg R A 2011 Hallmarks of cancer: the next generation *Cell* **144** 646–74
[3] Baker S G 2013 Paradox-driven cancer research *Disruptive Science and Technology* **1** 143–8
[4] Soto A M and Sonnenschein C 2004 The somatic mutation theory of cancer: growing problems with the paradigm? *Bioessays* **26** 1097–107
[5] Tarin D 2013 Role of the host stroma in cancer and its therapeutic significance *Cancer metastasis reviews*
[6] Tarin D 2012 Inappropriate gene expression in human cancer and its far-reaching biological and clinical significance *Cancer Metastasis Reviews* **31** 21–39
[7] Tarin D 2012 Clinical and Biological Implications of the Tumor Microenvironment *Cancer Microenviron.*
[8] Tarin D 2011 Cell and tissue interactions in carcinogenesis and metastasis and their clinical significance *Semin. Cancer Biol.* **21** 72–82
[9] Potter J D 2007 Morphogens, morphostats, microarchitecture and malignancy *Nat Rev Cancer* **7** 464–74
[10] Potter J D 2001 Morphostats: a missing concept in cancer biology *Cancer Epidemiol. Biomarkers Prev.* **10** 161–70
[11] Capp J P 2005 Stochastic gene expression, disruption of tissue averaging effects and cancer as a disease of development *BioEssays* **27** 1277–85

[12] Soto A M and Sonnenschein C 2011 The tissue organization field theory of cancer: a testable replacement for the somatic mutation theory *BioEssays* **33** 332–40
[13] Soto A M and Sonnenschein C 2005 Emergentism as a default: cancer as a problem of tissue organization *J. Biosci.* **30** 103–18
[14] Levin M 2012 Morphogenetic fields in embryogenesis, regeneration, and cancer: non-local control of complex patterning *Bio. Systems* **109** 243–61
[15] Burr H S and Northrop F 1935 The electrodynamic theory of life *Quarterly Review of Biology* **10** 322–33
[16] Plankar M, Jerman I and Krasovec R 2011 On the origin of cancer: Can we ignore coherence? *Prog. Biophys. Mol. Biol.*
[17] Dinicola S, D'Anselmi F, Pasqualato A, Proietti S, Lisi E, Cucina A and Bizzarri M 2011 A systems biology approach to cancer: fractals, attractors, and nonlinear dynamics *OMICS* **15** 93–104
[18] Bizzarri M, Cucina A, Biava P M, Proietti S, D'Anselmi F, Dinicola S, Pasqualato A and Lisi E 2011 Embryonic morphogenetic field induces phenotypic reversion in cancer cells. Review article *Curr. Pharm. Biotechnol.* **12** 243–53
[19] Bizzarri M, Cucina A, Conti F and D'Anselmi F 2008 Beyond the oncogene paradigm: understanding complexity in cancerogenesis *Acta. Biotheor.* **56** 173–96
[20] Rubin H 2007 Ordered heterogeneity and its decline in cancer and aging *Adv. Cancer. Res.* **98** 117–47
[21] Rubin H 2006 What keeps cells in tissues behaving normally in the face of myriad mutations? *BioEssays* **28** 515–24
[22] Bissell M J and Radisky D 2001 Putting tumours in context *Nat. Rev. Cancer* **1** 46–54
[23] Ingber D E 2008 Can cancer be reversed by engineering the tumor microenvironment? *Semin. Cancer Biol.* **18** 356–64
[24] Weaver V M and Gilbert P 2004 Watch thy neighbor: cancer is a communal affair *J. Cell Sci.* **117** 1287–90
[25] Prehn R T 1994 Cancers beget mutations versus mutations beget cancers *Cancer Res.* **54** 5296–300
[26] Rowlatt C 1994 Some consequences of defining the neoplasm as focal self-perpetuating tissue disorganization *New Frontiers in Cancer Causation* ed O H Iversen (Washington, DC: Taylor & Francis) 45–58
[27] Tsonis P A 1987 Embryogenesis and carcinogenesis: order and disorder *Anticancer Res.* **7** 617–23
[28] Rubin H 1985 Cancer as a dynamic developmental disorder *Cancer Res.* **45** 2935–42
[29] Baker S G, Soto A M, Sonnenschein C, Cappuccio A, Potter J D and Kramer B S 2009 Plausibility of stromal initiation of epithelial cancers without a mutation in the epithelium: a computer simulation of morphostats *BMC Cancer* **9** 89
[30] Needham J 1963 *Chemical embryology* (New York: Hafner Pub. Co)
[31] Needham J 1936 New Advances in the Chemistry and Biology of Organized Growth: (Section of Pathology) *Proc. R. Soc. Med.* **29** 1577–626
[32] Waddington C H 1935 Cancer and the theory of organisers *Nature* **135** 606–8
[33] Fan Y and Bergmann A 2008 Apoptosis-induced compensatory proliferation. The Cell is dead. Long live the Cell! *Trends Cell Biol.* **18** 467–73
[34] Huang S, Ernberg I and Kauffman S 2009 Cancer attractors: a systems view of tumors from a gene network dynamics and developmental perspective *Semin. Cell Dev. Biol.* **20** 869–76

Influence of Regenerative Capacity and Innervation on Oncogenesis in the Adult Frog (*Rana pipiens*)[1,2]

H. C. Outzen, R. P. Custer, and R. T. Prehn[3,4]

ABSTRACT—Twenty-two sarcomas were induced in 19 adult frogs (*Rana pipiens*) treated with 3-methylcholanthrene pellets. Thirteen of these tumors arose first in a denervated forelimb, and only 2 arose first in normal or nerve-supplemented control fore-limbs ($P = 0.004$). The remaining tumors developed either as a second tumor in a tumor-bearing frog or in hindlimbs. The critical role of innervation in regenerative capacity suggests that the predilection to tumor formation in the denervated limbs may have resulted from their lessened regenerative capacity.—J Natl Cancer Inst 57: 79–84, 1976.

The view that an inverse association may exist between the regenerative capacity of an animal and its susceptibility to neoplasia is not new. Waddington *(1)* and others *(2, 3)* proposed that tumor formation occurs when an animal has an incomplete or defective regenerative capacity. A more recent refinement of the hypothesis suggests that the neoplasm is actually an aberrant attempt at blastema formation in an animal that has defective regenerative capability *(3)*.

One of the early measurable responses to a chemical oncogen is the proliferation of fine nerves into the treated area *(4)*, like that found at the outset of limb regeneration *(5)*. Mescher and Tassava *(6)*, in their studies on amputated newt limbs, emphasized the importance of such nerve entry as a trigger for blastemal proliferation and subsequent cell division. Perhaps the proliferation of fine nerves in response to a chemical oncogen is actually the preliminary stage of blastema formation.

By surgically increasing or decreasing the nerve supply to the forelimbs of *Rana pipiens*, we were able, respectively, to increase *(7)* or decrease *(8)* the regenerative capacity of those limbs. Using this surgical approach to manipulate such ability in the forelimbs, we observed whether tumor induction by a chemical oncogen correlated with the relative differences in regenerative capacity in these forelimbs.

MATERIALS AND METHODS

Animal husbandry.—Adult *R. pipiens* (J. M. Hazen Co., Alburg, Vt.), were housed in plastic aquaria. Chlorinated city water maintained at 25° C by a hot and cold water-mixing valve (Hydroguard #420; Powers Regulator Co., Skokie, Ill.) was kept constantly flowing through the aquaria at a depth of about 2–5 cm. Several wooden

---

[1] ABBREVIATION USED; MCA = 3-methylcholanthrene.
Received August 26, 1975; accepted January 5, 1976.
[2] Supported by Public Health Service grants CA08856, CA06927, CA05255, and CA13456 from the National Cancer Institute, RR05539 from the Division of Research Resources, National Cancer Institute, and by an appropriation from the Commonwealth of Pennsylvania.
[3] The Institute for Cancer Research, The Fox Chase Cancer Center, Philadelphia, Pa. 19111.
[4] We thank Ellen Lawler, Sharon Howard, Kathy Tierney, and Carol Thomas for technical assistance.

blocks were placed in each aquarium to provide an out-of-water perch for the frogs. Fluorescent lamps (Vitalite #3028; Durotest Corp., North Bergen, N.J.) were placed about 30 cm above the water level. These lights have a low UV component and helped to maintain the frogs' good health. The frogs were force-fed 2- to 5-day-old mice twice a week. We placed a small drop of a multivitamin preparation (Linatone; Lambert-Kay Div., Carter Wallace, Inc., Los Angeles, Calif.) on the skin of each baby mouse to supplement the diet.

Denervation and nerve supplementation.—Animals were anesthetized by ip injection of 0.25 ml of a 2-g tricaine methanesulfonate (#C5431; Gallard-Schlesinger, Corle Place, N.Y.)/100 ml solution in water. Denervation was performed by the dissection of the brachial nerves free of the blood vessels from the point of their emergence from the vertebral column for a distance of 15–20 mm. A 10-mm segment of the brachial plexus was removed, and the proximal ends of the nerves (2–3 mm) were directed away from the arm *(8)*.

The opposite forelimb was then sham denervated, following the same procedure used for denervation, except that the brachial nerves were not actually cut.

In experiments 2 and 3, nerve supplementation of the sham-denervated forelimb was also done at the time the contralateral forelimb was denervated. The sciatic nerve was dissected free and deviated into the forelimb. The blood vessels to this rear limb were tied off and the leg was amputated *(7)*.

Carcinogen treatment.—Strips of Millipore filter material (Millipore Corp., Bedford, Mass.) were impregnated with a 20% solution of MCA in molten paraffin. Pellets 3 mm in diameter were cut from these strips with a cork-borer and were implanted in the frogs when they were denervated.

In experiments 1 and 2, we inserted MCA pellets im in denervated and sham-denervated forelimbs and also implanted a third pellet im in the remaining rear limb of each frog in the second experiment.

In experiment 3, we placed two small pieces of mouse liver on each side of the MCA pellet. Mouse liver can metabolize MCA to the proximate carcinogen *(9)*, and there is some evidence that embryo-derived frog cell lines cannot readily metabolize polycyclic hydrocarbon oncogens *(10)*. We added murine liver tissue to the MCA pellets hoping that some of the MCA would be metabolized to the active carcinogen before the liver tissue was immunologically rejected, and that the tumor incidence would consequently be increased. These MCA-liver sandwiches were implanted im in both the denervated and the nerve-supplemented, sham-denervated forelimbs.

Paraffin control pellets were used in two separate sets of frogs (expts PC-1 and PC-2) in the same fashion as in experiments 1 and 2, except that the strips of Millipore filter and material were impregnated only with molten paraffin.

The frogs were examined weekly after carcinogen treatment, and any swelling around implanted pellets was recorded. Generally, whenever a mass 5 mm or greater in diameter was observed at the implantation site, the frog was killed. The implantation sites were excised, fixed in neutral buffered formalin, and sectioned for histologic examination. All pellet sites were examined and the presence or absence of a neoplasm was histologically confirmed.

Frogs that died from extraneous causes were likewise examined histologically for evidence of incipient tumor formation.

Statistical analysis.—Sign test: These experiments were designed to test the hypothesis that tumors arise earlier in denervated than in control forelimbs. The pairing design requirements of the sign test were met by the incorporation in each animal of both an experimental and a control treatment *(11)*. Once a tumor occurred in either the experimental or the control forelimb, that animal could be eliminated from the experiment because the question of where the tumor first arose had been answered for that frog.

Fisher's Exact Probability Test: This *(11)* was used to evaluate the following hypotheses: *a)* The greatest tumor incidence would occur in the denervated forelimbs of all carcinogen-treated frogs, and *b)* the greatest number of tumors would occur in the denervated limbs of tumor-bearing frogs.

RESULTS

Frogs that survived the first 2 months after their surgical manipulation generally remained healthy during these experiments (text-fig. 1).

General Tumor Incidence

In 3 experiments with a total of 105 frogs, 22 sarcomas arose at the implantation sites of MCA pellets (table 1). In the 105 denervated forelimbs, we observed 14 tumors at the implantation sites, and in 105 control forelimbs, only 5 induced tumors were seen. These results were significantly different ($P < 0.02$) when we used a one-tailed Fisher Exact Probability Test. Comparison of the 16 tumor-bearing frogs demonstrated that tumors occurred more frequently in the denervated forelimbs (14/16) than in the control forelimbs (5/16) ($P < 0.002$, one-tailed Fisher Exact Probability Test).

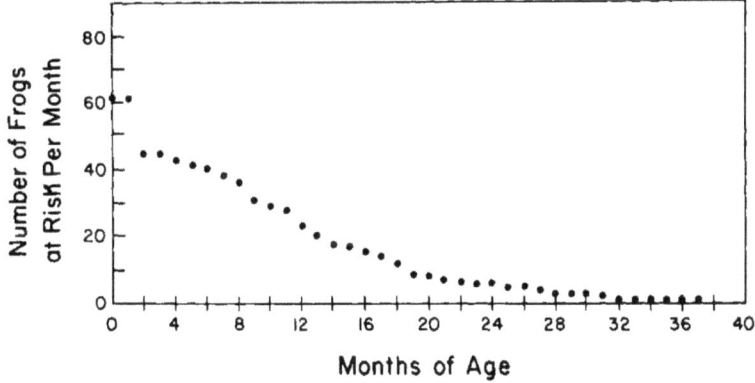

Text-figure 1.—Cumulative numbers of frogs-at-risk per month of age after surgical manipulation and carcinogen pellet implantation in experiments 1 and 2.

Table 1.—. Occurrence of sarcomas in denervated and control forelimbs and in normal hindlimbs that received an im implant of either a 20% MCA pellet or a paraffin control pellet

Experiment No. (No of frogs/ percent MCA)	Tumor-bearing frog No.	Tumor latent period, mo			Sign-test direction of difference in time of tumor appearance in forelimbs
		Forelimb		Hindlimb	
		Denervated	Control[a]		
1 (25/20)	1-A	10.7			+
	1-B	11.0			+
	1-C	11.6	13.3		+
	1-D	19.2			+
	1-E		13.0		—
2 (37/20)	2-A	6.5			+
	2-B	17.1			+
	2-C	24.8			+
	2-D		6.4		—
	2-E			4.2	[b]
	2-F			7.8	[b]
	2-G			8.1	[b]
3 (43/20)	3-A	2.2			+
	3-B	6.0	8.0		+
	3-C	9.0			+
	3-D	9.0			+
	3-E	10.0			+
	3-F	11.0			+
	3-G	12.0	12.0		-0- (tie)
PC-1 and PC-2 (19/0)[c]					

[a] Controls in expt 1 were sham denervated and in expts 2 and 3 were sham denervated and nerve supplemented.
[b] Tumors arose in hindlimbs; not incorporated in sign test.
[c] No tumors were observed in these paraffin-treated control frogs.

The time from carcinogen application until detection of the first tumor was 2.2 months; the longest latent period has been 24.8 months. The tumor incidence curve for experiments 1 and 2 suggested that the time when 50% of the carcinogen-treated animals would be expected to develop tumors would occur 23 months after the application of the carcinogen (text-fig. 2). Thus 23 frogs in experiments 1 and 2 survived more than 1 year after MCA implantation. Experiment 3 is still in progress 17 months after implantation of the carcinogen.

Two additional sets of control frogs had paraffin pellets without MCA implanted im. All remained free of tumor for an average observation period of 10 months, several in excess of 16 months (table 1).

Comparison of Tumor Induction in Denervated and Control Forelimbs

In the first experiment, 6 sarcomas were observed in 5 frogs; 4 arose first in the denervated limb, the fifth in the sham-denervated limb. The sixth sarcoma occurred in the sham-denervated forelimb of a frog about 2 months after a tumor appeared in its denervated forelimb (table 1).

The second experiment consisted of 37 frogs and differed from the first in that the sham-denervated control forelimb was also nerve supplemented. Three tumors arose in the denervated forelimbs, 1 in the control forelimb, and 3 in the hindlimb (table 1). In this experiment, the trend of more tumors arising first in the denervated forelimbs was continued.

The third experiment was like the second except that mouse liver was placed on both sides of the MCA pellets. Seven of 43 frogs developed 9 sarcomas during the first 17 months. Of these, 6 arose first in the denervated forelimbs (table 1). In the nerve-supplemented forelimbs, tumors developed at 8 and 12 months (table 1). The first sarcoma appeared in the frog that 2 months previously had developed a tumor in its denervated forelimb. The tumors found at 12 months appeared in both

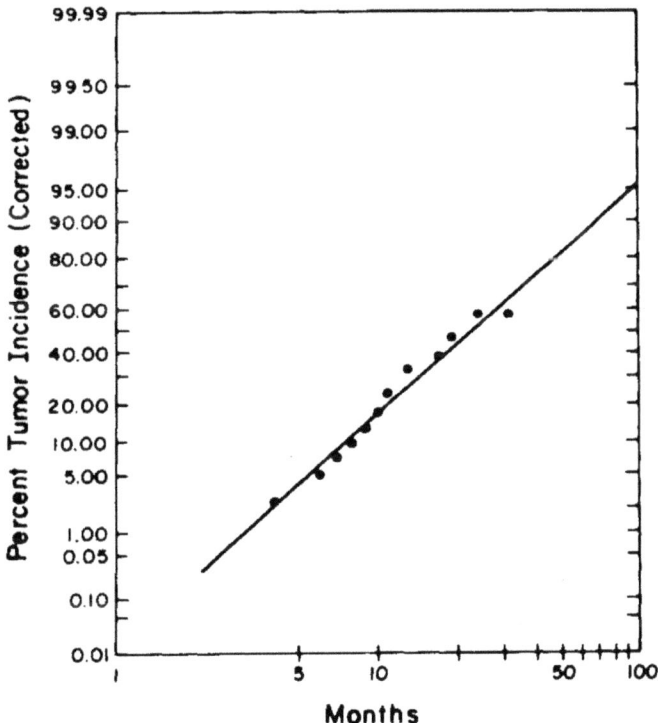

Text-figure 2. —Percent tumor incidence (corrected) in all frogs from expts 1 and 2 plotted against months of age. Twelve frogs with 13 tumors induced in these 2 experiments were used to compute this tumor incidence curve. Curve was corrected to compensate for loss of animals due to extraneous non-tumor-related deaths *(12)*.

the denervated and control forelimbs of a single frog. The present data permit no conclusion as to any influence of mouse liver on MCA metabolism.

Since each frog received an experimental and a concomitant control treatment, it was possible to consider each animal as a separate and individual experiment, which could then be pooled for analysis *(11)*.

The first appearance of tumors at the MCA implantation site in the 3 experiments occurred earlier in the denervated limbs than in the control limbs. Sixteen of these frogs developed tumors in their forelimbs; 13 arose first in the denervated forelimbs, only 2 in the controls (table 1). The probability of this distribution occurring by chance was 0.004 by the sign test *(11)*.

Influence of Nerve Supplementation to the Forelimb on the Regenerative Response

Three frogs were nerve supplemented by transplantation of the sciatic nerve from the ipsilateral hindlimb to the forelimb as previously described. The nerve-supplemented forelimbs were amputated. The time at which the first regenerative response in these limbs occurred was recorded and the length of the regenerate measured. Forelimbs of 3 normal frogs were amputated as controls. No regenerative response of any kind was seen in the control limbs. Regeneration in the nerve-supplemented forelimbs produced a limb stump of about 4–6 mm.

Pathology

The tumors were readily detectable, forming a hemispherical cap over the MCA pellet, expanding locally for the most part (fig. 4), and occasionally infiltrating the entire limb (fig. 1). They did not metastasize during the observation period. All tumors were confirmed histologically when the frogs were killed.

The same basic histologic pattern was found in each instance; the tumors were composed of wavy spindled cells usually arranged in interlacing fasciculi to form short radial whorls, palisades, and so-called herringbone stripes (figs. 2,3). These characteristics, plus the delicate, fibrillar cytoplasmic laciness, indicated a perineural histogenesis rather than the usual fascial fibrosarcoma or rhabdomyosarcoma induced by MCA in rodents. Varying degrees of anaplasia were encountered (figs. 5, 6), where the spindled tumor cells were vastly larger and more pleomorphic than in figures 2 and 3, and the interstices were considerably edematous. The aggressive invasiveness of this poorly differentiated tumor is shown in figure 5 in which the voluntary muscle fibers are seen pushed widely apart by the infiltrative growth.

The paraffin control pellets implanted im in both denervated and sham-denervated forelimbs, and in the remaining rear limb, evoked little or no inflammatory reaction (fig. 9). This was in striking contrast to early stages of the MCA-induced tumor shown in figure 10.

Transplantability of Tumors

Two tumors that arose in the denervated forelimbs were tested for transplantability. One was transplanted into the hindleg of the autochthonous host where it grew

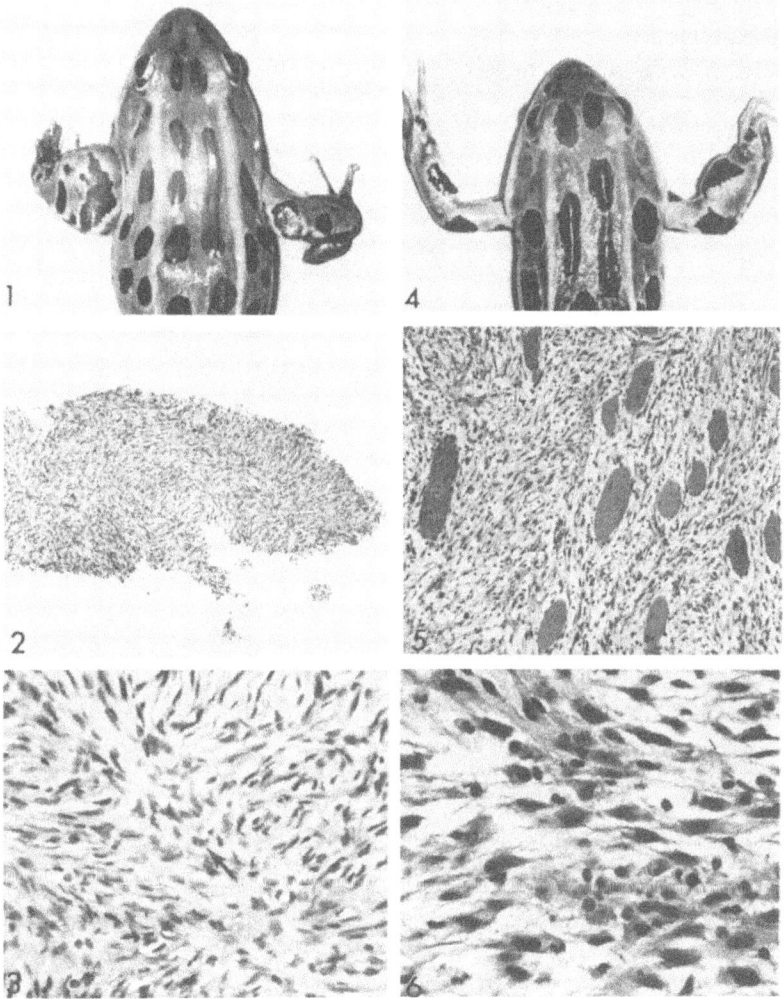

Figure 1. —MCA-induced tumor involving entire denervated left forelimb, 20 months after implantation of pellet. Frog No. 1-D. Figure 2. —Biopsy of tumor from frog No. 1-D showing compact growth of spindled cells arranged in short radial whorls and herringbone stripes, × 100. Figure 3. —Detail of figure 2 emphasizing the slender, wavy spindled cells with tendency toward nuclear palisading. A mitosis is indicated *(arrow)*, × 400. Figure 4. —MCA-induced tumor of denervated right foreleg forming hemispheric cap over implanted pellet after 12 months. Frog No. 1-C. Figure 5. —Tumor from frog No. 1-C showing invasion of host's voluntary muscle, × 100. Figure 6. —Detail of figure 5 showing marked dedifferentiation when contrasted with the tumor in figure 3. Many giant cells are present. Tumor also appears edematous, × 400

successfully. The other was propagated in cell culture for 7 passage generations and then implanted into the anterior chamber of the eye and into the brain of allogeneic hosts. Growth was apparent at both sites. There was considerable pleomorphism in the anterior chamber graft (figs. 7, 8).

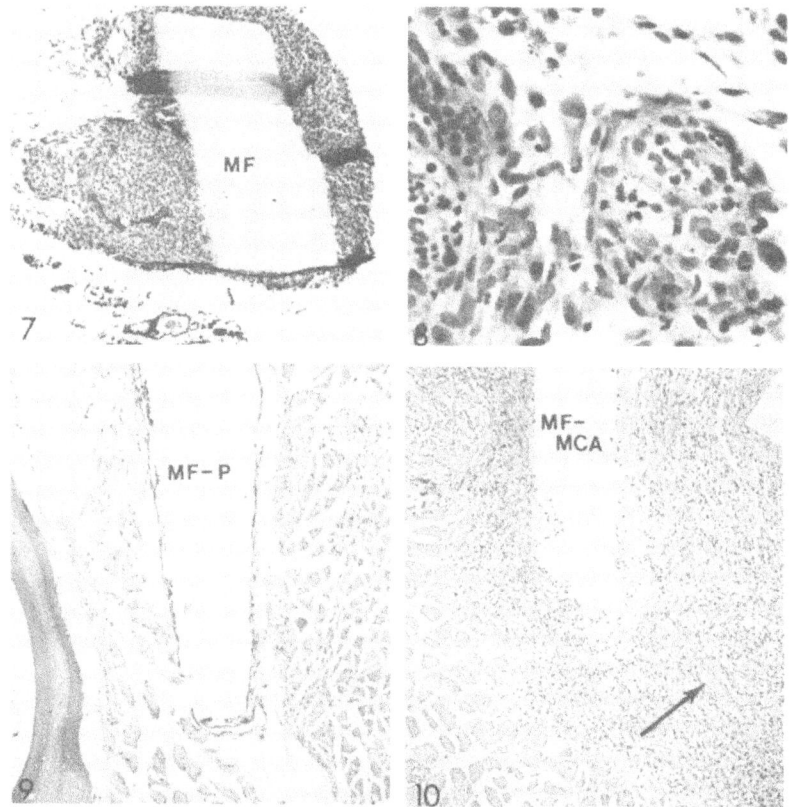

Figure 7. —Tumor graft in anterior chamber of eye, implanted on a Millipore filter (MF). Tumor had previously undergone 7 transfers through tissue culture, × 100 Figure 8. —Detail of anterior chamber transplant. Tumor shows considerable pleomorphism. × 400 Figure 9. —Control Millipore filter impregnated with paraffin alone (MF-P) shows virtually no inflammatory reaction along its margin after 13 months' implantation, × 100 Figure 10. —In contrast to figure 9, an MCA pellet (MF-MCA) is bordered by tumor that is forming fasciculi and whorls *(arrow)* 9 months after implantation, × 100

DISCUSSION

Reports of the successful induction of malignant mesenchymal tumors in Amphibia are limited *(13)*. Briggs *(14)* observed 3 tumors in 154 *R. pipiens* tadpoles treated with MCA, only 1 of which was malignant. After treatment of newts with several carcinogenic hydrocarbons, Breedis *(15)* found 2 sarcomas in 500, and Ingram *(16)* observed 1 sarcoma in 200. Other studies dealt with the increase in incidence of lymphosarcoma in *Xenopus (17)* and the induction of hepatomas in newts *(18)*. Thus our report of 22 carcinogen-induced sarcomas in the frog is unprecedented.

The conceptual relationship linking regeneration to oncogenesis is old *(1–3)*, and it has been reassessed repeatedly without much success *(13)*. Denervation of a limb allowed us to depress its regenerative capacity *(8)*, thus testing the hypothesis that the regenerative potential of a limb was inversely linked to its tendency toward neoplasia. Such an inverse relationship has been vigorously championed by Seilern-Aspang and Kratochwil *(19)*. They reported an increased incidence of malignant epithelial tumors in the less regenerative areas (i.e., dorsal body wall) of newts. Tumors that arose in the more regenerative areas (i.e., limbs and tail) differentiated into normal epithelial components. These epithelial tumors were generally limited to the integument, arose after short latent periods (8–20 days), then promptly regressed, and Balls and Ruben *(13)* questioned whether they were truly neoplastic.

Our data further support the hypothesis that there is an inverse relationship between regenerative ability and the capacity for *true neoplasia* after treatment with a chemical carcinogen. Whether the increased incidence of neoplasia in the denervated area is directly influenced by the inability to regenerate cannot yet be determined. There is evidence, however, that mammalian tumor susceptibility is increased in a surgically denervated area *(20, 21)*, though a much earlier paper was contradictory *(22)*. That regeneration will neither begin nor proceed unless an adequate nerve supply is present is well known *(5, 8)*, except in the case of newts made aneurogenic at the tailbud stage *(23)*. Apparently the dependence of regeneration on the presence of nervous tissue is a learned response.

The increased number of tumors that developed in the denervated limbs in our experiments implies that an intact nerve supply provides a defense against tumor induction in the frog. Some nerve regrowth into the denervated limbs was highly probable, though the denervated forelimbs never regained normal mobility. Furthermore, hydrocarbon carcinogens have a demonstrable neurotropic activity *(4)* where many fine nerve fibers proliferate into the carcinogen-treated area. This activity may have been important in providing a receptive milieu for tumorigenesis in the denervated fore-limb, i.e., one with the requisite small amount of regenerative capacity, insufficient to allow for morphogenetic control but enought to permit cell division and early blastemal events to occur. The greater regenerative potential in the normal or nerve-supplemented limbs aided by the neurotropic action of the carcinogen may have raised the level of morphogenetic control in these limbs far above the threshold at which a tumorous response would become likely.

That normal innervation is an important element in the early stages of blastemal formation *(5, 6)* and in the response to chemical carcinogens *(4)* has been clearly demonstrated. We can thereby postulate a direct causal link between innervation of a limb, its regenerative capacity, and whether or not morphogenetic control over neoplastic transformation will occur.

REFERENCES

1. WADDINGTON CH: Cancer and the theory of organizers. Nature 135:606–608, 1935
2. NEEDHAM J: New advances in the chemistry and biology of organized growth. Proc R Soc Med 29:1577–1626, 1936
3. PREHN RT: Immunosurveillance, regeneration and oncogenesis. Prog Exp Tumor Res 14:1–24, 1971
4. FITZGERALD MJ, LAVELLE SM: Response of murine cutaneous nerves to skin painting with methylcholanthrene. Anat Rec 154:617–634, 1966

5. SINGER M: The influence of the nerve in regeneration of the amphibian extremity. Q Rev Biol 27:169–200, 1952
6. MESCHER AL, TASSAVA RA: Denervation effects on DNA replication and mitosis during the initiation of limb regeneration in adult newts. Dev Biol 44:187–197, 1975
7. SINGER M: Induction of regeneration of the forelimb of the post-metamorphic frog by augmentation of the nerve supply. J Exp Zool 126:419–471, 1954
8. SINGER M, KAMRIN RP, ASHBAUGH A: The influence of denervation upon trauma induced regenerates of the forelimb of the post-metamorphic frog. J Exp Zool 136:36–51, 1957
9. SIMS P, GROVER PL: Epoxides in polycyclic aromatic hydrocarbon metabolism and carcinogenesis. Adv Cancer Res 20:165–274, 1974
10. DIAMOND L, CLARK HF: Comparative studies on the interaction of benzo[*a*]pyrene with cells derived from poikilothermic and homeothermic vertebrates. I. Metabolism of benzo[*a*]pyrene. J Natl Cancer Inst 45:1005–1012, 1970
11. SIEGEL S: *In* Nonparametric Statistics for the Behavioral Sciences (Harlow HF, ed.). New York, McGraw-Hill, 1956, pp 68–75
12. PILGRIM HI, DOWD JE: Correcting for extraneous death in the evaluation or morbidity or mortality from tumor. Cancer Res 23:45–48, 1963
13. BALLS M, RUBEN LN: A review of the chemical induction of neoplasms in Amphibia. Experientia 20:241–247, 1964
14. BRIGGS RW: Tumour induction in *Rana pipiens* tadpoles. Nature 146:29, 1940
15. BREEDIS C: Induction of accessory limbs and of sarcoma in the newt (*Triturus viridescens*) with carcinogenic substances. Cancer Res 12:861–873, 1952
16. INGRAM AJ: The reactions of carcinogens in the axolotl (*Ambystoma mexicanum*) in relation to the "regeneration field control" hypothesis. J Embryol Exp Morphol 26:425–441, 1971
17. BALLS M: Methylcholanthrene-induced lymphosarcoma in *Xenopus laevis*. Nature 196:1327–1328, 1962
18. INGRAM AJ: The lethal and hepatocarcinogenic effects of dimethylnitrosamine injection in the newt *Triturus helveticus*. Br J Cancer 26:206–215, 1972
19. SEILERN-ASPANG F, KRATOCHWIL K: Induction and differentiation of an epithelial tumour in the newt (*Triturus cristatus*). J Embryol Exp Morphol 10:337–356, 1962
20. PAWLOWSKI A, WEDDELL G: Induction of tumours in denervated skin. Nature 213:1234–1235, 1967
21. HASSON J: The effect of methylcholanthrene on the denervated skin of strain A mice. Cancer Res 18:267–273, 1958
22. CRAMER W: Innervation as a factor in the experimental production of cancer. Br J Exp Pathol 6:71–74, 1925
23. YNTEMA CL: Regeneration in sparsely innervated and aneurogenic forelimbs or *Ambystoma* larvae. J Exp Zool 140:101–123, 1959

Chapter 5

Mathematics and modeling

Case Study 14: Top-Down Causation: Not All The Work Is Done By Molecules

Perspective on *Quantifying causal emergence shows that macro can beat micro*

Sara Imari Walker,
Arizona State University

More is Not Just Different, its Causal

The reductionist paradigm in science has worked remarkably well for almost four centuries, ever since Newton consolidated much of our understanding of motion into a few concise equations. In the ensuing centuries, reductionism has given us insights into the structure of the atom, the fundamental forces of nature, and the inner workings of the cell among many others. Nonetheless there remain hints that reductionism qua reductionism cannot be the entire story. Fundamental components of reality—including life and mind—are not easily reconciled with the idea that all of reality may be reduced to description solely in terms of those fundamental laws that operate at its most microscopic layers. In the words of the Nobel Laureate Philip Anderson in his famous essay *More is Different* [1]:

'The ability to reduce everything to simple fundamental laws does not imply the ability to start from those laws and reconstruct the universe.'

As we move up in scale, the universe appears to become more and more complex, and it seems clear that as complexity increases entirely new properties emerge. For example, *How is it that the collection of atoms in a human brain is organized to give rise to thoughts, feelings and urges?* Perhaps more mysterious is that our thoughts appear to matter to the world. A simple thought experiment (no pun intended) of intending to lift your left arm and then executing that action provides the evidence. There is intense debate about whether this kind of example represents a conscious or subconscious action (it may not matter for this debate), and specifically whether one can really consider the action to be *caused* by a thought in your brain. A thought is an informational abstraction—so the debate really centers on whether *information* can have causal power. If all of reality can be described by mechanical actions at the micro-level only, there is no 'room at the top' for your thoughts to be causal. Stated succinctly, the challenge we are faced with is to determine what is doing the causal work—*is it you or your atoms?*

Settling the debate requires quantifying causal emergence, such that we can identify where the 'causal work' actually lies. There has been a long-held assumption that the bottom-up picture of reality excludes higher-levels from playing a role. This is refuted by the work of Hoel *et al* who explicitly quantify *cause* and *effect* information and show that more causal power can exist at macroscopic levels in some systems, depending on how they are networked together [2]. The key result is that even though there may exist a complete lower-level description of a physical system, higher levels can still be causal above and beyond the causal work of lower levels. Emergence and reductionism are not at odds. In fact, reductionism is what makes emergence possible [3]: if we could not reduce reality to the study of a few component building blocks, we would not be able to describe how those building blocks come together to create more complex 'higher-levels'.

This is significant since there are many examples across the sciences where 'top-down' causal effects are essential to building a coherent explanatory narrative [4]. Our conscious experience is only the most blatant example. Others come from quantum physics, thermodynamics, complex systems science and biology. For the latter, a causal role for macroscopic entities may even play a critical role in defining its life's origins [5]. In short, these same issues emerge (pun!) anywhere the term 'emergence' is cast about. We need a theory of causal emergence to understand the hierarchical organization of our universe, and why 'higher-levels' not only appear so much richer and more complex than the lower-levels they emerge from, but why they seem to matter at all.

This story is just beginning. While the idea of causal emergence has been demonstrated as a proof of concept, it is unclear if the formalism as presented will be the ultimate one science settles on. Important open questions remain. For example, causal emergence occurs in networks with particular architectures that are highly non-linear and include feedback loops and irreversible logic operations. Clearly these kinds of structures exist at macroscopic levels in the human brain, but how are these compatible with our lowest level descriptions of reality in physics, which are cast in terms of initial states and local, reversible laws of motion? Can causal emergence still occur if we consider the deepest layers of reality and do not already start from a macroscopic description? There also remains the open question of whether the macro could actually *intervene* on the micro (and not just beat it), such that top-down causation might be said to truly occur. These questions represent the next frontier for 21st century science to explain the emergent layers of reality as we experience them.

References

[1] Anderson P W 1972 More is Different *Science, New Series* Vol. 177 No. 4047 393–6
[2] Hoel E P, Albantakis L and Tononi G 2013 Quantifying causal emergence shows that macro can beat micro *Proc. Natl Acad. Sci.* **110** 19790–5
[3] Smith E and Morowitz H J 2016 *The Origin and Nature of Life on Earth: The Emergence of the Fourth Geosphere* (Cambridge University Press)
[4] Ellis G F R "Top-down causation and emergence: some comments on mechanisms." *Interface Focus* 2011 rsfs20110062
[5] Walker S I and Davies P C W 2013 The Algorithmic Origins of Life *J. Roy. Soc. Interface* **6** 20120860

Quantifying causal emergence shows that macro can beat micro

Erik P. Hoel, Larissa Albantakis, and Giulio Tononi[1]

Department of Psychiatry, University of Wisconsin, Madison, WI 53719

Edited by Michael S. Gazzaniga, University of California, Santa Barbara, CA, and approved October 22, 2013 (received for review August 6, 2013)

Causal interactions within complex systems can be analyzed at multiple spatial and temporal scales. For example, the brain can be analyzed at the level of neurons, neuronal groups, and areas, over tens, hundreds, or thousands of milliseconds. It is widely assumed that, once a micro level is fixed, macro levels are fixed too, a relation called supervenience. It is also assumed that, although macro descriptions may be convenient, only the micro level is causally complete, because it includes every detail, thus leaving no room for causation at the macro level. However, this assumption can only be evaluated under a proper measure of causation. Here, we use a measure [effective information (EI)] that depends on both the effectiveness of a system's mechanisms and the size of its state space: EI is higher the more the mechanisms constrain the system's possible past and future states. By measuring EI at micro and macro levels in simple systems whose micro mechanisms are fixed, we show that for certain causal architectures EI can peak at a macro level in space and/or time. This happens when coarse-grained macro mechanisms are more effective (more deterministic and/or less degenerate) than the underlying micro mechanisms, to an extent that overcomes the smaller state space. Thus, although the macro level supervenes upon the micro, it can supersede it causally, leading to genuine causal emergence—the gain in EI when moving from a micro to a macro level of analysis.

In science, it is usually assumed that, the better one can characterize the detailed causal mechanisms of a complex system, the more one can understand how the system works. At times, it may be convenient to resort to a "macro"-level description, either because not all of the "micro"-level data are available, or because a rough model may suffice for one's purposes. However, a complete understanding of how a system functions, and the ability to predict its behavior precisely, would seem to require the full knowledge of causal interactions at the micro level. For example, the brain can be characterized at a macro scale of brain regions and pathways, a meso scale of local populations of neurons such as minicolumns and their connectivity, and a micro scale of neurons and their synapses (1). With the goal

---

[1] To whom correspondence should be addressed. E-mail: gtononi@wisc.edu.
Author contributions: E.P.H., L.A., and G.T. designed research; E.P.H. and L.A. performed research; E.P.H., L.A., and G.T. analyzed data; and E.P.H., L.A., and G.T. wrote the paper.
The authors declare no conflict of interest.
This article is a PNAS Direct Submission.
This article contains supporting information online at www.pnas.org/lookup/suppl/doi:10.1073/pnas.1314922110/-/DCSupplemental.

of a complete mechanistic understanding of the brain, ambitious programs have been launched with the aim of modeling its micro scale (2).

The reductionist approach common in science has been successful not only in practice, but has also been supported by strong theoretical arguments. The chief argument starts from the intuitive notion that, when the properties of micro-level physical mechanisms of a system are fixed, so are the properties of all its macro levels—a relation called "supervenience" (3). In turn, this relation is usually taken to imply that the micro mechanisms do all of the causal work, i.e., the micro level is causally complete. This leaves no room for any causal contribution at the macro level; otherwise, there would be "multiple causation" (4). This "causal exclusion" argument is often applied to argue against the possibility for mental causation above and beyond physical causation (5), but it can be extended to all cases of supervenience, including the hierarchy of the sciences (6).

Some have nevertheless argued for the possibility that genuine emergence can occur. Purported examples go all of the way from the behavior of flocks of organisms (7) to that of ant colonies (8), brains (9), and human societies (10). Unfortunately, it remains unclear what would qualify some systems as truly emergent and others as reducible to their micro elements. Also, most arguments in favor of emergence have been qualitative (11). A convincing case for emergence must demonstrate that higher levels can be causal above and beyond lower levels ["causal emergence" (CE)]. So far, the few attempts to characterize emergence quantitatively (12) have not been based on causal models.

Here, we make use of simple simulated systems, including neural- like ones, to show quantitatively that the macro level can causally supersede the micro level, i.e., causal emergence can occur. We do so by perturbing each system through its entire repertoire of possible causal states ("counterfactuals," in the general sense of alternative possibilities) and evaluating the resulting effects using "effective information" (EI) (13). EI is a general measure for causal interactions because it uses perturbations to capture the effectiveness/selectivity of the mechanisms of a system in relation to the size of its state space. As will be pointed out, EI is maximal for systems that are deterministic and not degenerate, and decreases with noise (causal divergence) and/or degeneracy (causal convergence).

For each system, we completely characterize the causal mechanisms at the micro level, fixing what can happen at any macro level (supervenience). Macro levels are defined by coarse graining the micro elements in space and/or time, and this mapping defines the repertoire of possible causes and effects at each level. By comparing EI at different levels, we show that, depending on how a system is organized, causal interactions can peak at a macro rather than at a micro spatiotemporal scale. Thus, the macro may be causally superior to the micro even though it supervenes upon it. Evaluating the changes in EI that arise from coarse or fine graining a system provides a straightforward way of quantifying both emergence and reduction.

Theory

In what follows, we consider discrete systems S of connected binary micro elements that implement logical functions (mechanisms) over their inputs. We first introduce a state-dependent measure of causation, the "cause" and "effect information" of a single system state s_0, before we describe the state-independent *EI* of the system *S*.

> **Significance**
>
> Properly characterizing emergence requires a causal approach. Here, we construct causal models of simple systems at micro and macro spatiotemporal scales and measure their causal effectiveness using a general measure of causation [effective information (*EI*)]. *EI* is dependent on the size of the system's state space and reflects key properties of causation (selectivity, determinism, and degeneracy). Although in the example systems the macro mechanisms are completely specified by their underlying micro mechanisms, *EI* can nevertheless peak at a macro spatiotemporal scale. This approach leads to a straightforward way of quantifying causal emergence as the supersedence of a macro causal model over a micro one.

State-Dependent Causal Analysis. The micro mechanisms of *S* specify its state-to-state transition probability matrix (TPM) at a micro time step *t*. Building upon the perturbational framework of causal analysis developed by Judea Pearl (14; see also ref. 18), the TPM can be obtained by perturbing *S* at t_0 (13) into all possible *n* initial states with equal probability $1/n$ [do($S = s_i$) $\forall i \in ...n$]. Perturbing the system in this way corresponds to the unconstrained repertoire (probability distribution) of possible causes U^C and determines the probability of the resulting states at t_{+1}, corresponding to the unconstrained repertoire of possible effects U^E. Although U^C is thus identical to the uniform distribution *U* [with $p(s) = 1/n$, $\forall s \in S$], U^E is typically not uniform. A current system state $S = s_0$ is associated with the probability distribution of past states that could have caused it ("cause repertoire $S_P|s_0$," obtained by Bayes' rule), and the probability distribution of future states that could be its effects ("effect repertoire $S_F|s_0$"). A system's mechanisms and current state thus constrain both the repertoire of possible causes U^C and that of possible effects U^E. An informational measure of the causal interactions in the system (15) can then be defined as the difference [here Kullback–Leibler divergence (D_{KL}) (16)] between the constrained and unconstrained distributions:

$$\text{Cause information}(s_0) = D_{KL}((S_P|S_0), U^C),$$
$$\text{Effect information}(s_0) = D_{KL}((S_F|s_0), U^E).$$

Cause/effect information depends on two properties: (*i*) the size of the system's state space (repertoire of alternatives), because both are bounded by $\log_2(n)$; (*ii*) the effectiveness of the system's mechanisms in specifying past and future states. To isolate effectiveness from size, we define the following normalized coefficients:

$$\text{Cause coefficient}(s_0) = \frac{\text{Cause Information}(s_0)}{\log_2(n)},$$
$$\text{Effect coefficient}(s_0) = \frac{\text{Effect Information}(s_0)}{\log_2(n)}.$$

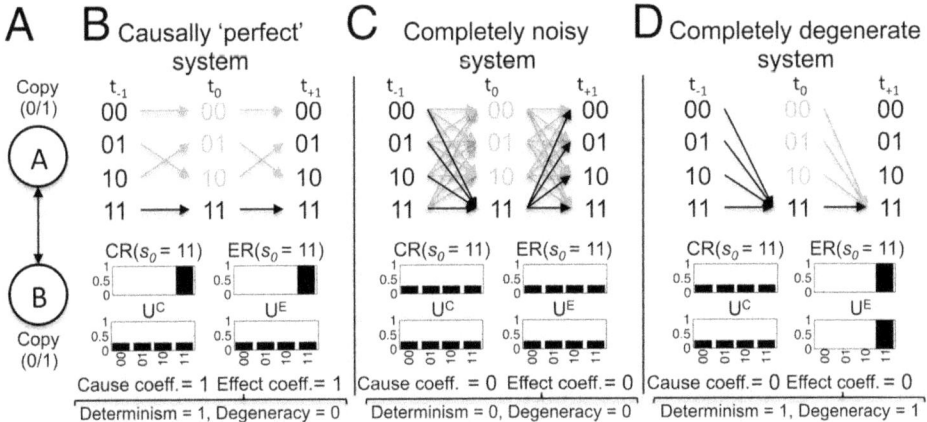

Fig. 1. Cause and effect coefficients in example systems with different causal architectures. (*A*) The systems consist of two interconnected binary COPY gates with possible states 0 and 1. (*B*) A causally perfect system, in which each state has one cause and one effect. Thus, $s_0 = [11]$ has a cause and effect coefficient (coef.) of 1. Moreover, there is no divergence (determinism coef. = 1) and no convergence (degeneracy coef. = 0). (*C* and *D*) In both the completely indeterministic and completely degenerate systems, state $s_0 = [11]$ is completely insufficient to specify past system states and completely unnecessary to specify future states (cause and effect coefficient = 0). Note that the degeneracy coef. is 0 in the completely noisy system, because all convergence is due to noise alone.

The "cause coefficient" describes to what extent a state is sufficient to specify its past causes, and the "effect coefficient" indicates how necessary a state is to specify its future effects (Fig. 1*B*). In turn, the effect coefficient itself is a function of two terms, "determinism" and "degeneracy" (see *Effect Coefficient and Effectiveness (Eff) Expressed as Determinism and Degeneracy* for derivation):

Effect coefficient = Determinism coefficceint (s_0) − Degeneracy coefficient (s_0)

$$= \frac{1}{\log_2(n)} \sum_{S_F \in U^E} p(S_F|S_0) \log_2(n \cdot p(S_F|S_0))$$

$$- \frac{1}{\log_2(n)} \sum_{S_F \in U^E} p(S_F|S_0) \log_2(n \cdot p(S_F|S_0)).$$

The determinism coef. is the difference $D_{KL}((S_F|s_0), U)$ between the effect repertoire and the uniform distribution (U) of system states, divided by $\log 2(n)$, and measures how deterministically (reliably) s_0 leads to the future state of the system: it is "1" (complete determinism) when the current state leads to a single future state with probability $p = 1$, and is "0" (complete in-determinism or noise) if it could be followed by every future state with $p = 1/n$. The degeneracy coef. measures to what degree there is deterministic convergence (not due to noise) from other states onto the future states specified by s_0. In broad terms, degeneracy refers to multiple ways of deterministically achieving the same effect or function (17, 18). The degeneracy coef. is 1 (complete degeneracy) when s_0 specifies the same future state as all other states, and 0 when s_0 specifies a unique future state (no degeneracy).

Both cause and effect coefficients are minimal (0) in a completely noisy or completely degenerate system (Fig. 1 *C* and *D*) and maximal (1) in a deterministic, nondegenerate system (*Bounds of Cause and Effect Coefficients and Effectiveness Eff (S)*). The contribution of a single state to the system's determinism and degeneracy are best demonstrated by decomposing the effect coefficient. Although the cause coefficient also reflects the degeneracy and determinism of the system, it is not subdivided further here.

State-Independent Causal Analysis. A state-independent informational measure of a system's causal architecture can be obtained by taking the expected value of cause or effect information over all system states, a quantity called effective information (*EI*):

$$EI(s) = \langle \text{Cause Information}(s_0) \rangle = \sum_{s_0 \in U^E} p(s_0) D_{KL}((S_P|s_0), U^C)$$

$$= \langle \text{Effect Information}(s_0) \rangle = \frac{1}{n} \sum_{s_0 \in U^E} D_{KL}((S_F|s_0), U^E).$$

The two terms are identical, because the system is assumed to be time invariant ($\langle t_{-1} \rightarrow t_0 \rangle = \langle t_0 \rightarrow t_{+1} \rangle$), and cause and effect information are related via Bayes' rule. *EI* is also the mutual information (*MI*) between all possible causes and their effects, $MI(U^C; U^E)$ (Effective Information EI(S) Expressed in Terms of Cause and Effect Information and Mutual Information MI).

As a measure of causation, EI captures how effectively (deterministically and uniquely) causes produce effects in the system, and how selectively causes can be identified from effects. As with the state-dependent measures, the effectiveness (Eff) of the causal interactions within a system can be captured by normalizing EI by the system's size: $Eff(S) = EI(S)/\log_2(n)$. Also as in the state-independent case, effectiveness can be split into two components, determinism and degeneracy:

$$Eff(S) = \langle \text{Determinism coefficient}(s_0) \rangle$$
$$- \langle \text{Degeneracy coefficient}(s_0) \rangle$$
$$= \langle D_{KL}((S_F|s_0), U) \rangle / \log_2(n) - D_{KL}(U^E|U)/\log_2(n).$$

Thus, *Eff*(S) = 1 if *EI* is maximal for a given system size, and decreases with indeterminism (divergence due to noise) or degeneracy (deterministic convergence), with *Eff*(S) = 0 for completely noisy or degenerate systems (Fig. 1 C and D). In a system with perfect effectiveness (Fig. 1*B*), each cause has a unique effect, and each effect has a unique cause. Thus, such a system [where *Eff*(S) = 1] is perfectly retrodictive/predictive, in the sense that not only the unique future trajectory, but also the unique past trajectory of all states can be deduced from the TPM (complete causal reversibility).

Levels of Analysis. A finite, discrete system *S* can be considered at various levels, from the most fine-grained micro causal model S_m through various coarse-grained causal models S_M. All macro levels S_M are assumed to be "supervenient" on the micro level S_m: given the micro elements of S_m and the causal relationships between them, all other members of {**S**}—the set of all possible causal models of system *S*—are fixed as well (19). Although S_m fixes S_M, any S_M may be fixed by a number of different lower level descriptions, a property known as "multiple realizability" (20).

Groupings. Micro elements are binary and labeled by Latin letters {A, B, C ...}, macro elements by Greek letters {α, β, γ ...}. Micro states are labeled {1, 0} and macro states {"on," "bursting," "quiet"...). Micro elements can be grouped into macro elements spatially, temporally, or both. Micro states are grouped into macro states through a mapping $M : S_m \to S_M$. The mapping must be exhaustive and disjunctive over micro elements (all of the states of one micro element must be mapped to the states of the same macro element; note that a macro element can consist of a single micro element as long as the state space of the system is reduced). Moreover, the mapping must be such that no micro-level information is available at the macro level (the identity of the micro elements grouped into a macro element is lost). For example, the grouping of the four states of two micro elements into the two states of one macro element as [[00, 01, 10] = off, [11] = on] is permitted, whereas the grouping [[00, 01], [10, 11]] is not, because distinguishing 01 from 10 requires knowing the identity of the micro elements.

Level-Specific Perturbations. Causal analysis at the micro level S_m, requires setting S into all possible micro states with equal probability (i.e., testing all micro alternatives) and determining the resulting effects. When moving to a macro level S_M, S must similarly be set into all possible macro states with equal probability (i.e., testing all macro alternatives). To causally assess any macro state, then, one must set S into all of the n_{micro} micro states $\{s_m\}$ that are grouped into the corresponding macro state s_M, and average over the effects. This is done using a "macro perturbation": $do(S_M = s_M) = \frac{1}{n_{micro}} \sum_{s_{m,i} \in S_M} do(S_M = s_{m,i})$. Using such macro perturbations, one can obtain cause/effect information and EI for every coarse grain of S_m. EI at each macro level is then equivalent to the MI between the set of macro causes and their macro effects.

Causal Emergence/Reduction. Finally, by assessing $EI(S)$ over all coarse grains of S_m, one can ask at which level of $\{S\}$ causation reaches a maximum. This provides an analytical definition of causal emergence, expressed in bits: $CE = \underline{EI(S_M)} - EI(S_m)$.

Thus, if $EI(S)$ is maximal for a macro-level S_M rather than the micro-level S_m, then $CE > 0$ and causal emergence occurs. If for every macro-level $CE < 0$, causal reduction holds. Although the focus here is on emergence/reduction relative to the micro-level S_m, the above measure can of course be used to compare different macro levels.

As mentioned above, $EI(S)$ depends on both the size of the system's repertoire of states and on the effectiveness of its mechanisms. When moving from one system level to another, both terms change as the state space becomes smaller or larger, and the individual states become more or less selective with respect to the past, and more or less determined or degenerate with respect to the future. The respective informational contributions of repertoire size and effectiveness to $\Delta EI(S)$ can be expressed separately as follows: $\Delta I_{Eff} = (Eff(S_M) - Eff(S_m)) \cdot log_2(n_M)$, $\Delta I_{Size} = Eff(S_m) \cdot (log_2(n_M) - log_2(n_m))$, where $n_{m/M}$ is the state repertoire size of $S_{m/M}$. It follows that $\Delta EI = \Delta I_{Eff} + \Delta I_{Size} = CE$. A positive ΔI_{Eff} can thus be due to the macro reducing the degeneracy of the micro level, increasing the determinism of the micro level, or both. Notably, coarse graining the micro-level S_m into macro-level S_M implies that ΔI_{Size} is always negative. Hence, for causal emergence to occur [$EI(S_M) > EI(S_m)$], the increase in effectiveness ΔI_{Eff} must outweigh the decrease in ΔI_{Size}.

Results

Causal analysis was performed across all coarse grains of a system [only the S_M with maximal $EI(S)$ is shown in the figures] with a custom-made Python program. Data plots were created using MATLAB. Below, we consider examples of spatial, temporal, and spatiotemporal emergence (see Fig. S1 for an example of spatial reduction).

Spatial Causal Emergence. As a proof-of-principle example, consider a system of four binary elements S_m = {ABCD} (Fig. 2A). Each micro mechanism is an AND-gate (two inputs) operating over some intrinsic noise. The 16 × 16 S_m TPM was constructed by setting the system into all possible micro states from [0000] to [1111] with equal probability (Fig. 2B). At the micro level S_m, effective information $EI(S)$ = 1.15 bits, out of maximally 4 bits, with effectiveness $Eff(S_m)$ = 0.29. The macro level S_M (Fig. 2D), composed of two elements {α, β}, each with states {"on," "off"}, is a coarse graining of S_m as defined by the mapping M in Fig. 2C. The 4 × 4 S_M TPM was obtained by setting the system into all possible macro states from [off, off] to [on, on] with equal probability (Fig. 2E). For the macro level, $EI(S_M)$ = 1.55 bits, higher than $EI(S_m)$ = 1.15 bits. Thus, $CE(S)$ = 0.40 bits, demonstrating that in

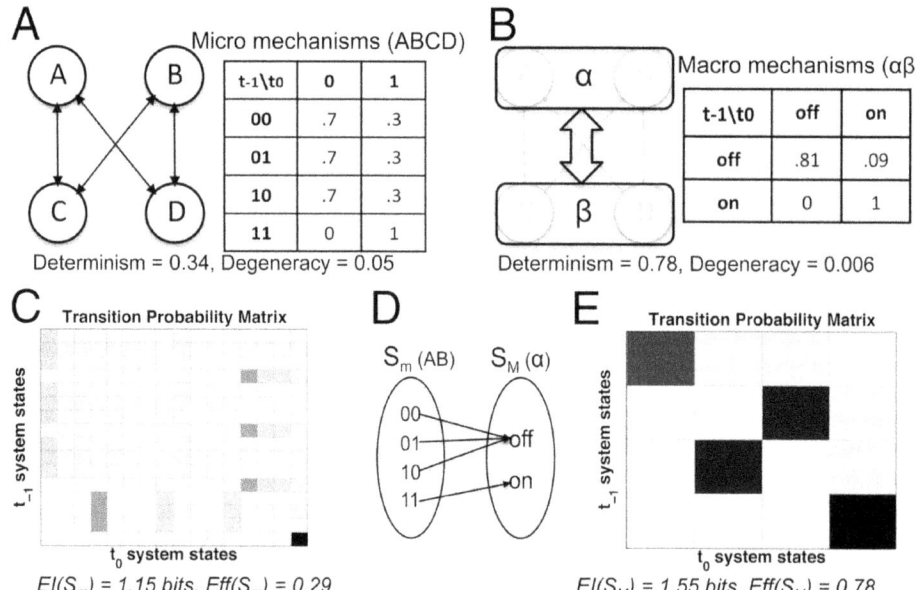

Fig. 2. Spatial causal emergence (counteracting indeterminism). (*A*) The micro level S_m of system S is composed of identical noisy micro mechanisms. (*B*) The micro TPM. (*C*) A macro causal level S_M and its TPM are defined by the mapping M (shown for AB to α, CD to β is symmetric). (*D*) S_M and its macro mechanisms. (*E*) By reducing indeterminism and increasing effectiveness *Eff*, the macro beats the micro in terms of *EI* despite the reduced repertoire size (CE = 0.40 bits).

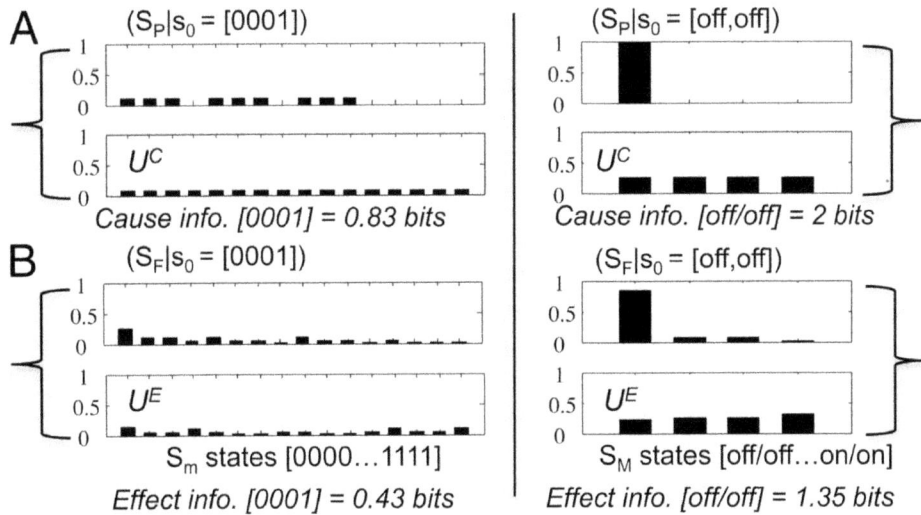

Fig. 3. State-dependent cause/effect information. (*A*) The cause information of S_m in micro state {ABCD} = [0001] is calculated as the difference (D_{KL}) between the cause repertoire of state [0001] and the unconstrained micro repertoire U^C (*Left*). The cause information of S_M in the supervening macro state {αβ} = [off/off] (*Right*) is the difference (D_{KL}), between the cause repertoire of [off/off] and the unconstrained macro repertoire U^C. (*B*) Effect information. The higher cause and effect information at the macro level is due to an increase in determinism and decrease in degeneracy, reflecting higher selectivity.

this case the macro S_M beats the micro S_m and constitutes the optimal causal model of system S. This is because the TPM for S_M is much closer to perfect effectiveness [$Eff(SM) = 0.78$] and the increase in effectiveness gained by grouping $\Delta I_{Eff} = 0.97$ bits outweighs the loss in size $\Delta I_{Size} = -0.57$ bits. In this example, the gain in effectiveness ΔI_{Eff} at the macro level comes primarily (91%) from counteracting noise [determinism coef. $(Sm) = 0.34$; $(S_M) = 0.78$] and less so (9%) from reducing degeneracy [degeneracy coef. $(S_m) = 0.05$; $(S_M) = 0.006$].

The higher effectiveness of the macro level is also evident comparing S_m and S_M in a state-dependent manner. As an example, the cause/effect distributions for S_m in state {ABCD} = [0001] are compared with the corresponding S_M state {αβ} = [off, off] in Fig. 3. Comparing the cause/effect distributions of S_m = [0001] against the unconstrained repertoires (using D_{KL}) yields 0.83 bits of cause information and 0.43 bits of effect information. For the macro S_M, cause information is 2 bits and effect information 1.35 bits. The macro beats the micro because {αβ} = [off, off] is both more selective and more reliable than {ABCD} = [0001].

Causal emergence may arise not only from macro gains in determinism (as above), but also from reducing degeneracy. In Fig. 4, micro elements A-F are deterministic AND gates connected in away that ensures high degeneracy (Fig. 4A, determinism coef. = 1; degeneracy coef. = 0.6), resulting in $Eff(S_m) = 0.4$ and $EI(S_m) = 2.43$ bits (Fig. 4C). The optimal macro groups the six micro AND gates into three macro COPY gates (αβγ) (Fig. 4B). Both macro and micro are deterministic, but by eliminating degeneracy $\Delta I_{Eff} = 1.79$ bits > $-\Delta I_{Size} = 1.22$ bits. As a result, $Eff(S_M) = 1$, $EI(S_M) = 3$ bits, and the macro emerges over the micro ($CE = 0.57$ bits).

Temporal Causal Emergence. The same principles allowing for emergence through spatial groupings hold for temporal groupings, which coarse grain micro time steps (t_x) into macro time steps (T_x). The example in Fig. 5 shows micro elements that, upon receiving an input "burst" of two spikes, respond with an output burst of two spikes. Thus, elements implement second-order Markov mechanisms over both inputs and outputs (Fig. 5A). Fig. 5B shows that causal interactions assessed over one micro time step are weak [$EI(S_m) = 0.16$ bits; $Eff(S_m) = 0.03$] because they fail to capture the second-order mechanisms. By contrast, causal analysis over two micro time steps (Fig. 5C) gives $EI = 1.38$ bits and $Eff(S_m) = 0.34$. The temporal grouping of micro into macro states $\alpha = \{A_t, A_{t+1}\}$ and $\beta = \{B_t, B_{t+1}\}$ (Fig. 5D) is analogous to the spatial grouping in Fig. 2: {00, 01, 10} = {off} and {11} = {on}. Over macro time steps, the system becomes fully deterministic and nondegenerate, $EI(S_M) = 2$ bits, $Eff(S_M) = 1$, and $CE(S) = 0.62$ bits (Fig. 5 E and F).

Spatiotemporal Causal Emergence. In general, emergence may occur simultaneously over space and time (Fig. 6). As in Fig.5, the nine neural-like micro elements in Fig. 6A are second-order Markov mechanisms, integrating inputs and outputs over two micro time steps, t_{-2} t_{-1}, and t_0 t_{+1}, respectively [compare to longer time constants of NMDA receptors (21)]. Moreover, in the examples above, the micro elements within a macro element were not connected and were causally equivalent. To demonstrate that this is not a requisite for causal emergence, in Fig. 6, the micro elements are fully connected and causally heterogeneous (self-connections not drawn). All elements are spontaneously active (1) with heterogeneous probabilities: $p(A/D/G) = 0.45$; $p(B/E/H) = 0.5$; $p(C/F/I) = 0.55$. The elements are structured into three groups {ABC, DEF, GHI} due to different intra- group and intergroup mechanisms: within each group, if the sum of intragroup connections $\Sigma(intra) = 0$ (for two time steps), all elements stay 0 (for the next two time steps). However, if the sum of intergroup connections $\Sigma(inter) = 6$ from one or both of the other two groups over two time steps (burst of synchronous activity), $p(1)$ is raised by 0.5 for the next two time steps (see Fig. S2 for macro and micro TPMs of a spatial system with equivalent rules). At the macro-level S_M (Fig. 6B), the three groups of neurons become macro elements, and two micro time steps (t_x) are grouped into one macro time step (T_x). In neural terms, these macro elements could represent "minicolumns" having three states: "inhibited" (all minicolumn neurons silent at T_x), "receptive" (some firing at T_x), or "bursting" (all firing at T_x). Macro causal interactions can be summarized as follows: if a macro element is inhibited, only receiving a burst can move it to the receptive or (more unlikely) the bursting state; otherwise, it stays inhibited. As in previous examples, the coarse-grained S_M has higher $EI(SM) = 3.51$ bits and $Eff(SM) = 0.74$ than S_m [$EI(Sm) = 0.59$ bits; $Eff(S_m) = 0.033$]. In this case, spatiotemporal causal emergence [$CE(S) = 2.92$ bits] is due to an increase in determinism that far outweighs a slight increase in degeneracy and the decrease in size.

Discussion
This paper provides a principled way of assessing at which spatio-temporal grain size the causal interactions within a system reach a maximum. Causal interactions

are evaluated by effective information (*EI*), a measure that is sensitive both to the effectiveness of the system's mechanisms and to the size of its state space. Examples with simulated systems demonstrate that, after coarse graining the micro mechanisms in both space and time, *EI* can be higher at a macro level than at a micro level. In these cases, the macro mechanisms, rather than the micro ones, can be said to be doing the causal work within a system.

Effective Information, Effectiveness, and Emergence. As shown here, *EI* corresponds to the "effectiveness" of a system's mechanisms multiplied by repertoire size, expressed in bits. Effectiveness *Eff(S)* is the average of the effect coefficients over all system states. The effect coefficient measures to what extent the current system state is necessary to specify the system's future state. This, in turn, is a function of determinism minus degeneracy. On the cause side, the equivalent to the effect coefficient is the cause coefficient, which measures to what extent the current state is sufficient to specify the system's past state. For a particular current state, cause and effect coefficients may differ: for example, a state may have many causes but only one effect. However, the average of the effect coefficients over system states, i.e., effectiveness, corresponds to the average of the cause coefficients (weighted by the probability of the effects). In other words, within a time-invariant system the average selectivity of the causes corresponds to the average selectivity of the effects. Note that, in principle, other measures of causation that, like *EI*, reflect causal

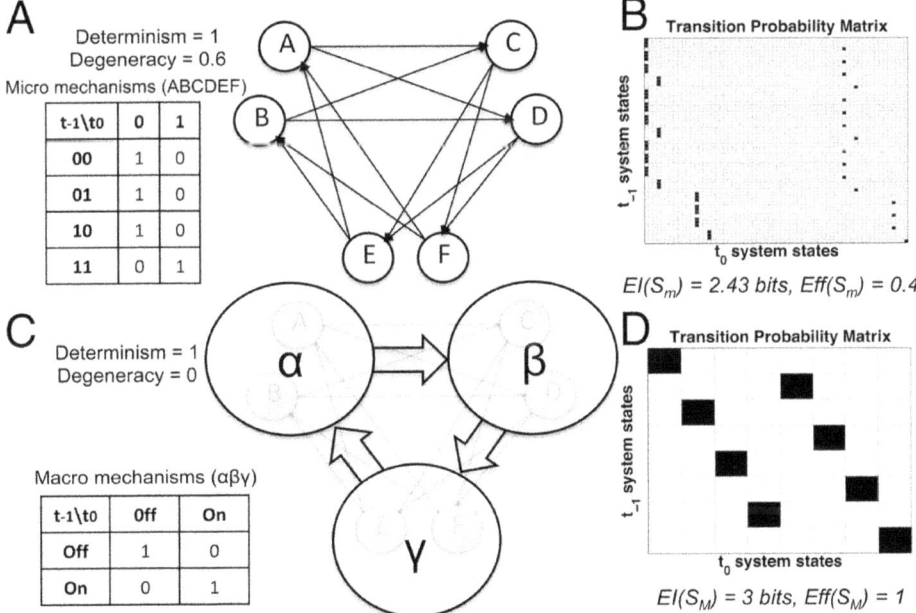

Fig. 4. Spatial causal emergence (counteracting degeneracy). (*A*) A degenerate S_m with deterministic AND gates. (*B*) The cycle of AND gates is mapped onto a cycle of COPY gates at the macro level. (*C*) The deterministic but degenerate micro TPM. (*D*) The deterministic macro TPM with zero degeneracy. By eliminating degeneracy and achieving perfect effectiveness, the macro beats the micro (*CE* = 0.57 bits).

structure (selectivity, determinism, degeneracy) and system size, should demonstrate causal emergence as well.

The main result obtained in the simulations is that coarse graining, both in space and in time, can yield a higher value of *EI*. This happens even though the micro has, by definition, a larger state space than the macro—an advantage with respect to *EI*. Given this inherent advantage of the micro, it is understandable why the default scientific strategy for analyzing systems has been one of reduction (*Causal Reduction*). However, the examples presented above show that the inherent loss in *EI* due to the macro's smaller repertoire size can be offset if the macro achieves a greater gain in effectiveness. In turn, greater effectiveness stems from macro mechanisms constructed from their constituting micro mechanisms in such a way that, at the macro level, determinism is increased and/or degeneracy is decreased. Genuine causal emergence can then be said to occur whenever there is a gain in *EI* (*CE* > 0) at the optimal macro level. If instead there is a loss in *EI* (*CE* < 0), causal reduction is appropriate, and the micro level is the optimal level of causal analysis. The causal approach pursued here suggests that qualitative or noncausal accounts of

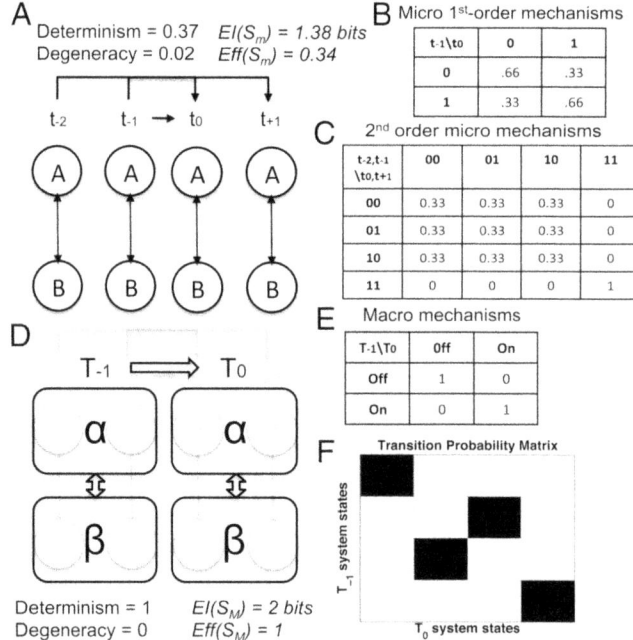

Fig. 5. Temporal causal emergence. (*A*) S_m is composed of second-order Markov mechanisms A and B: at t_0, each mechanism responds based on the inputs at t_{-2} and t_{-1}, and outputs over t_0 and t_{+1}. (*B*) Causal analysis over one micro time step gives an incomplete view of the system. (*C*) A causal analysis over two micro time steps reveals the second-order Markov mechanisms. (*D*) The optimal macro system S_M groups two micro time steps into one macro time step for macro elements {α,β}. (*E*) Each coarse grained macro mechanism effectively corresponds to a deterministic COPY gate. (*F*) The macro one-time step TPM S_M has $Eff(S_M) = 1$, and the micro two-time step TPM has $Eff(S_m) = 0.34$; $CE = 0.62$ bits.

emergence may have been hindered by not being able to characterize how and why a macro level can actually have greater causal effectiveness than a micro level (22, 23).

Micro Macro Mappings and Repertoires of Alternatives. The present approach makes it possible to compare causation at the micro and macro levels in a fair manner. First, the simulated examples are such that the macro supervenes strictly upon the micro: once the micro is defined, all macro levels are fixed. Specifically, no extra causal ingredients are added at the macro level, such as rules that apply to the macro only (24). Furthermore, the mapping of micro into macro elements is such that the identity of micro elements is lost; otherwise, the macro level would have access to micro-level information that could offset its reduced repertoire size. Finally, when causation is evaluated a uniform distribution of alternatives is imposed independently at the micro and macro levels. For this uniform distribution of perturbations to be imposed at the macro level, the probability of the underlying micro perturbations must be modified by averaging the micro states that map into the same macro state. The modified distribution of micro perturbations yielding a uniform distribution of macro perturbations makes EI sensitive to the causal structure at each level, ultimately allowing the supervening macro EI to exceed the micro EI.

Emergence as an Intrinsic Property of a System. EI is a causal measure, because it requires perturbing the system in all possible ways and evaluating the resulting effects on the system. It is also an informational measure, because its value depends on the size of the repertoire of alternatives. Indeed, in the present approach, causation and information are necessarily linked (25), hence the term "effective information." Finally, measuring EI reveals an "intrinsic" property of the system, namely the average effectiveness/selectivity of all possible system states with respect to the system itself. Effectiveness/selectivity can be assessed at multiple spatiotemporal grains, and the particular spatiotemporal grain at which EI reaches a maximum is again an intrinsic property of the system. This in no way precludes an observer from profitably investigating the system's properties at other macro levels, at the micro level, or at multiple levels at once (e.g., neuroscientists studying the brain at the level of ion channels, individual neurons, local field potentials, or functional magnetic resonance signals). However, causal emergence implies that the macro level with highest EI is the one that is optimal to characterize, predict, and retrodict the behavior of the system—the one that "carves nature at its joints" (26).

The search for the macro level at which EI is maximal has a parallel in information theory: channel capacity is an intrinsic property defined as the maximal amount of information that can be transmitted along the channel at a certain rate, found by searching over all possible input distributions (27). Finding the optimal level of coarse graining for causal emergence is based on a similar search, with several differences. First, EI is evaluated using perturbations over the system itself, rather than across a channel (the system is its own input and output). Second, the probability distributions over micro states that can be considered must conform to a proper mapping of micro into macro elements (or time intervals). Additional

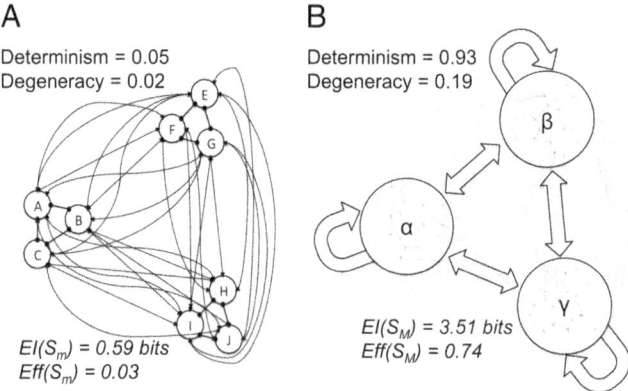

Fig. 6. Spatiotemporal causal emergence. (*A*) A "neuronal" system merging the temporal characteristics of the system in Fig. 5 with a differentiated spatial structure (Fig. S2). Regular and rounded arrows indicate intergroup and intragroup connections, respectively. (*B*) Each macro element receives inputs from itself and the other macro element. The macro level beats the micro level, leading to spatiotemporal emergence [$CE(S) = 2.92$ bits].

connections of causal emergence to established measures, such as reversibility and lumping in Markov processes (28), or epsilon machines (29), are a potential subject for future work.

Causal Exclusion and Its Implications. Causal analysis as presented here endorses both supervenience (no extra causal ingredients at the macro level) and causal exclusion [for a given system at a given time, causation occurs at one level only, otherwise causes would be double counted (4)]. However, causal analysis also demonstrates that *EI* can actually be maximal at a macro level, depending on the system's architecture. In such cases, causal exclusion turns the reductionist assumption on its head, because to avoid double-counting causes, optimal macro causation must exclude micro causation. In other words, macro mechanisms can always be decomposed to their constituting micro mechanisms (supervenience); however, if there is emergence, macro causation does not reduce to micro causation, in which case the macro wins causally against the micro and takes its place (supersedence). The notion of irreducibility among levels (does the macro beat the micro?) is complemented by the notion of irreducibility among subsets of elements within a level [is the whole more than its parts (15, 25)?]. From the perspective of a system, emergence ($CE > 0$) implies causal "self-definition" at the optimal macro level—the one at which its causal interactions "come into focus" (30) and "the action happens."

Applicability to Real Systems. Measuring *EI* exhaustively, across all micro/macro levels, is not feasible for complex physical or biological systems (*Applicability—Network Motifs as Indicators of Emergence*). However, some useful guidelines can be derived from the above analysis: (*i*) if $Eff(S_m) \geqslant Eff(S_M)$, then causal emergence is impossible and causal reduction holds; (*ii*) if $EI(S_m) > \log_2(n_M)$, where n_M is the state repertoire size of S_M, causal reduction holds; (*iii*) if for some coarse graining, *Eff* increases drastically, causal emergence is to be suspected (as $\Delta I_{Eff} \gg -\Delta I_{Size}$). Therefore, systems that already are close to maximal effectiveness at the micro level (Fig. S1) indicate causal reduction. By contrast, heavily interconnected groups of elements with spontaneous activity and the ability to distinguish between intragroup and intergroup connections, such as the simplified neural system of Fig. 6, are more suitable for emergence.

In real neural systems, one could compare the respective effective information at the micro scale of single neurons over millisecond intervals, the meso scale of neuronal groups over hundreds of milliseconds, and the macro scale of brain regions over several seconds (using tools such as optogenetics and calcium imaging). In this way, classic notions, such that cortical minicolumns may constitute the fundamental units of brain function (31), or that the cortex works by population coding in space (32) or rate coding in time (33) in the face of high intertrial variability (34), could then be tested rigorously using a measure of effectiveness. Examining small motifs that are overrepresented in complex networks [such as brains (35)] could determine whether the network as a whole is biased toward emergence or reduction. Heuristic assessments of the likelihood of emergence could also rely on the analysis of wiring diagrams, which can offer an estimate of degeneracy, combined with knowledge of the amount of intrinsic noise in a system, which can provide an estimate of determinism.

Conclusions

The approach to emergence investigated here provides theoretical support for the intuitive idea that, to find out how a system works, one should find the "differences that make [most of] a difference" to the system itself (25) (cf. ref. 36). It also suggests that complex, multilevel systems such as brains are likely to "work" at a macro level because, in biological systems, selectional processes must deal with unpredictability and lead to degeneracy (18). This may also apply to some engineered systems designed to compensate for noise and degeneracy. More broadly, this view of causal emergence suggests that the hierarchy of the sciences, from microphysics to macroeconomics, may not just be a matter of convenience but a genuine reflection of causal gains at the relevant levels of organization.

ACKNOWLEDGMENTS. We thank M. Boly, C. Cirelli, A. Hashmi, C. Koch, L. Morton, A. Nere, M. Oizumi, and L. Shapiro for helpful discussions, and P. Rana for assisting with the Python software. This work has been supported by Defense Advanced Research Planning Agency Grant HR 0011-10-C-0052 and The Paul G. Allen Family Foundation.

1. Sporns O, Tononi G, Kotter R (2005) The human connectome: A structural description of the human brain. *PLoS Comput Biol* 1(4):e42.
2. Markram H (2006) The blue brain project. *Nat Rev Neurosci* 7(2):153–160.
3. Davidson D (1980) Mental events. *Readings in Philosophy of Psychology,* ed Block N (Harvard Univ Press, Cambridge, MA), Vol 1, pp 107–119.

4. Kim J (1993) *Supervenience and Mind: Selected Philosophical Essays* (Cambridge Univ Press, Cambridge, UK).
5. Kim J (2000) *Mind in a Physical World: An Essay on the Mind-Body Problem and Mental Causation* (MIT Press, Cambridge, MA).
6. Bontly T (2002) The supervenience argument generalizes. *Philos Stud* 109:75–96.
7. Seth A (2008) Measuring emergence via nonlinear Granger causality. *ALIFE* 2008:545–552.
8. Hölldobler B, Wilson E (2009) *The Superorganism: The Beauty, Elegance, and Strangeness of Insect Societies* (W. W. Norton, New York).
9. Sperry R (1983) *Science and Moral Priority: Merging Mind, Brain, and Human Values* (Columbia Univ Press, New York).
10. Sawyer R (2005) *Social Emergence: Societies as Complex Systems* (Cambridge Univ Press, Cambridge, UK).
11. Broad C (1925) *The Mind and Its Place in Nature* (Routledge & Kegan Paul, London).
12. Bar-Yam Y (2004) A mathematical theory of strong emergence using multiscale variety. *Complexity* 9:15–24.
13. Tononi G, Sporns O (2003) Measuring information integration. *BMC Neurosci* 4:31.
14. Pearl J (2000) *Causality: Models, Reasoning and Inference* (Cambridge Univ Press, Cambridge, UK).
15. Albantakis L, Hoel EP, Koch C, Tononi G (2013) Intrinsic Causation and Consciousness. *Association for the scientific study of consciousness conference (ASSC17)*. Available at www.theassc.org/files/assc/docs/ASSC17-PB-070113-online-version-with-Addendum.pdf. Accessed November 2, 2013.
16. Kullback S (1997) *Information Theory and Statistics* (Dover Publications, New York).
17. Edelman GM (1987) *Neural Darwinism: The Theory of Neuronal Group Selection* (Basic Books, New York).
18. Tononi G, Sporns O, Edelman GM (1999) Measures of degeneracy and redundancy in biological networks. *Proc Natl Acad Sci USA* 96(6):3257–3262.
19. Stalnaker R (1996) Varieties of supervenience. *Philos Perspect* 10:221–241.
20. Fodor J (1974) Special sciences (or: The disunity of science as a working hypothesis). *Synthese* 28:97–115.
21. Jahr CE, Stevens CF (1990) Voltage dependence of NMDA-activated macroscopic conductances predicted by single-channel kinetics. *J Neurosci* 10(9):3178–3182.
22. Bedau M (1997) Weak emergence. *Nous* 31:375–399.
23. Chalmers D (2006) Strong and weak emergence. *The Reemergence of Emergence,* eds Clayton P, Davies P (Oxford Univ Press, Oxford), pp 244–256.
24. Butterfield J (2012) Laws, causation and dynamics at different levels. *Interface Focus* 2(1):101–114.
25. Tononi G (2012) Integrated information theory of consciousness: An updated account. *Arch Ital Biol* 150(2-3):56–90.

26. Hamilton E, Cairns H (1961) *The Collected Dialogues of Plato: Including the Letters* (Pantheon Books, New York).
27. Shannon CE (1997) The mathematical theory of communication. 1963. *MD Comput* 14(4):306–317.
28. Kemeny J, Snell J (1976) *Finite Markov Chains* (Springer, New York).
29. Shalizi C, Crutchfield J (2001) Computational mechanics: Pattern and prediction, structure and simplicity. *J Stat Phys* 104:817–879.
30. Alexander S (1920) *Space, Time, and Deity: The Gifford Lectures at Glasgow, 1916–1918* (Macmillan, London).
31. Buxhoeveden DP, Casanova MF (2002) The minicolumn hypothesis in neuroscience. *Brain* 125(Pt 5):935–951.
32. Georgopoulos AP, Schwartz AB, Kettner RE (1986) Neuronal population coding of movement direction. *Science* 233(4771):1416–1419.
33. London M, Roth A, Beeren L, Hausser M, Latham PE (2010) Sensitivity to perturbations in vivo implies high noise and suggests rate coding in cortex. *Nature* 466(7302):123–127.
34. Knoblauch A, Palm G (2005) What is signal and what is noise in the brain? *Biosystems* 79(1-3):83–90.
35. Sporns O (2010) *Networks of the Brain* (MIT Press, Cambridge, MA).
36. Fitelson B, Hitchcock C (2010) *Probabilistic Measures of Causal Strength* (Oxford Univ Press, Oxford, UK).

Supporting Information

Hoel etal. 10.1073/pnas.1314922110

Effect Coefficient and Effectiveness *(Eff)* Expressed as Determinism and Degeneracy

The state-dependent effect coefficient $(s_0) = \dfrac{\text{effect information}(s_0)}{\log_2(n)}$ can be described as a function of two terms, the determinism and degeneracy coefficient. To derive these two terms, the effect information (s_0), the distance between the effect repertoire $(S_F|s_0)$ and the unconstrained repertoire of effects U^E, is split into the distance between $(S_F|s_0)$ and the uniform distribution U with $p(s_U) = 1/n$, and a residual term:

$$\text{Effect Information}(s_0) = D_{KL}((S_F|s_0), U^E)$$

$$= \sum_{s_F \in U^E} p(S_F|s_0) \log_2\left(\frac{p(S_F|s_0)}{p(s_F)}\right) \tag{S1}$$

$$= \sum_{s_F \in U^E} p(s_F|s_0) \log_2\left(\frac{p(s_F|s_0)}{p(s_U)} + \frac{p(s_U)}{p(s_F)}\right) \tag{S2}$$

$$= \sum_{s_F \in U^E} p(s_F|s_0)\left(\log_2\left(\frac{p(s_F|s_0)}{p(s_U)}\right) - \log_2\left(\frac{p(s_F)}{p(s_U)}\right)\right) \tag{S3}$$

$$= \sum_{s_F \in U^E} p(s_F|s_0)\log_2\left(\frac{p(s_F|s_0)}{p(s_U)}\right) - \sum_{s_F \in U^E} p(s_F|s_0)\log_2\left(\frac{p(s_F)}{p(s_U)}\right) \tag{S4}$$

$$(\text{using } p(s_U) = 1/n) = \sum_{s_F \in U^E} p(s_F|s_0)\log_2(n \cdot p(s_F|s_0)) - \sum_{s_F \in U^E} p(s_F|s_0)\log_2(n \cdot p(s_F)) \tag{S5}$$

$$= D_{KL}((S_F|s_0), U) - \sum_{s_F \in U^E} p(s_F|s_0)\log_2(n \cdot p(s_F)), \tag{S6}$$

where s_F denotes a state of the system S_F at t_{+1} with probability $p(s_F)$ according to the unconstrained distribution of effects U^E. s_0 is the present system state. The determinism coefficient is then the left term in lines S5 and S6 divided by $\log 2(n)$:

$$\text{Degeneracy coefficient}(S_0) = \frac{\sum_{s_F \in U^E} p(S_F|s_0) \log_2(n \cdot p(S_F|s_0))}{\log_2(n)} \tag{S7}$$

the degeneracy coefficient the right term:

$$\text{Degeneracy coefficient}(s_0) = \frac{\sum_{s_F \in U^E} p(s_F|s_0)\log_2(n \cdot p(s_F|s_0))}{\log_2(n)} \quad (S8)$$

as defined in the main article.

The effectiveness (*Eff*) of a system assesses the causal relations in a system in a state-independent manner, irrespective of the size of the system's state space:

$$Eff(S) = \frac{EI(S)}{\log_2(n)} = \frac{\langle \text{Effect Information}(s_0) \rangle}{\log_2(n)}$$
$$= \frac{\sum_{s_0 \in U^C} p(s_0) D_{KL}((S_F|s_0), U^E)}{\log_2(n)}, \quad (S9)$$

where the effective information $EI(S)$ is the average effect information of all system states s_0, distributed according to U^C, the unconstrained repertoire of causes, which is identical to the uniform distribution U; thus, here $p(s_0) = 1/n$. $EI(S)$ can then be divided in the same way as the state-dependent effect information:

$$EI(S) = \langle \text{Effect Information}(s_0) \rangle, \quad (S10)$$

$$= \left\langle D_{KL}((S_F|s_0), U) - \sum_{s_F \in U^E} p(S_F|s_0) \log_2\left(\frac{p(s_F)}{p(s_U)}\right) \right\rangle, \quad (S11)$$

$$= \langle D_{KL}((S_F|s_0), U) \rangle - \left\langle \sum_{s_F \in U^E} p(S_F|s_0) \log_2\left(\frac{p(s_F)}{p(s_U)}\right) \right\rangle, \quad (S12)$$

$$= \langle D_{KL}((S_F|s_0), U) \rangle - \sum_{s_0 \in U^C} p(s_0) \sum_{s_F \in U^E} p(s_F|s_0)\log_2\left(\frac{p(s_F)}{p(s_U)}\right), \quad (S13)$$

$$= \langle D_{KL}((S_F|s_0), U) \rangle - \sum_{s_F \in U^E} p(s_F)\log_2\left(\frac{p(s_F)}{p(s_U)}\right), \quad (S14)$$

$$= \langle D_{KL}((S_F|s_0), U) \rangle - D_{KL}(U^E, U). \quad (S15)$$

The last equality is due to the fact that $p(s_F)$ is the probability of state s_F to occur at t_{+1} following U^E, the unconstrained distribution of effects (future states) obtained by setting the system S at t_0 into all possible states s_0 with equal probability $p(s_0) = 1/n$.

Both, indeterminism and degeneracy at the micro level may be indicative of causal emergence (*Discussion*, main text). Note that, in previous work, it was suggested that a convergence of two causes onto the same effect—an instance of

degeneracy—may actually disqualify the micro level from causation (1, 2) (although see ref. 3).
1. Yablo S (1992) Mental causation. *Philos Rev* 101:245–280.
2. List C, Menzies P (2009) Non-reductive physicalism and the limits of the exclusion principle. *J Philos* CVI(9):475–502.
3. Shapiro L, Sober E (2012) Against proportionality. *Analysis* 72:89–93.

Effective Information *EI(S)* Expressed in Terms of Cause and Effect Information and Mutual Information *MI*

The effective information of a system, $EI(S)$, can be obtained as the expected value of the cause or effect information. Moreover, $EI(S)$ is identical to the mutual information $MI(U^C; U^E)$: the MI between the system S set to all possible counterfactuals (system states) with equal probability (unconstrained repertoire of causes, U^C) and the resulting distribution of system states at the next time step (unconstrained repertoire of effects, U^E). Note that EI was originally introduced as a measure of causal influence of one subset of a system over another (1), whereas here it captures the overall effectiveness of system S onto itself (see refs. 2 and 3 for related measures).

In the following derivation, we start from the definition of $EI(s)$ as the average effect information of all system states s_0 as counterfactual causes [distributed according to U^C with equal probability $p(s_0) = 1/n$ for all system states]:

$$EI(S) = \langle \text{Effect Information}(s_0) \rangle = \sum_{s_0 \in U^C} p(s_0) D_{KL}((S_F|s_0), U^E) = \quad (S1)$$

$$(\text{using } p(s_0) = 1/n \;\forall\; s_0) = \frac{1}{n} \sum_{s_0 \in U^C} D_{KL}((S_F|s_0), U^E). \quad (S2)$$

Using Bayes' rule and time invariance, we then show that the average effect information is indeed equivalent to the mutual information $MI(U^C; U^E)$ and to the expected value of the cause information, which is the average cause information of each accessible state at t_0, weighted by $p(s_0)$ according to U^E:

$$EI(S) = \langle \text{Effect Information}(s_0) \rangle = MI(U^C; U^E) \\ = \langle \text{Cause Information}(s_0) \rangle. \quad (S3)$$

In detail:

$$EI(S) = \langle \text{Effect Information}(s_0) \rangle = \sum_{s_0 \in U^C} p(s_0) D_{KL}((S_F|s_0), U^E) = \quad (S4)$$

$$= \sum_{s_0 \in U^C} p(s_0) \sum_{s_F \in U^E} p(s_F|s_0) \log_2 \left(\frac{p(s_F|s_0)}{p(s_F)} \right) = \quad (S5)$$

$$= \sum_{s_0 \in U^C} \sum_{s_F \in S^F} p(s_0)p(s_F|s_0)\log_2\left(\frac{p(s_F|s_0)}{p(s_F)}\right) = \tag{S6}$$

$$\text{(Bayes' rule)} = \sum_{s_0 \in U^C} \sum_{s_F \in U^E} p(s_0|s_F)\log_2\left(\frac{p(s_0,s_F)}{p(s_0)p(s_F)}\right) = \tag{S7}$$

$$= MI(U^C; U^E) = \tag{S8}$$

$$\text{(time invariance)} = \sum_{s_P \in U^C} \sum_{s_0 \in U^E} p(s_P|s_0)\log_2\left(\frac{p(s_P,s_0)}{p(s_P)p(s_0)}\right) = \tag{S9}$$

$$\text{(Bayes' rule)} = \sum_{s_P \in U^C} \sum_{s_0 \in U^E} p(s_0)p(s_P|s_0)\log_2\left(\frac{p(s_P,s_0)}{p(s_P)}\right) = \tag{S10}$$

$$= \sum_{s_0 \in U^E} p(s_0) \sum_{s_P \in U^C} p(s_P|s_0)\log_2\left(\frac{p(s_P|s_0)}{p(s_P)}\right) = \tag{S11}$$

$$= \sum_{s_0 \in U^E} p(s_0) D_{KL}((s_P|s_0), U^C) = \langle \text{Cause Information}(s_0) \rangle. \tag{S12}$$

MI is originally a statistical measure of how much information is shared between a source and a target (4). In the present context, *MI* is applied between two time steps of a system that is first perturbed into all counterfactuals (alternative states) with equal probability and then observed at the next time step. Because of the system perturbations, *MI* here is a causal measure. In other words, *EI(S)* is the *MI* between the set of all possible causes U^C and the set of all their effects U^E. Usually, however, *MI* is calculated for observed distributions of system states and thus not a causal measure, but a statistical measure of correlation.

1. Tononi G, Sporns O (2003) Measuring information integration. *BMC Neurosci* 4:31.
2. Ay N, Polani D (2008) Information flows in causal networks. *Adv Complex Syst* 11(1): 17–41.
3. Korb KB, Nyberg EP, Hope L (2011) *Causality in the Sciences,* eds Illari P, Russo F, Williamson J (Oxford Univ Press, Oxford), pp 628–652.
4. Cover TM, Thomas JA (2006) *Elements of Information Theory* (Wiley-Interscience, Hoboken, NJ).

Bounds of Cause and Effect Coefficients and Effectiveness *Eff(S)*
In the following, we will show that the cause and effect coefficients, as well as the effectiveness *Eff(S)*, are bounded between 0 and 1 ($\in [0 \ldots 1]$):

$$\text{Cause coefficient}(s_0) = \frac{\text{Cause information}(s_0)}{\log_2(n)}$$
$$= \frac{D_{KL}((S_P|s_0), U^C)}{\log_2(n)}, \quad \text{(S1)}$$

$$\text{Effect coefficient}(s_0) = \frac{\text{Effect information}(s_0)}{\log_2(n)}$$
$$= \frac{D_{KL}((S_F|s_0), U^E)}{\log_2(n)}, \quad \text{(S2)}$$

$$\textit{Eff}(S) = \frac{EI(S)}{\log_2(n)} = \frac{\frac{1}{n}\sum_{s_0 \in U^C} D_{KL}((S_F|s_0), U^E)}{\log_2(n)} \quad \text{(S3)}$$
$$= \langle \text{Effect cofficient}(s_0) \rangle.$$

The lower bound (0) is given by the fact that the Kullback–Leibler divergence (D_{KL}) is always nonnegative (Gibbs' inequality). Because the cause and effect information are expressed in terms of D_{KL} and the state-independent effective information $EI(S)$ is just an average of the state-dependent values, neither of the three coefficients can be negative. It thus remains to show that cause and effect coefficients cannot exceed 1.

The cause information (s_0) is the D_{KL} between the cause repertoire ($S_p|s_0$) and U^C, the unconstrained cause repertoire, which is identical to the uniform distribution with $p(s_p) = 1/n \; \forall s_P$. It follows that

$$\text{Cause information}(s_0) = D_{KL}((S_P|s_0), U^C)$$
$$= \sum_{s_P \in U^C} p(s_P|s_0) \log_2\left(\frac{p(s_P|s_0)}{p(s_P)}\right) = \quad \text{(S4)}$$

$$= \sum_{s_P \in U^C} p(s_P|s_0) \log_2(n \cdot p(s_P|s_0)) = \quad \text{(S5)}$$

$$(\text{since } p(s_P|s_0) \leq 1) \leq \sum_{s_P \in U^C} p(s_P|s_0) \log_2(n) = \log_2(n), \quad \text{(S6)}$$

and thus
$$\text{Cause coefficient}(s_0) \leq 1. \quad \text{(S7)}$$

The effect information (s_0) is the D_{KL} between the effect repertoire ($S_F|s_0$) and U^E, the unconstrained effect repertoire. U^E is in general not identical to the uniform distribution. However,

$$p(s_F) = \sum_{s_0 \in U^C} p(s_F|s_0) \cdot p(s_0), \quad \text{(S8)}$$

where $p(s_0) = 1/n$ $\forall s_0$ and thus:

$$p(s_F|s_0) \leq n \cdot p(s_F), \forall s_F. \tag{S9}$$

Using Eq. **S9**, if follows that:

$$\text{Effect information}(s_0) = D_{\text{KL}}((S_F|s_0), U^E)$$
$$= \sum_{s_F \in U^E} p(s_F|s_0) \log_2\left(\frac{p(s_F|s_0)}{p(s_F)}\right) = \tag{S10}$$

$$(\text{using Eq. S9}) \leq \sum_{s_F \in U^E} p(s_F|s_0) \log_2\left(\frac{n \cdot p(s_F)}{p(s_F)}\right) = \sum_{s_F \in U^E} p(s_F|s_0) \log_2(n) \tag{S11}$$

$$= \log_2(n), \tag{S12}$$

and thus

$$\text{Effect coefficient}(s_0) \leq 1. \tag{S13}$$

Finally, because the effect coefficient $(s_0) \in [0 \ldots 1]$ $\forall s_0$, also its average over all system states, the state-independent effectiveness $\mathit{Eff}(S) \in [0 \ldots 1]$.

Causal Reduction

To complement the examples of causal emergence in the main text, we here provide an example in which causal reduction is called for. In Fig. S1, a macro mechanism works as an XOR logic gate (as an isolated part of a larger circuit board) with inputs X, Y, and output Z. (Fig. S1A). At the macro level, the system (XOR,X,Y,Z) generates 2 bits of *EI* over one macro time step T_x (the XOR operates after a "decision" period where it processes the input) and $\mathit{Eff}(S_M) = 0.5$. The macro XOR gate is actually composed of (supervenes upon) nine deterministic micro logic gates (COPY, NOT, AND, OR). In this case, however, causal interactions are stronger at the micro level and over a single micro time step $t_x[EI(S_M) = 7.43$ bits and $\mathit{Eff}(s_M) = 0.83$]. Thus, $CE = -5.43$ bits, corresponding to negative causal emergence, i.e., reduction. Note that in this case the micro circuit is deterministic and minimally degenerate (0.17), so the macro cannot offset the loss of effective information due to its reduced size by a gain in determinism or a reduction in degeneracy.

To demonstrate this case of causal reduction, we have assumed that a deterministic micro circuit underlies the above macro circuit. In general, however, real digital circuits are often built from many stochastic analog micro elements in a highly degenerate manner, to compensate for noise at the lower level and to create deterministic macro elements. In this way, digital circuits and other engineered systems follow similar design principles as the more physiological examples presented in the main text. Consequently, there is the potential for either causal emergence or reduction in digital circuits, depending on the underlying micro level, just as in physiological systems.

More generally, the notion of causal reduction ($CE < 0$) stands in contrast to previous accounts of reduction that focused on the relationship between scientific theories and whether or not they are reducible to one another (1). In the present account based on causal analysis, the focus is instead on the relationship between micro and macro levels of mechanisms. This account reveals why there is a bias in favor of reductionism in mechanistic scientific explanations. The bias is understandable given that, everything else being equal, the micro would always beat the macro: being more detailed by definition, the micro has an inherent advantage in how informative its causal mechanisms are. This inherent advantage is captured quantitatively in causal analysis because the micro can benefit from both ΔI_{Eff} and ΔI_{Size}, whereas the macro can only gain from ΔI_{Eff}.

1. Nagel E (1961) *The structure of science: problems in the logic of scientific explanation* (Harcourt, Brace & World, New York).

Causal Emergence in a System with Causally Heterogeneous Elements

Although the examples in the main text (with the exception of Fig. 6) all have macro elements with underlying unconnected and causally equivalent micro elements, this is not a necessity for causal emergence. In Fig. S2A, the six micro elements are fully interconnected and causally heterogeneous. The elements are structured into two groups {ABC, DEF} due to different intra-group and intergroup mechanisms: within each group, if the sum of intragroup connections Σ(intra) = 0, all elements stay 0 (inactive) the next time step. However, if the sum of intergroup connections Σ(inter) = 3 (synchronous activity from the other group), all elements turn 1, unless they are all 0, in which case they become spontaneously active (1) with probabilities: p(A/D) = 0.45; p(B/ E) = 0.5; p(C/F) = 0.55. Because the micro transition probability matrix (TPM) is noisy, $EI(S_m)$ = 1.13 bits and $Eff(S_m)$ = 0.19 (Fig. S2B). The optimal macro grouping S_M (Fig. S2C) has a more deterministic TPM (Fig. S2D), $EI(S_M)$ = 1.84 bits and $Eff(S_M)$ = 0.58. Thus, the macro supersedes the micro [$CE(S)$ = 0.72 bits] despite its reduced repertoire size, because it counteracts noise by responding almost deterministically to synchronous activity over intergroup connections.

The neural-like system of Fig. 6 in the main text has equivalent spatial properties to the example system of Fig. S2 (fully connected, causally heterogeneous elements, sensitive to differences in intraconnections and interconnections). In addition, it has the same temporal properties as the system shown in Fig. 5 (main text), with second-order Markov mechanisms at the micro level. The system's states space at the micro level thus contains 2^{18} states, which prohibited an exhaustive search for the optimal macro level. Nevertheless, the spatiotemporally emergent macro grouping shown in Fig. 6B (main text) is assumed to be the optimal macro grouping based on the results obtained from the examples of Fig. S2 and Fig. 5 (main text).

Applicability—Network Motifs as Indicators of Emergence

Measuring *EI* exhaustively, across all micro/macro levels, is not feasible for large systems. This is because, assuming N binary elements, $B_N - 1$ (Nth Bell number) possible groupings of those micro elements into macro elements exist, each of which entails $\prod_{j=1}^{k} (B_{m(j)+1} - 1)$ possible groupings of micro into macro states, where k is the number of macro elements with $m(j)$ micro elements each. The number of *EI* computations to determine the spatiotemporal grain with maximal *EI* thus increases dramatically with N ($N = 1, 1; N = 2, 5; N = 3, 27; N = 4, 180$ computations, etc.) if calculated exhaustively.

In large, complex networks where an exhaustive causal analysis is unfeasible, overrepresented network motifs could already indicate whether the network as a whole is biased toward emergence or reduction. For example, the two most common network motifs shared by the gene networks in *Escherichia coli* and the brain of *Caenorhabditis elegans* are the feedforward loop and the bifan (1). Both these network motifs mimic in their connectivity precisely the micro element groups that made up the optimal (winning) macro elements in our chosen examples. In Fig. 2 (main text), the first spatial example, the macro elements are bifans, whereas in Fig. 6 (main text), the first temporal example, the macro elements are feedforward loops. These are perhaps the simplest possible functionally relevant macro elements. Both the bifan and the feedforward loop show causal convergence (degeneracy) in either space or time. A greater than random prevalence of these or similar network motifs, paired with some amount of intrinsic noise in the system, may indicate that the system operates at a macro level.

1. Milo R, et al. (2002) Network motifs: Simple building blocks of complex networks. *Science* 298(5594):824–827.

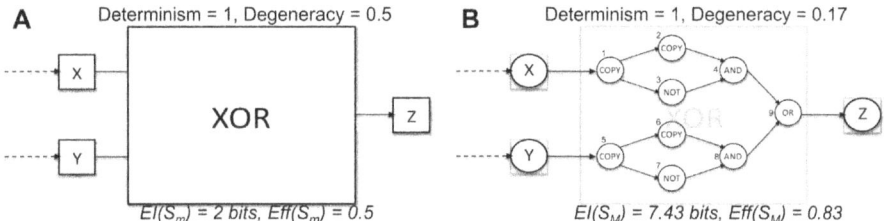

Fig. S1. Causal reduction. (*A*) A part of a larger circuit is presented, which performs a macro XOR logic function over its inputs X, Y, and outputs to Z. (*B*) At the micro level, the XOR consists of nine deterministic logic gates. The system is deterministic at both the micro and the macro level. Moreover, the degeneracy coefficient at the micro level is lower than at the macro level. Therefore, in this case, the micro beats the macro, leading to causal reduction. $CE(S) = -5.43$ bits.

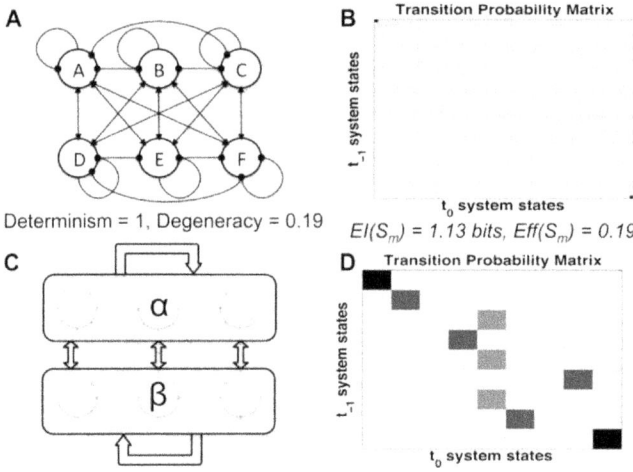

Fig. S2. Causal emergence in a system with differentiated connectivity. (*A*) Micro system S_m with six elements. Regular and rounded arrows indicate intergroup and intragroup connections, respectively. (*B*) Noisy micro-level TPM. (*C*) Macro system S_M. Each macro element receives inputs from itself and the other macro element. (*D*) More deterministic macro-level TPM. $CE(S) = 0.72$ bits.

Case Study 15: Standard Deviation not S.E.M.

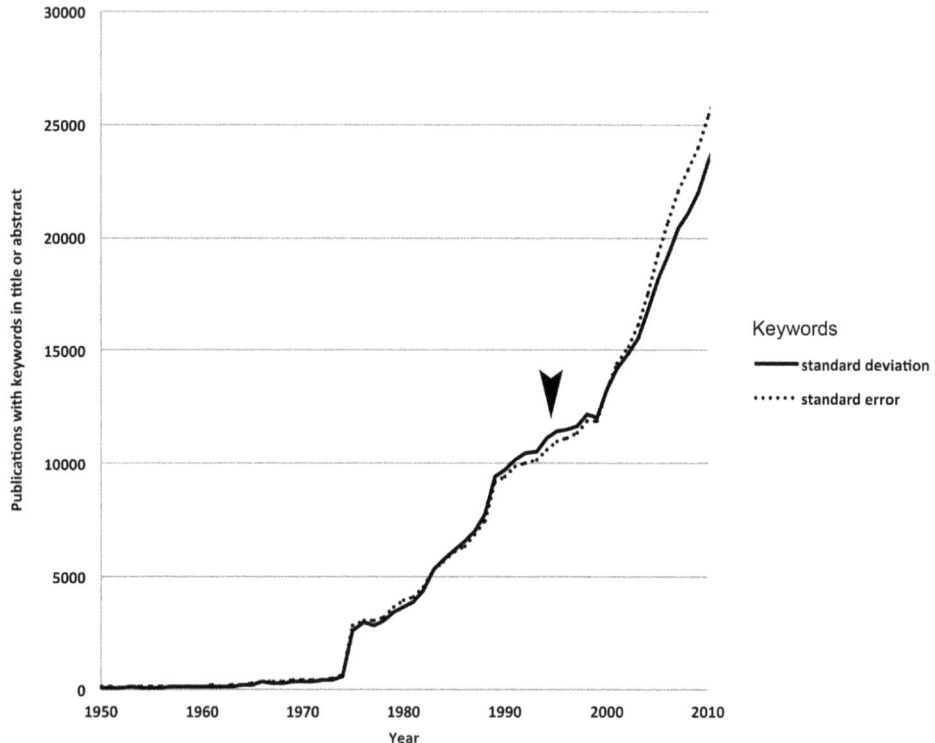

Streiner (1996) Maintaining standards: differences between the standard deviation and the standard error, and when to use each.

Perspective on *Maintaining Standards: Differences between the Standard Deviation and Standard Error, and When to Use Each*
Can J. Psychiatry 41:498–502

Dany Spencer Adams
Research Professor
Tufts University

> **'...[U]ncertainty and its components should be expressed in the form of 'standard deviations.'** Recommendation INC-1 (1980) by the Working Group on the Statement of Uncertainties convened by the [International Bureau of Weights and Measures], a recommendation that the [International Committee for Weights and Measures] approved in 1981 and reaffirmed in 1986 via its own Recommendations 1 (CI-1981) and 1 (CI-1986).

> **'The SE reflects the variability of the mean values, as if the study were repeated a large number of times. By itself, the SE is not particularly useful...'** (Streiner 1996).

This is not the first paper to introduce or even review these ideas. A PubMed search for articles similar to this one finds 122 papers published since 1946, including 16 that were published prior to 1996, and there are surely more that were not recognized by this search technique. This one was chosen for this collection because it is exceptionally clear. The rules it explains were taught to us all but are, nevertheless, often disregarded, probably because dividing by the square root of the sample size makes the uncertainty look smaller; that is not a good reason. The trend graph shows how SE and SD are neck and neck when it comes to use. This article explains why that is unfortunate.

Maintaining Standards: Differences between the Standard Deviation and Standard Error, and When to Use Each

David L Streiner, PhD[1]

Many people confuse the standard deviation (SD) and the standard error of the mean (SE) and are unsure which, if either, to use in presenting data in graphical or tabular form. The SD is an index of the variability of the original data points and should be reported in all studies. The SE reflects the variability of the mean values, as if the study were repeated a large number of times. By itself, the SE is not particularly useful; however, it is used in constructing 95% and 99% confidence intervals (CIs), which indicate a range of values within which the "true" value lies. The CI shows the reader how accurate the estimates of the population values actually are. If graphs are used, error bars equal to plus and minus 2 SEs (which show the 95% CI) should be drawn around mean values. Both statistical significance testing and CIs are useful because they assist the reader in determining the meaning of the findings.

(Can J Psychiatry 1996;41:498-502)

Key Words: *statistics, standard deviation, standard error, confidence intervals, graphing*

Imagine that you've just discovered a new brain protein that causes otherwise rational people to continuously mutter words like "reengineer," "operational visioning," and "mission statements." You suspect that this new chemical, which you call LDE for Language Destroying Enzyme, would be found in higher concentrations in the cerebrospinal fluid (CSF) of administrators than that of other people. Difficult as it is to find volunteers, you eventually get samples from 25 administrators and an equal number of controls and find the results shown in Table I. Because you feel that these data would be more compelling if you showed them visually, you prepare your paper using a bar graph. Just before you mail it off, though, you vaguely remember something about error bars, but can't quite recall what they are; you check with a few of your colleagues. The first one tells you to draw a line above and below the top of each bar so that each part is equal to the standard deviation. The second person disagrees, saying that the lines should reflect the standard errors, while the third person has yet another opinion—the lines should be plus and minus 2 standard errors, that is, 2 standard errors above and 2 below the mean. As you can see in Figure 1, these methods result in very different pictures of

---

Manuscript received March 1996.
This article is the eleventh in the series on Research Methods in Psychiatry.
For previous articles please see Can J Psychiatry 1990;35:616–20, 1991; 36:357–62, 1993;38:9–13, 1993;38:140–8, 1994;39:135–40, 1994; 39:191–6, 1995;40:60–6, 1995;40:439–44, 1996;41:137–43, and 1996;41:491–7.
[1]Professor, Departments of Clinical Epidemiology and Biostatistics and of Psychiatry, McMaster University, Hamilton, Ontario.
Address reprint requests to: Dr David L Streiner, Department of Clinical Epidemiology and Biostatistics, McMaster University, 1200 Main Street West, Hamilton, ON L8N 3Z5
e-mail: streiner@fhs.csu.mcmaster.ca

what's going on. So, now you have 2 problems: first, what is the difference between the standard error and the standard deviation, and second, which should you draw?

Standard Deviation
The standard deviation, which is abbreviated variously as S.D., SD, or s (just to confuse people), is an index of how closely the individual data points cluster around the mean. If we call each point "X_i," so that "X_1" indicates the first value, "X_2" the second value, and so on, and call the mean "M," then it may seem that an index of the dispersion of the points would be simply $\Sigma(X_i - M)$, which means to sum (that's what the Σ indicates) how much each value of X deviates from M; in other words, an index of dispersion would be the *Sum of (Individual Data Points—Mean of the Data Points)*.

Logical as this may seem, it has 2 drawbacks. The first difficulty is that the answer will be zero—not just in this situation, but in every case. By definition, the sum of the values above the mean is always equal to the sum of the values below it, and thus they'll cancel each other out. We can get around this problem by taking the absolute value of each difference (that is, we can ignore the sign whenever it's negative), but for a number of arcane reasons, statisticians don't like to use absolute numbers. Another way to eliminate negative values is to square them, since the square of any number—negative or positive—is always positive. So, what we now have is $\Sigma(X_i - M)^2$.

The second problem is that the result of this equation will increase as we add more subjects. Let's imagine that we have a sample of 25 values, with an SD of 10. If we now add another 25 subjects who look exactly the same, it makes intuitive sense that the dispersion of these 50 points should stay the same. Yet the formula as it now reads can result only in a larger sum as we add more data points. We can compensate for this by dividing by the number of subjects, N, so that the equation now reads $\Sigma(X_i - M)^2/N$.

In the true spirit of Murphy's Law, what we've done in solving these 2 difficulties is to create 2 new ones. The first (or should we say third, so we can keep track of our problems) is that now we are expressing the deviation in squared units; that is, if we were measuring IQs in children with autism, for instance, we may find that their mean IQ is 75 and their dispersion is 100 squared IQ points. But what in heaven's name is a squared IQ point? At least this problem is easy to cure: we simply take the square root of the answer, and we'll end up with a number that is in the original units of measurement, so in this example, the dispersion will be 10 IQ points, which is much easier to understand.

The last problem (yes, it really is the last one) is that the results of the formula as it exists so far produce a *biased* estimate, that is, one that is consistently either higher or (as in this case) lower than the "true" value. The explanation of this is a bit more complicated and requires somewhat of a detour. Most of the time when we do research, we are not interested so much in the samples we study as in the populations they come from. That is, if we look at the level of expressed emotion (EE) in the families of young schizophrenic males, our interest is in the families of all people who meet the criteria (the population), not just those in our study. What we do is *estimate* the population mean and SD from our sample. Because all we are studying is a sample, however, these estimates will deviate by some unknown amount from the population values. In calculating the SD, we would ideally see how much each person's score deviates from the population mean, but all we have available to us is the sample mean. By definition, scores deviate less from their own mean than from any other number. So, when we do the calculation and subtract each score from the sample mean, the result will be smaller than if we subtracted each score from the population mean (which we don't know); hence, the result is biased downwards. To correct for this, we divide by N − 1 instead of N. Putting all of this together, we finally arrive at the formula for the standard deviation, which is:

$$SD = \sqrt{\frac{\Sigma(X_i - M)^2}{N - 1}}$$

(By the way, don't use this equation if, for whatever bizarre reason, you want to calculate the SD by hand, because it leads to too much rounding error. There is another formula, mathematically equivalent and found in any statistics book, which yields a more precise figure.)

Now that we've gone through all this work, what does it all mean? If we assume that the data are normally distributed, then knowing the mean and SD tells us everything we need to know about the distribution of scores. In any normal distribution, roughly two-thirds (actually, 68.2%) of the scores fall between −1 and +1 SD, and 95.4% between −2 and +2 SD. For example, most of the tests used for admission to graduate or professional schools (the GRE, MCAT, LSAT, and other instruments of torture) were originally designed to have a mean of 500 and an SD of 100. That means that 68% of people get scores between 400 and 600, and just over 95% between 300 and 700. Using a table of the normal curve (found in most statistics books), we can figure out exactly what proportion of people get scores above or below any given value. Conversely, if we want to fail the lowest 5% of test takers (as is done with the LMCCs), then knowing the mean and SD of this year's class and armed with the table, we can work out what the cut-off point should be.

So, to summarize, the SD tells us the distribution of *individual scores* around the mean. Now, let's turn our attention to the standard error.

Table I. Levels of LDE in the CSF of Administrators and Controls

Group	Number	Mean	SD
Administrators	25	25.83	5.72
Controls	25	17.25	4.36

Standard Error

I mentioned previously that the purpose of most studies is to estimate some population parameter, such as the mean, the SD, a correlation, or a proportion. Once we have that estimate, another question then arises: How accurate is our estimate? This may seem an unanswerable question; if we don't know what the

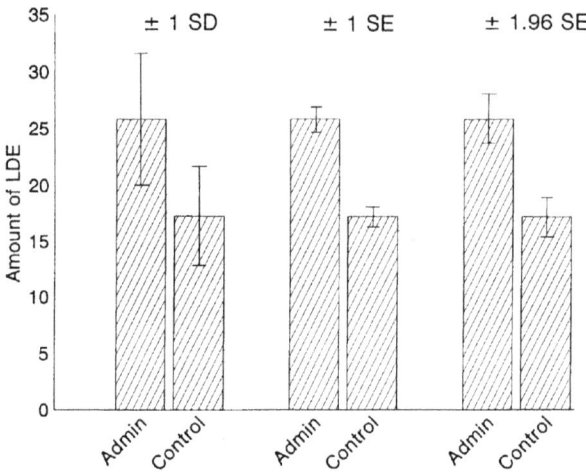

Figure 1. Data from Table I, plotted with different types of error bars.

population value is, how can we evaluate how close we are to it? Mere logic, however, has never stopped statisticians in the past, and it won't stop us now. What we can do is resort to probabilities: What is the probability (P) that the true (population) mean falls within a certain range of values? (To cite one of our mottos, "Statistics means you never have to say you're certain.")

One way to answer the question is to repeat the study a few hundred times, which will give us many estimates of the mean. We can then take the mean of these means, as well as figure out what the distribution of means is; that is, we can get the standard deviation of the mean values. Then, using the same table of the normal curve that we used previously, we can estimate what range of values would encompass 90% or 95% of the means. If each sample had been drawn from the population at random, we would be fairly safe in concluding that the true mean also falls within this range 90% or 95% of the time. We assign a new name to the standard deviation of the means: we call it the *standard error of the mean* (abbreviated as SEM, or, if there is no ambiguity that we're talking about the mean, SE).

But first, let's deal with one slight problem—replicating the study a few hundred times. Nowadays, it's hard enough to get money to do a study once, much less replicate it this many times (even assuming you would actually want to spend the rest of your life doing the same study over and over). Ever helpful, statisticians have figured out a way to determine the SE based on the results of a single study. Let's approach this first from an intuitive standpoint: What would make us more or less confident that our estimate of the population mean, based on our study, is accurate? One obvious thing would be the size of the study; the larger the sample size, N, the less chance that one or two aberrant values are distorting the results and the more likely it is that our estimate is close to the true value. So, some index of N should be in the denominator of SE, since the larger N is, the smaller SE would become. Second, and for similar reasons, the smaller the variability in the data, the more confident we are that one value (the mean) accurately reflects them. Thus, the SD should be in the numerator: the larger it is, the larger SE will be, and we end up with the equation:

$$SE = \frac{SD}{\sqrt{N}}$$

(Why does the denominator read $\sqrt{N}$ instead of just N? Because we are really dividing the variance, which is SD^2, by N, but we end up again with squared units, so we take the square root of everything. Aren't you sorry you asked?)

So, the SD reflects the variability of *individual data points*, and the SE is the variability of *means*.

Confidence Intervals

In the previous section, on the SE, we spoke of a range of values in which we were 95% or 99% confident that the true value of the mean fell. Not surprisingly, this range is called the confidence interval, or CI. Let's see how it's calculated. If we turn again to our table of the normal curve, we'll find that 95% of the area falls between −1.96 and +1.96 SDs. Going back to our example of GREs and MCATs, which have a mean of 500 and an SD of 100, 95% of scores fall between 304 and 696. How did we get those figures? First, we multiplied the SD by 1.96, subtracted it from the mean to find the lower bound, and added it to the mean for the upper bound. The CI is calculated in the same way, except that we use the SE instead of the SD. So, the 95% CI is:

$$95\% \ CI = M \pm (1.96 \times SE)$$

For the 90% CI, we would use the value 1.65 instead of 1.96, and for the 99% CI, 2.58. Using the data from Table I, the SE for administrators is $5.72/\sqrt{25}$, or 1.14, and thus the 95% CI would be $25.83 \pm (1.96 \times 1.14)$, or 23.59 to 28.07. We would interpret this to mean that we are 95% confident that the value of LDE in the population of administrators is somewhere within this interval. If we wanted to be more confident, we would multiply 1.14 by 2.58; the penalty we pay for our increased confidence is a wider CI, so that we are less sure of the exact value.

The Choice of Units

Now we have the SD, the SE, and any one of a number of CIs, and the question becomes, which do we use, and when? Obviously, when we are describing the results of any study we've done, it is imperative that we report the SD. Just as obviously, armed with this and the sample size, it is a simple matter for the reader to figure out the SE and any CI. Do we gain anything by adding them? The answer, as usual, is yes and no.

Essentially, we want to convey to the reader that there will always be sample-to-sample variation and that the answers we get from one study wouldn't be exactly the same if the study were replicated. What we would like to show is how much of a difference in findings we can expect: just a few points either way, but not enough to substantially alter our conclusions, or so much that the next study is as likely to show results going in the opposite direction as to replicate the findings. To some degree, this is what significance testing does—the lower the P level, the less likely the results are due simply to chance and the greater the probability that they will be repeated the next time around. Significance tests, however, are usually interpreted in an all-or-nothing manner: either the result was statistically significant or it wasn't, and a difference between group means that just barely squeaked under the $P < 0.05$ wire is often given as much credence as one that is highly unlikely to be due to chance.

If we used CIs, either in a table or a graph, it would be much easier for the reader to determine how much variation in results to expect from sample to sample. But which CI should we use? We could draw the error bars on a graph or show in a table a CI that is equal to exactly one SE. This has the advantages that we don't have to choose between the SE or the CI (they're identical) and that not much calculation is involved. Unfortunately, this choice of an interval conveys very little useful information. An error bar of plus and minus one SE is the same as the 68% CI; we would be 68% sure that the true mean (or difference between 2 means) fell within this range. The trouble is, we're more used to being 95% or 99% sure, not 68%. So, to begin with, let's forget about showing the SE: it tells us little that is useful, and its sole purpose is in calculating CIs.

What about the advice to use plus and minus 2 SEs in the graph? This makes more sense; 2 is a good approximation of 1.96, at least to the degree that graphics programs can display the value and our eyes discern it. The advantages are twofold. First, this method shows the 95% CI, which is more meaningful than 68%. Second, it allows us to do an "eyeball" test of significance, at least in the 2-group situation. If the top of the lower bar (the controls in Figure 1) and the bottom of the higher bar (the administrators) do not overlap, then the difference between the groups is significant at the 5% level or better. Thus we would say that, in this example, the 2 groups were significantly different from one another. If we actually did a t test, we would find this to be true: $t(48) = 2.668$, $P < 0.05$. This doesn't work too accurately if there are more than 2 groups, since we have the issue of multiple tests to deal with (for example, Group 1 versus Group 2, Group 2 versus 3, and Group 1 versus 3), but it gives a rough indication of where the differences lie. Needless to say, when presenting the CI in a table, you should give the exact values (multiply by 1.96, not 2).

Wrapping Up
The SD indicates the dispersion of individual data values around their mean, and should be given any time we report data. The SE is an index of the variability of the means that would be expected if the study were exactly replicated a large number of times. By itself, this measure doesn't convey much useful information. Its main function is to help construct 95% and 99% CIs, which can supplement statistical significance testing and indicate the range within which the true mean or difference between means may be found. Some journals have dropped significance testing entirely and replaced it with the reporting of CIs; this is probably going too far, since both have advantages, and both can be misused to equal degrees. For example, a study using a small sample size may report that the difference between the control

and experimental group is significant at the 0.05 level. Had the study indicated the CIs, however, it would be more apparent to the reader that the CI is very wide and the estimate of the difference is crude, at best. By contrast, the much-touted figure of the number of people affected by second-hand smoke is actually not the estimate of the mean. The best estimate of the mean is zero, and it has a very broad CI; what is reported is the upper end of that CI.

To sum up, SDs, significance testing, and 95% or 99% CIs should be reported to help the reader; all are informative and complement, rather than replace, each other. Conversely, "naked" SEs don't tell us much by themselves, and more or less just take up space in a report. Conducting our studies with these guidelines in mind may help us to maintain the standards in psychiatric research.

Résumé

Beaucoup de gens confondent l'écart-type et l'erreur-type de la moyenne et ne savent pas lequel utiliser pour présenter les données sous forme graphique ou tabulaire. L'ecart-type indique la variabilité des données originales et devrait être mentionné pour toutes les etudes. L'erreur-type montre la variabilité des valeurs moyennes, comme si l'étude avait été reprise de nombreuses fois. En soi, l'erreur-type n'a pas d'utilité particulière; toutefois, on s'en sert pour créer les intervalles de confiance à 95% et à 99% utilises pour établir la fourchette de valeurs dans laquelle se situe la valeur «réelle». Les intervalles de confiance signalent au lecteur la précision des estimations des valeurs démographiques. Lorsqu'on se sert de graphiques, la barre d'erreur représente un intervalle de plus à moins 2 écarts-types (ce qui correspond à l'intervalle de confiance de 95%). Elle devrait entourer la valeur moyenne. Les épreuves de signification statistique et les intervalles de confiance présentent une grande utilité, car ils aident le lecteur à établir l'importance des constatations.

Case Study 16: Field Models of Pattern Formation

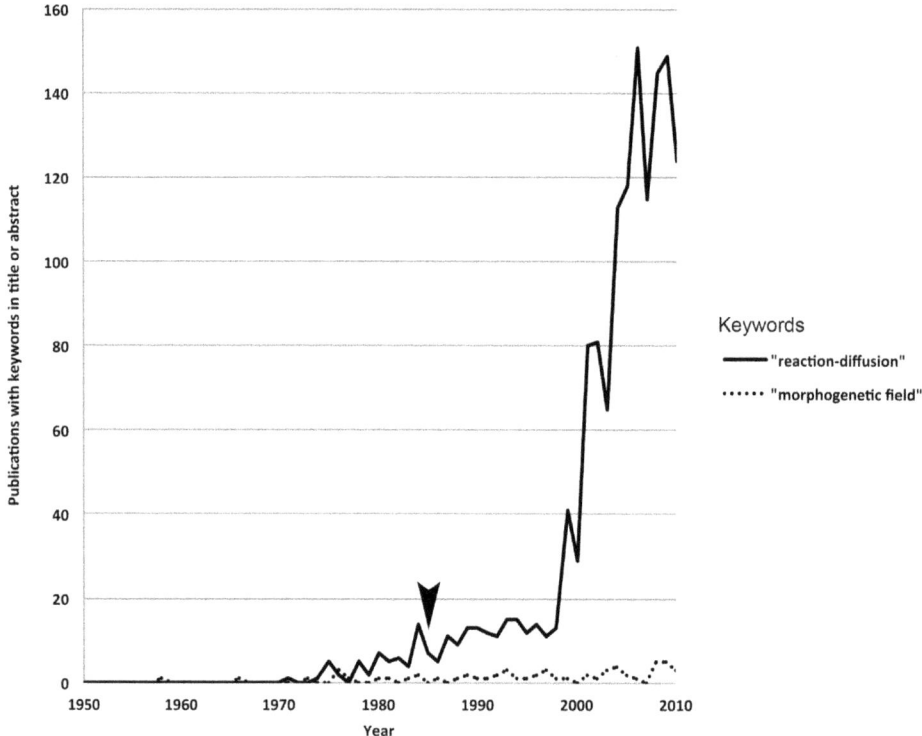

Goodwin (1985) Developing organisms as self-organizing fields. In this graphic we have compared this theoretical approach to the frequently cited reaction-diffusion theory.

Perspective on *Developing Organisms as Self-Organizing Fields*

Wendy Brandts, PhD
University of Ottawa,

Understanding complex systems top-down

I. Introduction

Against the tide of fashion dictating that biological systems are to be understood primarily from their molecular- and cellular-level behaviours, this ground-breaking paper makes the case that understanding biological systems requires structural laws at multiple levels. Drawing on the deep biological understanding and unique perspective of Goodwin, and the insight and knowledge of quantitative structural laws of L.E.H. Trainor, a mathematical framework is proposed for structural laws of biological form.

After all, biological systems are also physical systems, with some special additional properties. The historical success of physics to describe physical systems (e.g. quarks, nucleons, atoms, molecules) via structural laws at many levels suggests carrying out a similar program at higher levels of organization and complexity.

II. Structuralist Approach

Goodwin takes the structuralist point of view, going back to Hans Driesch and Darcy Thompson, that seek laws of form showing regularities of morphology over many taxonomic groups. He builds on the approach of the Clockface Model (French *et al* 1976) which are rules of morphogenesis making no reference to composition or inheritance. In particular, Goodwin's morphogenetic field serves to inform different parts of an organism 'where in the whole it is'.

In this interpretation, the morphogenetic field is the self-organizing entity. The organism, not the cell, is the fundamental biological entity. Spatial organization of the whole derives from the global morphogenetic field plus the constraints of the parts.

Goodwin and Trainor (1980) provide a mathematical formalism for this morphogenetic field, which was not attained by the creators of the Clockface Model. The field is a structure: it belongs to an invariant set defined by internal relations (the field equations). An analogy from classical mechanics is the trajectories of motion of an object in a central force field, where the set of possible solutions to Newton'ss laws are restricted to the conic sections.

The formalism of this morphogenetic field is akin to electrodynamic field equations of physics. The concept of 'selection rules', also borrowed from physics, determines which possible solutions of the field equations will actually be manifested in the system, subject to other constraints. The resemblance to physical theory in this approach echoes the background of the late L.E.H. Trainor, originally a nuclear physicist, who developed the new field of theoretical biophysics at the University of

Toronto with inspiration from Goodwin, C.J. Lumsden and others (see various authors, [Physical Theory in Biology: Foundations and Explorations (1997)]).

The use of physical fields is both convenient and natural because the proposed biological fields actually demonstrate properties similar to physical fields, e.g. smoothness and continuity (one cell is more similar to a neighbouring cell than to a distant cell). The holistic nature of physical fields which allows, for example, a unique electric field to be determined everywhere in a region given sufficient information about the field on the boundary, is reminiscent of the ability of a morphogenetic field to interpolate in regeneration, and thus to reform a whole organism from the information contained in some of the parts.

With this formulation of the morphogenetic field, Goodwin describes early cleavage processes by applying the Laplace equation over a sphere (representing the blastula) and selecting solutions of spherical harmonics whose null lines then indicate cleavage planes.

He concludes that constraints determining specific forms arise from other organizational levels than simple molecular composition and therefore top-down models are required for understanding. Gene products act within the context of fields (electric, visco-elastic, etc) which generate morphology. He argues the inadequacy of any theory of biological form based only on genes and molecules.

III. Models based on Goodwin's Morphogenetic Field Approach.

In the years since Goodwin's pioneering papers, the morphogenetic field formalism (Goodwin and Trainor 1980) has formed the starting point for a number of more detailed applications, which in turn confirm the usefulness of the fundamental concepts.

1. Limb regeneration in amphibians

Valuable predictions can be made, such as the outcome of transplant experiments, without needing to know all the underlying details of structure and composition. For example, a field model of limb regeneration based on fields with solutions of specific symmetry can predict handedness of supernumerary growth (Tevlin and Trainor 1985, Totafurno and Trainor 1987).

2. Regulation in a single cell

Analogous to multi-cellular pattern formation, the growth is governed by global patterning fields, not based on genes or molecules. The field model describes the organism at a 'high level' or macroscopic level of the structural hierarchy, organizing large and various collections of experimental data and predicting developmental configurations and pathways (Brandts and Trainor 1990a, Brandts and Trainor 1990b, Brandts 1993).

3. Electrical signals (dynamics of voltage potentials in gap junction-coupled bioelectric network within non-neural tissues) as global patterning fields (Levin *et al* 2015).

These applications all demonstrate the power of top-down models.

While morphogenetic field models stand on their own as useful tools, it may nevertheless be informative to investigate what underlies the morphogenetic field. Many different lower-level factors could contribute to the expression and maintenance of the field.

Some examples are: order parameters relating to the organization of microfilaments in the cell surface (Goodwin 1980); viscoelastic forces (Hart et al 1989); reaction-diffusion of a binary fluid (Brandts 1995); electrical potentials (Levin et al 2015).

IV. Recent ideas that support Goodwin's perspective.

The structuralist approach embraced by Goodwin has been largely underappreciated due to fashions and successes in microbiology. Scientific progress in recent years in the new field of complexity has brought several fundamental issues to the fore that should encourage a more general acceptance of Goodwin's approach. Here I touch on three of these issues:

1. Simulation *vs* duplication

A simulation of the behavior of a system does not necessarily duplicate the way a system itself actually operates. Nevertheless, simulating or modeling the behaviour of a complex system through top-down processes can aid understanding. After all, no one questions whether a particle calculates its velocity while deciding how to move according to laws of motion; the laws are valuable for predicting behavior, and making control decisions, regardless of the interpretation.

Complex system behaviour can often be viewed as dynamics or as computation. [Brandts (1997). 'Complexity: A Pluralistic Perspective'.].

The point is, when systems behave 'as if' they were calculating, they can fruitfully be modelled as such. Regardless of the philosophical interpretations, such as those entertained by Einstein and Bohr on Quantum Mechanics, in the end, the predictive power of a model wins out.

Recent concerns about goal-directed behaviour and the apparent role of teleology in top-down models may thus be dismissed.

2. Emergence and reducibility

One definition of emergence (from physics) is that emergence is a new behaviour that arises at a higher level and which can not be predicted from the lower level behavoirs. For example, in physics the exchange symmetry of fermions, which influences the structure of the periodic table, is a rule which applies to multi-fermion systems but cannot be derived from the behaviour of a single fermion. See also the two proposed broad and distinctive classes of emergence, *reducible* and *a prior irreducible* (Trainor 1997, Trainor 1985).

In a different case, consider Thermodynamics, TD, the higher-level macroscopic model of the behaviour of the gaseous state of matter. The concepts and laws of TD relate observable quantities, namely temperature and pressure, and provided essential understanding to power the industrial revolution. But TD laws do not apply to single molecules, or even require any particular molecular structure. The lower-level model, Statistical Mechanics, which was developed later, provides the microscopic theoretical underpinnings of TD; nevertheless, TD concepts and laws are still useful and applicable today.

In addition, it is important to discriminate between reduction of *theories* and reduction of *phenomenon* [Brandts (1997). 'Complexity: A Pluralistic Perspective'.]. If the high level processes are irreducible, then they will certainly not be explicable by a lower-level theory—i.e. a higher level or top-down theory will be *essential* for understanding. But even if the *processes* are reducible, the *theories* may not be.

Moreover, while the biological distinction between local and global interactions can often be made clear, and tested in a laboratory situation, in the mathematics of modeling, the distinction between global or local control can be ambiguous. This is because dynamical systems often have alternative mathematical representations which may be entirely equivalent but lead to seemingly different interpretations. An example from classical physics is mechanics, in which a particle can be described as moving according to Newton's laws, which are local, or so as to minimize action over any segment of its future path, which is non-local in both space and time.

In the particular case of field models, the mathematical expression of pattern control is formulated in terms of minimization of total energy, a global property. However, the model could in principle have been formulated in terms of differential equations which express field values and interactions locally. In fact, there are many different dynamical processes (equations) which could give rise to the same steady state minimum energy solutions of the global field.

3. Complexity

Complexity exists on many levels and scales in biological systems. Even defining a quantitative measure of complexity is a daunting task, demanding that many different levels of explanation be employed.

Recent developments in the science of complexity and consideration of biological systems as complex systems has lead to the realization multiple levels of modeling will be required to understand complex biological systems (Brandts 1997). Top-down modeling is essential for a pluralistic approach; no single level of analysis has priority. Each description or model naturally emphasizes its own variables and should be considered a legitimate method of modeling without requiring descriptions and their language to translate into each other in a reductionist fashion (although it may sometimes be possible to do so).

Laws and models are simplifications of the world. For example, even position and momentum, in Newton's classical mechanics, are properties of the formalism, and do not necessarily reflect the real world because position and momentum do not exist simultaneously in the lower-level Quantum Mechanical formulation. That is, Newton's formalism is only applicable to the real world over a restricted range of velocities, timescales, and sizes; at other levels, new phenomenology and models are required. Real systems map to many formal systems, each one capturing different aspects of behaviour. There are distinctive ideas in biology not translatable into physics.

In complex systems such as biological ones, incorporating all lower-level details is unwieldy or impossible, and top-down models like the field approach are required in order to make predictions about higher-level biological behaviours. High level models, with all their simplifications, are needed because when models contain 'everything'—field data, a plethora of parameters, a patchwork of dynamics—and simulations are run as a black box, little in the way of strategic relationships between variables, or the effect of the numerous parameters on the dynamics, can be uncovered (Brandts 2002). We may simulate nature, but we will gain little insight, or control.

Or as J.L. Borges put it, *Of what use is a map as big as the world?*

References

Brandts W A M and Trainor L E H 1990a A nonlinear field model of pattern formation -- intercalation in morphalactic regulation *J. Theor. Biol.* **146** 37–57

Brandts W A M and Trainor L E H 1990b A nonlinear field model of pattern formation -- application to intracellular pattern reversal in tetrahymena *J. Theor. Biol.* **146** 57–87

Brandts W A 1993 A field model of left-right asymmetries in the pattern regulation of a cell *IMA J. Math. Appl. Med. Biol.* **10** 31–50

Brandts W A M 1995 Relevance of field models to global patterning in ciliates *Interplay of Genetic and Physical Processes in the Development of Biological Form* ed E Beysens, G Forgacs and F Gaill (Singapore: World Scientific)

Brandts W A 1997 Complexity: A Pluralistic Perspective *Physical Theory in Biology: Foundations and Explorations* ed C J Lumsden, W A Brandts and L E H Trainor (London: World Scientific Publishing)

Brandts and others 1997 *Physical Theory in Biology: Foundations and Explorations* ed C J Lumsden, W A Brandts and L E H Trainor (London: World Scientific Publishing)

Brandts W A M 2002 Tetrahymena and Ants -- Simple Models of Complex Systems *Formal Descriptions of Developing Systems. Nato Science Series* v. 121 ed J Nation, I Trofimova, J D Rand and W Sulis (Kluwer Academic Publishers)

French V, Bryant P J and Bryant S V 1976 Pattern Regulation in Epimorphic Fields *Science* **193** 969–81

Goodwin B C and Trainor L E H 1980 A field description of the cleavage process in embryogenesis *J. Theor. Biol.* **85** 757–70

Levin M, Pezzulo G and Levin M 2015 Re-membering the body: applications of computational neuroscience to the top-down control of regeneration of limbs and other complex organs *Integrative Biology* **7** 1487–517

Tevlin P and Trainor L E H 1985 A two vector field model of limb regeneration and transplant phenomena *J. Theor. Biol.* **115** 495–513

Totafurno J and Trainor L E H 1987 A non-linear vector field model of supernumerary limb production in salamanders *J. Theor. Biol.* **124**

Trainor L E H 1997 *Physical Theory in Biology: Foundations and Explorations* ed C J Lumsden, W A Brandts and L E H Trainor (London: World Scientific Publishing)

Trainor L E H 1985 Remarks on emergence in physics and biology *Mathematical Essays on Growth and the Emergence of Form* ed P Antonelli (Edmonton: University of Alberta Press)

DEVELOPING ORGANISMS AS SELF-ORGANIZING FIELDS

B.C. *GOODWIN*

5.1 Introduction

Self-organization is a property of a structurally and functionally integrated entity which is considered to be made up of, or to have, "parts." There are essentially two ways of relating the whole and the parts: either the parts are regarded as given (with or without some temporal sequence of presentation or production) and the entity is generated by their interaction; or the entity is regarded as a whole defined by certain invariant relations, the "parts" coming into being as a result of systematic transformations which preserve invariance while generating heterogeneity ("parts") within a functional and structural unity. The most extreme form of the first description is atomism, which assumes that all the information necessary for generating the entity is resident in the parts, so that spontaneous assembly occurs simply as a result of their interaction. This hypothesis, in various forms and with various modifications, is a dominant theme in contemporary models of evolutionary, developmental, behavioral, and even cognitive processes. What characterizes this conceptualization is the assumption that there are no laws of self- organization other than those governing the interaction of the parts or constituents so that the whole is reducible to these parts and their interactions. In developmental biology, the constituents are usually considered to be, ultimately, molecules, although there are theories in which cells are used as intermediate "atoms" in the analysis. Since genes are generally considered to be the determinants of which molecules are present in organisms, it follows that organisms are reducible to genes. The self-organizing process of embryo-genesis is then regarded as a consequence of two activities: the operation of the "genetic program" which determines the types of molecular constituents in the organism, the sequence and spatial location of their appearance; and the interaction of these constituents according to physical and chemical laws.

Such an approach to embryogenesis means that the specific forms generated by morphogenetic processes, defining different species of organism, are irreducibly complex because there are no laws or principles which constrain the "genetic program" which is the determinant of specific form. This program is the result of random permutation and natural selection, purely contingent processes as far as organisms are concerned, the only constraint being that the organism specified by the program must be able to survive and leave offspring in some environment. In such a view, there can be no general laws of biological form. Each species is, speaking more than metaphorically, a law unto itself. This is reflected in the primacy of the species concept in neo-Darwinism and the emphasis placed upon competition and survival as the expression of the individual species' success in establishing a unique, singular relationship of order and stability within itself and with its environments. Thus, as one might expect since cognitive constructs are themselves "self- j organizing," there is a clearly-defined continuity between the way organisms are conceptualized in neo-Darwinism as "survival machines" and the way in which they are considered to be generated from their molecular parts.

There is no *a priori* reason why such a description of self-organization should not be valid, and indeed it is clear that there are special cases in both the animate and the inanimate realms where atomistic explanations appear to be appropriate. As regards structure or form, these are the instances where a crystallization or "self-assembly" type of process leads to a unique morphology. However, even in inorganic chemisty one encounters many instances of polymorphism in which the same substance can crystallize into different forms, familiar examples being graphite and diamond, or rhombic and mono-clinic sulphur. Thus, in general, composition does not determine form. A,similar polymorphism is observed in biological structures at different levels of organization, the molecular (Oosawa et al., 1966), the cellular (Sonneborn, 1970), the tissue (Saunders et al., 1957) and so on. Much of embryology consists in fact of generating a variety of "abnormal" forms out of cells of identical genotype: for example, the induction of supernumerary limbs in an amphibian by a simple manipulation of tissue in the embryonic limb bud involving no addition or deletion of cells, but simply a change of relative position, and no change in the external environment. Furthermore, organisms containing specific mutant genes (e.g. homoeotics) may or may not express them; while organisms with wild-type genes can show the "mutant" phenotype (spontaneous homoeotic transformation). Hence there is no one-to-one relationship between genotype and phenotype, always assuming j a constant external environment, so that genes are not the specific determinants of morphology. That is to say, the form of an organism is not determined by its genome, with the consequence that self-assembly theories together with a genetic program are inadequate to provide a generative theory of biological self-organization (see Webster and Goodwin, 1982, for a more detailed argument leading to this conclusion).

We must now consider whether or not there is empirical evidence relating to general organizational principles, or laws of form in biology, manifesting as regularities of morphology over large taxonomic groups. We have seen above that neo-Darwinism, which takes the view that "the chief part of the organization of every being is simply cue to its inheritance" (Darwin, 1859; inheritance meaning, in contemporary usage, the genetic program), provides no basis for understanding any such regularities since biological form in this theory is determined by contingency, not by law. However, if ordering constraints do exist in the biological realm, then this must be taken into account of in any theory of biological self-organization. This leads us to the work of the pre-Darwinian rational morphologists, who were animated by a belief in the possibility of a rational, intelligible ordering or classification of organisms which would provide an insight into the laws of organic creation (i.e. generative rules). This tradition reached its peak in the work and insights of the great comparative morphologists of the late eighteenth and early nineteenth centuries such as Geoffroy St. Hilaire, Cuvier, Reichert, and Owen, who searched for and discovered empirical regularities of organismic structure. These regularities appeared as invariant structural relations or "typical forms" which were seen to define that which is common to a variety of particular realizations of the same type. Owen's demonstration of the structural homologies which exist between the great variety of vertebrate limbs, leading to the concept of the pentadactyl limbs as the typical form, is characteristic of this work. Each specific member of the invariant set, such as the limb of the horse, of the bat, of the frog, etc., can then be seen as equivalent to every other member under a transformation, so that a common plan is revealed which unifies the diversity of manifest forms. It is, in fact, straightforward to demonstrate the simple proposition that tetrapod limb morphogenesis may be understood in terms of some basic generative principles capable of producing a great variety of limb forms which are all transformable one into the other under modifications of the limb generating process (Goodwin and Trainor, 1982). This is analogous to the realization that the different forms of motion shown by bodies under the action of a central attracting force, obeying Newton's laws, all belong to the same invariant set known as the conic sections; and indeed the rational morphologists were inspired by the same vision as Newton, which was the Enlightenment Ideal of a mathematical natural science. Their conviction was, and they provided good evidence for the belief, that the morphological complexity of organisms is not irreducibility complex but that there exist rational principles or laws of form which render the diversity intelligible.

Despite the fact that this tradition was largely eclipsed by Darwinism, which adopted the diametrically opposite view that organismic form is determined not by rational law but by historical accident, by contingency, a few rather isolated and sometimes misunderstood biologists have pursued this approach further. Among these the embryologist Hans Driesch stands out and his work is very relevant to the view of self-organization which will be developed below. He introduced the field concept into embryology as a result of his demonstration that relative position in the whole embryo is an important determinant of cell fate. He used the concepts of wholeness, self-regulation, and transformation to define the properties of tissues which respond to a variety of disturbances (e.g., removal or addition of cells, or spatial reordering of parts) by a reorganization such that tire normal form is generated. Examples of such fields are the amphibian embryo from fertilization up to about the gastrula stage, the limb and eye primordia, and a variety of other tissues domains which define secondary fields. Within such domains, relative position is a primary determinant of cell fate and the parts which emerge during individuation and differentiation come into being as a result of local and global ordering principles, generating a structural and functional unity. Driesch assumed, like the rational morphologists, that there are organizing rules which operate within organisms to constrain or limit the forms which can be generated (Driesch, 1929) but, again like his predecessors, he failed to give them any mathematical formulation. The problem to be addressed now is what type of mathematical description may be appropriate for these organizing principles, for which there is clear biological evidence.

5.2 Organisms as Fields

The proposition which emerges from the above analysis is that living entities are wholes or *structures* defined by internal relations which remain invariant under certain categories of transformation, the latter limiting the possible generative processes which can result in organisms of specific form (species). Organisms are not, in this view, generated as a result of the interaction of "atomic" constituents, whatever these may be construed to be. Heterogeneity ("parts") arises as a result of systematic transformations of the organized whole, which may be described as the manifestation of states selected from a potential set which satisfies a primary property of invariance characteristic of organisms. Thus the organism is not so much a self- organizing system which generates an ordered state from disordered or less ordered parts; it is more a self-organized entity which can undergo transformations preserving this state. The problems faced by this conceptualization are those of making explicit the nature of the invariant internal relationships which define the whole; the type of transformation which it can undergo; and the relationship between whole and part which confers upon it the properties of generation (reproduction) and regeneration.

Following Driesch's insight that developing organisms have field properties, we may proceed to the question of what type of field and how it may be characterized mathematically. A very extensive body of experimental work in developmental biology since Driesch's pioneering studies has led recently to the observation that an appropriate arithmetic description of the spatial smoothness characteristic of developmental fields is a simple spatial averaging or intercalculation rule applied to field values (French, Bryant, and Bryant, 1976). This states simply that the field value at any point within the boundaries of a developmental field is the arithmetic mean of the values at equidistant neighboring points. Mathematically, this leads to the most general field equation used in physics, namely Laplace's equation. The question then naturally arises whether one can use solutions of this and related equations, known as harmonic functions, to describe developmental fields and hence biological form. Preliminary essays in this direction have been published (Goodwin, 1980; Goodwin and Trainor, 1980; Goodwin and Trainor, 1982). This approach will now be illustrated by an analysis of the earliest stage of amphibian embryogenesis, following the treatment of Goodwin and Trainor (1980), and then certain conclusions regarding the problem of self-organization in biology will be drawn.

5.3 A Field Description of the Typical Cleavage Process

The first five stages of the typical cleavage pattern is described by classical investigation as shown in Figure 5.3-I, starting from the two-cell stage after the first division of the egg. From the 32-cell stage, cell divisions continue to show an alternation between vertical and horizontal cleavage planes, but there is at some stage a loss of spatial and temporal order (synchrony) which differs between species. Since our interest is in the geometry of the cleavage planes, we project the typical pattern onto the original spherical egg to get the schematic sequence in Figure 5.3-II, which illustrates the first seven cleavages up to the 128-cell morula. This is immediately suggestive of a sequence of harmonic functions on the sphere, the

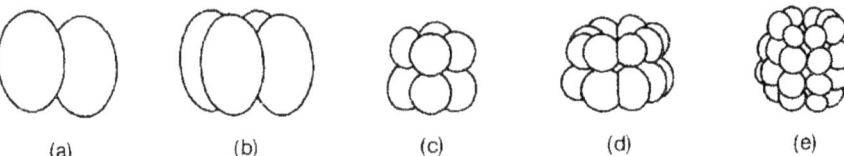

Figure 5.3-I. The holoblastic radial cleavage pattern.

cleavage lines corresponding to nodal lines of spherical harmonics. Accordingly, we develop an "eigenfunction" description of cleavage on the basis of a minimization principle wherein the eigenstates of a morphogenetic surface field describe the successive stages of the cleavage process in the early embryo. The cleavage process is then seen as a series of transformations to successively higher characteristic states of the morphogenetic field as metabolism proceeds. An analogy to this may be found in the electron density distributions of the hydrogen atom, in which the transitions to successively higher energy states results from the action of some external optical pumping field. A characteristic biological feature is that the cleavage transformations are "pumped" or induced internally, the system being self-generating.

5.4 A Variational Principle for Cleavage Planes

Proceeding with the analysis at a fairly abstract level, let us now introduce a field function $u(\theta, \phi)$ over the surface of the sphere and adopt the convention that its nodal lines represent lines of least resistance to a furrowing process preliminary to the development of cleavage planes. This function may be taken to be some kind of order-disorder parameter relating to the organization of microfilaments in the cell surface. The basic stability of the typical cleavage pattern suggests the use of a minimization principle on this field function. An appropriate surface density function, which in a physical problem would be the energy density, is

$$E(\theta, \phi) = A\left[\left(\frac{\partial u}{\partial \theta}\right)^2 + \frac{1}{\sin^2 \theta}\left(\frac{\partial u}{\partial f}\right)^2 + \beta u^2\right] \tag{5.4-1}$$

where the constants A and β incorporate relevant physiological units. Then suppose that the characteristic cleavage planes correspond to a minimum of the integral of this density function over the surface energy E,

$$\delta \int_0^{2\pi} \int_0^{\pi} E(\theta, \phi) \sin \theta \, d\theta \, d\phi = 0 \tag{5.4-2}$$

subject to a conservation law on u^2

$$\int_0^{2\pi} \int_0^{\pi} u^2(\theta, \phi) \sin \theta \, d\theta \, d\phi = 1 \tag{5.4-3}$$

which amounts to a normalization condition on the field variable u. Equations (5.4-2) and (5.4-3) require satisfaction of the Euler-Lagrange equation (see Trainor and Wise, 1979)

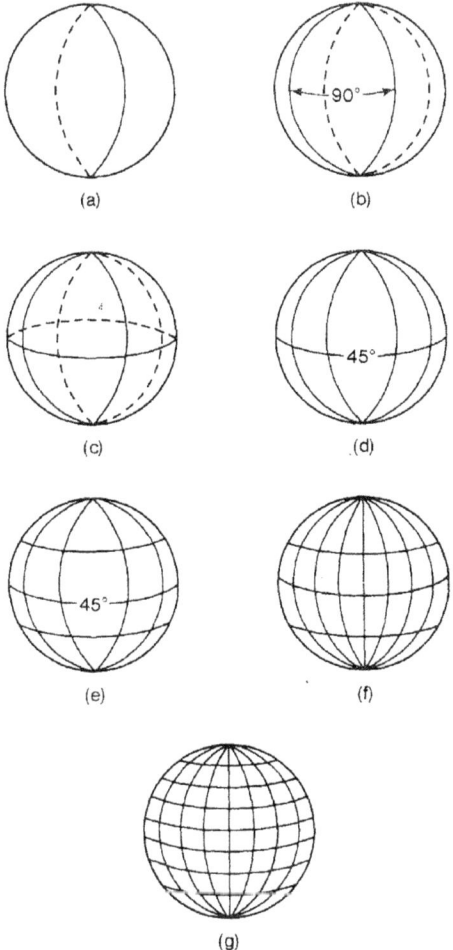

Figure 5.3-II. The geometry of typical cleavage planes up to the 128-cell stage.

$$\frac{1}{\sin\theta}\frac{\partial}{\partial\theta}\left(\sin\theta\frac{\partial u}{\partial\theta}\right) + \frac{1}{\sin^2\theta}\frac{\partial^2 u}{\partial\phi^e} - \alpha u = 0 \tag{5.4-4}$$

where α: incorporates the parameter β and an undetermined multiplier of equation (5.4-3).

The usual conditions on u, that it be finite, single-valued and continuous over the sphere restricts the possible solution u to a characteristic (eigen-function) set, viz. the spherical harmonics (real part taken):

$$u(\theta, \phi) \rightarrow Y_{\ell m}(\theta, \phi) = \sqrt{2}\,N_{\ell m}P_\ell^m(\cos\theta)\ m\phi \tag{5.4-5}$$

where ℓ takes on the integral values 0, 1, 2, etc., and for given ℓ, the m values are integers ranging from $-\ell$ to $+\ell$. The parameter α is restricted to the corresponding characteristic values (eigenvalues) $\ell(\ell + 1)$. In equation (5.4-5), the P_ℓ^m are associated

Legendre polynomials (Hobson, 1955) and the $N_{\ell m}$ are the normalization constants given by

$$N_{\ell m} = \left[\frac{(2\ell + 1)(\ell - |m|)!}{4\pi(\ell + |m|)!} \right]^{1/2} \qquad (5.4\text{-}6)$$

It is easy to calculate the surface "energy" corresponding to each characteristic cleavage state. The result is:

$$E_\ell = A\{\ell(\ell + 1) + \beta\}. \qquad (5.4\text{-}7)$$

However, not every characteristic state (5.4-5) is realized in the cleavage process since the mitotic apparatus imposes a biological constraint (somewhat analogous to superselection rules in physics: Wick, Wightman and Wigner, 1952) that the number of cells is doubled in each cleavage stage.

According to the ideas set out above, the nodal lines of the characteristic function $Y_{\ell m}(\theta, \phi)$ on the sphere are in correspondence with the furrow lines of a characteristic cleavage state, except that the set of characteristic states is limited by the requirement that the number of cells is given by 2^p where p is the number of cell divisions. It is easily shown that the number of cells in a state characterized by $Y_{\ell m}$ is $2m(\ell - m+1)$ unless m = 0 in which case the number is $\ell + 1$. (It is sufficient to choose the real part of $Y_{\ell m}$, i.e. the cos $m\phi$ solutions, so that we need consider only in m $\geq$ 0.) Hence the biological constraint requires that

$$\begin{aligned} 2^p &= 2m(\ell - m + 1) \quad \text{if} \quad m \neq 0, \quad \text{or} \\ &= \ell + 1 \qquad \qquad \text{if} \quad m = 0. \end{aligned} \qquad (5.4\text{-}8)$$

In general this equation, for a given characteristic cleavage state corresponding to p divisions, is satisfied by more than one set of (ℓ, m) values. It is natural to suppose that the choice is made primarily on the basis of lowest ℓ value, since according to equation (5.4-7), this minimizes the "energy." The choice is then nearly unique, except for a two-fold degeneracy every second division. It is assumed that the animal-vegetal polarity of the embryo defines a secondary polar field weaker than the primary field which removes this degeneracy in favor of the highest m value for a given ℓ, in much the same way as a magnetic field removes the $(2\ell + 1)$ fold degeneracy of magnetic- states in the hydrogen atom.

Table 5.4-1 shows the correspondence between cleavage states and characteristic functions $Y_{\ell m} = 2N_{\ell m} P_\ell^m \cos m\phi$ by listing the number of cells to be expected from each set of (ℓ, m) values up to $\ell = 7$. The appropriate (ℓ, m) pair is then selected out uniquely by the conditions expressed in equations (5.4-8), together with the minimization condition (5.4-7), except for the two-fold degeneracies at the first, second, fourth, etc., cell divisions. As remarked above a unique correspondence is achieved by assuming that a weaker polar field selects (1,1) over (1,0), (2,2) over (2,1) and (5,4) over (5,2), that is, it favors highest m value for a given ℓ.

In Table 5.4-1 the selected states for the 6th and 7th cleavages corresponding to the 64 and 128 cell stages have been included, without listing all of the rejected (ℓ, m) values. Figure 5.3-II illustrates the nodal lines for the successive stages in a typical cleavage process. (Note that in the model used here the furrowing process corresponds to an accumulative conjunction of surface harmonic patterns. In a variant of the model (Goodwin and LaCroix, 1982), the pattern is not cumulative and each stage is described by a single harmonic pattern of nodal lines.)

Table 5.4-1. Correspondence between $Y_{\ell m}(\theta, \phi)$ and cell number, defined by $\ell + 1$ if m = 0, otherwise by $2m(\ell - m + 1)$.

ℓ value	m value	cell number	(ℓ, m) pair selected
1	0	2	(1, 1)
	1	2	
	0	3	
2	1	4	(2, 2)
	2	4	
	0	4	
3	1	6	(3, 2)
	2	8	
	3	6	
	0	5	
	1	8	
4	2	12	
	3	12	
	4	8	
	0	6	
	1	10	
5	2	16	(5, 4)
	3	18	
	4	16	
	5	10	
	0	7	
	1	12	
	2	20	
6	3	24	
	4	24	
	5	20	
	6	12	
	0	8	
	1	14	
	2	24	
7	3	30	(7, 4)
	4	32	
	5	30	
	6	24	
	7	14	
11	8	64	(11, 8)
15	8	128	(15, 8)

5.5 Fields and Self-Organization

This example clarifies some of the abstract concepts introduced in the discussion on organisms as fields, and it is now of some interest to elaborate on these and their relationship to the concept of self-organization in biology. The first point to emerge is the primacy of the organism as the fundamental biological entity, replacing the usual definition of the cell as the unit of life. This follows from the fact that the field is the self-organizing entity, and this is coextensive in the above description with the organism. The orderly geometry of the cleavage planes is a reflection of this organization at the level of the developing embryo. This does not mean that the parts, which in this case are cells, have no properties of their own. On the contrary, the constraint of binary division, a cellular property, is a major source of the distinctive geometry of the cleavage process, since this defines one of the three selection rules which determine the harmonic functions allowable as descriptions of this process. What thus becomes apparent is that the spatial organization of the whole derives from principles relating to global field behavior together with constraints coming from the properties of the entities which are generated as parts. Such a description avoids both atomistic reduction and a holistic description which identifies the whole with some (conceptually or materially) isolable essence. The genome is such an isolable essence, traceable historically to its conceptual roots in idealistic holism via Weismann's conceptualization of the organism as separable into germ plasm (essence) and somatoplasm (expression of the essence; see Webster and Goodwin, 1982). A field description of developing organisms sees spatial organization as the expression of distributed influence, global order being constrained to give specific morphology as a result of autonomous properties of parts. This defines a "decentered structure."

The ambiguity between the concepts of "cell" and "organism" can be resolved in terms of the above description. The fertilized egg is both a cell and a developing organism. It is an organism insofar as it is a totality describable by a field; it is a cell insofar as it embodies the specific constraints (e.g., binary division) characteristic of such an entity. As cleavage proceeds, the organism continues to be identified with the whole field (the embryo), while cells are identified as parts which play specific roles within a transforming context. After gastrulation, more complex parts such as neural plate, limb fields, eye fields, etc., come into existence. These consist of aggregates of cells so that an integrated hierarchy of parts emerges within the context of the organism as the global field which continues to impose organizational constraints upon the whole, the parts imposing reciprocal constraints so that further specific form emerges.

This type of description allows one to make comparisons between developmental processes in a cellular (or unicellular) and in multicellular organisms and to understand them in the same terms, which is the classical view. If one adopts a position such as that of Wolpert (1971), that development is to be understood in terms of the responses of cells (really, of their genomes: Wolpert and Lewis, 1975) to "positional information" established over multicellular domains, then such comparisons become problematical, as discussed by Frankel (1974). The view of organisms as fields overcomes this difficulty and leads to the suggestion of some unexpected homologies between the morphology of ciliate protozoa and of amphibian gastrulae (Goodwin, 1980). This is very much in the tradition of rational morphology, since this makes clear that one must seek homologies at the level of "deep structure," i.e. at the level of generative principles such as those defining field properties, not in terms of "surface structure" such as whether or not an organism is partitioned into cells. Again, this is not to deny that surface structure imposes constraints which affect manifest form; it is simply that in comparing field effects at the level of the whole organism, these are secondary.

A very important problem in embryogenesis relates to the means whereby new or emergent aspects of morphogenesis are initiated at specific times in the process. For example, the relatively simple recursive process of cleavage in amphibian embryos is followed by the dramatic phenomenon of gastrulation, which transforms the hollow blastula into the three-layered late gastrula. The transition from cleavage to gastrulation has been described (Goodwin, 1980) as the expression of a cortical or surface field which is radially symmetric in the unfertilized egg but develops bilateral symmetry, normally as a result of sperm entry. The description of such a surface field in terms of harmonic functions reveals that with bilateral symmetry there appears a special point on the surface, a saddle point, which is identified with the future dorsal lip of the blastopore. However, it is assumed that the influence of this singularity on the morphogenetic process cannot become significant and be expressed until cleavage transforms the initially solid sphere of the egg into a hollow spherical shell, the blastula. Then the saddle point can make its presence felt as a point on the blastula where surface polarity is absent and a radial influence can manifest as the force causing bottle cell formation and the initiation of invagination. Thus cleavage is seen as a process which establishes a necessary condition for the emergence of a new phase of morphogenesis resulting from the interaction between a singularity in the surface or cortical field and the residual radial component of the cleavage field, which is described by solid harmonics.

5.6 Generation and Regeneration

There is another basic property of organisms relating the whole and the part, and this is the capacity of a part to generate or to regenerate the whole. So far, in discussing embryogenesis, the view has been developed that the fertilized egg is a whole which undergoes transformations resulting in the appearance of parts which have distinctive properties but are not, within the context of the whole organism, autonomous in the sense that atomistic theories would have them be. However, there was a time when the egg was a cell within the ovary, itself a part of the parent organism. The oocyte during its maturation develops the capacity to develop into a new whole. Such a transformation can be achieved also by parts (multicellular fragments) of hydroids such as *Hydra,* or parts (noncellular fragments) of ciliate protozoa such as *Stentor,* or a cell of a carrot, any of which can regenerate the whole organism. The capacity of plants to propagate from leaves and stems, of insects and urodeles to regenerate limbs from stumps, and of higher organisms to regenerate skin and liver, are other manifestations of this same property of parts to transform into wholes. It is evident that different organisms vary widely in their regenerative capacities, but it is true of all organisms that from particular parts, wholes can be produced. This defines reproduction or generation, as well as regeneration, and it is one of fundamental self-organizing properties which living creatures display. What does a field description of organisms have to say about such behavior?

There is an interesting property of harmonic functions which is suggestive of precisely this capacity. If such a function is described over any part of its domain of definition (e.g., part of the surface of a sphere), then the function can be recovered uniquely over the whole domain by analytic continuation. Thus, in a particular sense, the part contains the whole. This gives us a kind of existence theorem for the generative and regenerative properties of organisms, defined as fields describable by harmonic functions. It is just this type of property that needs to be embodied in a description of biological self-organization, although the specific property of harmonic functions described here is neither necessary nor sufficient to account for the actualities of generation and regeneration in the living realm.

5.7 Structuralist Biology

It may well have become apparent to the reader before this point that the general context within which this essay has been constructed is that of contemporary Structuralism, as defined by Levi-Strauss (1968) and Piaget (1968), and developed in another, more extended analysis by Webster and Goodwin (1982). A field as used above is an example of a *structure* in that it belongs to an invariant set (the harmonic functions) defined by internal relations (those defining the field equation over a domain), each member of which is a transformation of the other (by a change of boundary values). The particular field functions which have been used in this treatment are, in a general sense, not as important as the structuralist principles which inform the analysis. For these principles emphasize the necessity to identify that which is specific to a particular area of study, such as biology, before attempting to develop a theory to "explain" the phenomena. This is why, in approaching the problem of self-organization in embryogenesis, it was necessary first to identify any empirical regularities emerging from the study of biological form which might suggest the existence of principles of organization or invariant relationships in organisms. The evidence points clearly in this direction so that organisms, and hence embryos, are indeed structures in the technical sense: i.e., entities with the defining characteristics of wholeness and self-organization, capable of undergoing transformations which preserve these deep properties while changing manifest form. This is just a more elaborated description of a view which has been clearly articulated by those who insisted that the primary problem of biology is that of organization and form, not of composition or heredity, the latter finding their place within the context of the former (cf. Russell, 1930; Needham, 1936). The use of the field concept and the more specific description of biological form in terms of harmonic functions simply makes more explicit the implications of this view. Gene products can in certain instances select specific form, such as left- or right-handed spiralling in the third cleavage planes and, consequently, in the shells of the snail, *Limnaea*. However, in general we have seen from the field description of cleavage that the constraints determining specific form arise from other organizational levels than simple molecular composition, these being such processes as binary cleavage, animal-vegetal polarity, and a minimum "energy" condition. This emphasizes once again the primacy of organizational principles in biological process and the inadequacy of any theory based upon genes and molecules.

We may say that the role played by the genes is to specify the potential molecular composition of an organism and to define some temporal sequences in which molecular components are made. These constitute constraints which impose some limitations on the forms which organisms can assume, and in certain instances may actually determine higher-order form; but in general the relationship between "genotype" and "phenotype" is one of causal necessity, not sufficiency, since gene products act within the context of fields (electrical, visco-elastic, etc.) which generate morphology. A linguistic analogy would be that genes essentially determine the set of words out of which a text can be constructed. Words are clearly insufficient to define a text, which embodies higher-order syntactical, semantic, and; contextual constraints or rules. These are rules of organization which limit the set of allowed arrangements of the words within the text. Such organizational principles are described in this essay as field constraints, which limit the allowed range of biological forms. The generative principles of organismic morphogenesis then consist of organizational constraints or rules (laws of form) common to all organisms in the form of fields (thus defining biological universals) together with specific constraints characteristic of individual species (which then define particulars). Genes specify some of the latter but none of the former.

One of the most significant aspects of the structuralist approach is the deliberate avoidance of any *a priori* material reduction of the organism to parts such as cells, molecules, or genes. Once the problem of biological (self-) organization is clearly and explicitly described, and an appropriate description is available, then it *may* be possible to carry out a relevant material reduction. Certainly it will be possible to achieve an *abstract* reduction to laws and rules, such as those which have emerged from the analysis and description of cleavage given above. This description was itself inspired by a paper which was, in my opinion, a landmark in the discovery of general rules of morphogenesis expressed in terms which make no reference to composition or inheritance and are of a purely relational nature (French, Bryant, and Bryant, 1976). To start with the assumption that one knows what the basic parts of the organism are is to make a strategic error at the outset, which can lead one badly astray in seeking at once the most economical and the most rigorous analytical treatment of the problem. The description of organisms as fields which embody self-organizing properties and can undergo transformations preserving invariant relationships is, despite its limitations, a step towards a biological science of form and organization which relates part to whole in a manner which preserves organismic unity in the diversity of manifest morphology

Acknowledgements
I am indebted to my colleague, G.C. Webster, for the intellectual stimulus which has led to many of the ideas developed in this essay.

References

Darwin, C. 1859. *On the Origin of Species.* Reprint of the first edition, 1950. Watts & Co., London.

Driesch, H. 1929. *The Science and Philosophy of the Organism.* Black, London.

Frankel, J. 1974. Positional information in unicellular organisms. *J. Theoret. Biol,* 47: 439–81.

French, V., P.J. Bryant, and S.V. Bryant. 1976. Pattern regulation in epimorphic fields. *Science,* 193: 969–981.

Goodwin, B.C. 1980. Pattern formation and its regeneration in the protozoa. In G.W. Gooday, D. Lloyd, and A.P.J. Trinci, eds., *The Eukaryotic Microbial Cell* Symp. Soc. Gen. Microbial, 30: 377–404.

Goodwin, B.C., and N. LaCroix. 1984. A further study of the holoblastic cleavage field. *J. Theoret. Biol,* 109: 41–58.

Goodwin, B.C., and L.E.H. Trainor. 1980. A field description of the cleavage process in embryogenesis. *J. Theoret. Biol,* 85: 757–770.

Goodwin, B.C., and L.E.H. Trainor. 1982. The ontogeny and phylogeny of the pentadactyl limb. Brit. Soc. Dev. Biol., Symp. on Development and Evolution, p. 75.

Hobson, E.W. 1955. *The Theory of Spherical and Ellipsoidal Harmonics.* Chelsea Publishing Company, New York.

Levi-Strauss, C. 1968. *Structural Anthropology.* London.

Needham,J. 1936. *Order and Life.* Yale University Press.

Oosawa, F., H. Kasai, S. Hatano, and S. Asakura. 1966. Polymerization of actin and flagellin. In G.E.W. Wolstenholme and M. O'Connor, *eds., Principles of Biomolecular Organization,* pp. 273-286. Little, Brown and Co., Boston.

Piaget, J. 1968. *Structuralism.* Routledge and Kegan Paul, London.

Saunders, J.W., J.M. Cairns, and M.T. Gasseling. 1957. The role of the apical ridge of ectoderm in the differentiation of the morphological structure and inductive specificity of limb parts in the chick. *J. Morphol,* 101:57-87.

Trainor, L.E.H., and M.B. Wise. 1979. *From Physical Concept to Mathematical Structure.* University of Toronto Press.

Webster, G.C., and B.C. Goodwin. 1982. The origin of species : a structuralist view. *J. Soc. Biol. Struct.,* 5: 15-47.

Wick, G., A. Wightman, and E. Wigner. 1952. *Phys. Rev.,* 88: 101.

Wolpert, L. 1971. Positional information and pattern formation. *Curr. Top. Dev. Biol,* 6: 183-224.

Wolpert, L., and J. Lewis. 1975. Towards a theory of development. *Fed. Proc.,* 34: 14-20.

(a) 2 cells, (b) 4 cells, (c) 8 cells, (d) 16 cells, (e) 32 cells.

In (f) and (g) only, the silhouette forms one of the 8 longitudinal sections.

(a) 2 cells, (b) 4 cells, (c) 8 cells, (d) 16 cells, (e) 32 cells, (f) 64 cells, (g) 128 cells.

Case Study 17: Noise and Prediction in Biology

Perspective on *Irreproducible Results and the Breeding of Pigs (or Nondegenerate Limit Random Variables in Biology)*

Dany Spencer Adams
Research Professor
Tufts University
and
Michael Levin
Professor
Tufts University

'Bernoulli knew that life is lawful to the ensemble though chaos to the individual'

This paper is not an easy read, despite the fact that it mixes real-world examples with profound and fundamental mathematics concepts. Its applicability ranges among ecology (Ulanowicz 2007), developmental genetics (Ropers and Grimm 1977), cognitive neuroscience (Nakada 2011), the spread of disease in a population, and even cosmology. Indeed, any process where one of the variables studied is a 'non-degenerate limit random variable'. Random variable means just that. A non-degenerate limit means that chance can cause large differences in the results of replicated experiments even if they are performed identically. Flipping a fair coin is 'degenerate' because eventually, every experiment will converge on a probability of heads of 0.5. Non-degenerate means the probabilities will not converge to a single predictable number. Ignoring this can lead to false positives in all kinds of studies. Cohen gives a couple of examples, but an important point is summed up in this excerpt:

'I want to emphasize what almost sure convergence to a nondegenerate limit random variable looks like to people participating in a[n] experiment with this property. With increasing time, each [researcher's] experiment settles down to systematic, regular, and lawful behavior; his graphical plot of proportion [with phenotype X] as a function of time wiggles at first but smooths out gradually to a steady flat line [at a certain value]. However, if he repeats the experiment or gets a friend to do so under identical conditions, ... the curve of the replicate experiment levels out [at a value that] seems to bear no relation to the original. It is only after a change in the level of analysis—only after considering an ensemble of replications—that regularity and simplicity reappear.'

In other words, if you are dealing with a non-degenerate limit random variable, replicates of experiments (each of which comprises many samples) may be performed identically, but there *will* nonetheless be differences among results. To learn about the population, not just about sets of samples, requires examination of the results from 'ensembles' of replicated experiments.

More broadly, the paper explores the highly counter-intuitive dynamics of recursive processes, so prevalent in biology, where events are modified by the outcomes of prior events. Under those conditions, small initial differences can be amplified toward distinct outcomes. This not only makes prediction difficult, but can mislead one toward inferring the presence of deterministic causes where the role of chance is sufficient to explain strongly divergent results from identical experimental setups. Though not often cited, the concepts in this paper (and the even earlier discussion of Polya's Urn (Eggenberger and Polya 1923)) are fundamental to the understanding of chance, causation, and prediction in biology.

Cited:

Eggenberger F and Polya G 1923 Uber die statistik verketteter vorgange *Z Angew. Math. Mech.* **3** 279–89

Nakada T 2011 Brain science of the mind *Soc. Sci. Inform.* **50** 25–38

Ropers H H and Grimm T 1977 Variable composition of X chromosomal mosaics: due to asynchronous cell division during early embryogenesis? *Hum. Genet.* **39** 213–5

Ulanowicz R E 2007 Emergence, Naturally! *Zygon* **42** 945–60

Irreproducible Results and the Breeding of Pigs
(or Nondegenerate Limit Random Variables in Biology)

Joel E. Cohen

Many people believe that a coin in ordinary currency will come up heads nearly half the time it is tossed. Few people have reported a systematic experimental test of *that* belief.

During World War II an English statistician, J. E. Kerrich, was in Denmark when the Germans overran it. Interned under benevolent Danish supervision, he performed and recorded (Kerrich 1946) 10,000 spins of an ordinary coin. The proportions of trials which came up heads after ten, a hundred, a thousand, and ten thousand trials were, respectively, 0.400, 0.440, 0.502, and 0.507. If the ten thousand trials are broken into ten blocks of a thousand trials each, then the proportions of heads after each of the ten blocks were 0.502, 0.511, 0.497. 0.519, 0.504, 0.47 6, 0.507, 0.5I8, 0.504, and 0.529.

Viewed as a single long series, the data show that the proportion of heads tended toward and remained near one-half as the number of trials (tosses or spins) increased. Viewed as ten shorter series, the data suggest that the proportions of heads in independent experiments under the same conditions tended toward a single common value.

In 1713, Bernoulli constructed a mathematical idealization of the coin-tossing experiment as a sequence of independent trials each with a fixed probabilily p of coming up heads. Here p is some fraction near one-half. Imagine a very large number of copies of Kerrich all tossing copies of the same coin under the same conditions in perfect synchrony, but with the outcome of each coin toss independent of every other outcome. Bernoulli showed that, as the number of tosses increases, the proportion of all the copies of Kerrich for each of whom the fraction of his trials coming up heads differs from p by less than some arbitrarily small fixed amount approaches 100%. Mathematicians call this phenomenon convergence in probability to the constant limit p.

Two centuries later, in 1909, Borel proved that the same imaginary situation is even more lawful than Bernoulli had supposed. Bernoulli's result does not rule out the possibility that the proportion of heads in the trials of a particular copy of Kerrich could continue indefinitely to wander away from p by at least some fixed nonzero amount. Borel ruled out this possibility: for 100% of the copies of Kerrich, as the number of each man's tosses increases, the proportion of his trials coming up heads must approach and remain arbitrarily near the value p. Mathematicians call this phenomenon convergence with probability 1 or almost sure convergence to the constant limit p (see Loève 1963, pp. 14 and 19 for short proofs). Few people find these results a shock to their intuition.

The author is with Rockefeller University, New York, NY 10021.

ONE GOOD URN

Now consider an equally simple experiment. Suppose a very large box (whose capacity can be extended indefinitely by adjunction of similar boxes) initially contains one green ball and one blue ball. Choose one ball at random, look at its color, replace the ball in the box, and add to the box another ball of the same color as the one chosen. At each successive point in time, say once every second, choose one ball at random and then repeat exactly the above. The precise meaning of "at random" is that if there are n balls in the box when a drawing is made, each ball has an equal chance $1/n$ of being drawn.

The proportion of green balls in the box is the number of green balls divided by the total number of balls, whether blue or green. What will happen to the proportion of green balls as time increases?

Before reading further, please make a serious effort to guess. You have three guesses. When I proposed this problem to a very august mathematical ecologist in the course of a country march, he gave up after two wrong guesses. When I first heard the answer myself, I was astonished both by the general phenomenon it exemplifies and by the particular details. On reflection, I think the general phenomenon permeates population biology. My purpose here is to describe the phenomenon, give some biological examples of it, and suggest its consequences for the interpretation of biological data.

The experiment just described is a special case of what is known as "Polya's urn scheme." Eggenberger and Polya (1923) introduced the scheme in 1923 to model the spread of infection in a population. David Blackwell and David Kendall (1964) studied another generalization of this experiment and even mentioned its implications for stochastic population growth. But an overgrowth of related mathematical results obscured their message for biologists.

So, suppose a single Kerrich performs the experiment with blue and green balls. As time goes on, the proportion of green balls will converge to some limit p. (This fact alone is not obvious.) As in the Bernoulli model, "converge" means that if you pick some nonzero tolerance interval around p, then there is some time at which the proportion of green balls will be in that tolerance interval and after which it will never leave it; the proportion approaches and remains near p.

But what is p? For the first copy of Kerrich, call him $Kerrich_1$, all one can say is that his value of p, call it p_1, lies between 0 and 1 inclusive The chance that his p_1 is exactly equal to any particular fixed p between 0 and 1 is zero! However, the chance that his p_1 falls between 0.2 and 0.3 inclusive is exactly $0.3 - 0.2 = 0.1$.

At the next desk, $Kerrich_2$ is finding that the $proportion_2$ of green balls in his box_2 is getting and remaining closer and closer to a fixed number p_2. But whereas $Kerrich_1$'s proportion seems to be approaching $p_1 = 0.2435871 \ldots$, his $proportion_2$ is approaching $p_2 = 0.9342265 \ldots$ And on his other side, $Kerrich_3$'s proportion is approaching $p_3 = 0.59943312\ldots$. Each man's proportion of green balls converges to a limit, which is constant for each particular man but which varies from one man to another, even though all- are performing exactly the same experiment. In this case, the limiting proportion p is uniformly distributed over the interval from 0 to 1: that is, the chance that Kerrich$_{17}$'s limit p_{17} falls between a and b, where $0 \leqslant a \leqslant b \leqslant 1$, is $b - a$.

Blackwell and Kendall (1964) proved that if the box starts out with one ball of each of k different colors, where k may exceed 2, then the limiting distribution of proportions of each of the k colors is uniform over the set of all possible ways of dividing 100% into k proportions.

The behavior of this hypothetical experiment exemplifies what mathematicians call almost sure convergence to a nondegenerate limit random variable. "Nondegenerate" means that the limiting value of the proportion of green balls is not restricted to a single point. In Bernoulli's model of the real experiment which the real Kerrich performed, the limit random variable is degenerate because every such Kerrich would (in theory) have obtained the same limiting proportion p of heads.

This hypothetical experiment behaves identically to an apparently quite different experiment. Suppose each Kerrich has a box with one green and one blue ball. He receives a coin; one side is green, one blue. Though the coin looks fair, the real probability p that the coin will come up green is distributed uniformly between 0 and 1. For any given coin, p is constant in time. No man has any reason to suspect that his coin differs from any of the others; in particular, he does not know his coin's value of p. Once a second, each man flips his coin and adds to his box a ball of the color indicated. Then (since this experiment is just Bernoulli's model and Borel's theorem applies) each man's proportion of green balls converges with probability 1 to his coin's p.

Here each man's limiting proportion p is assigned first. The color of the next ball is chosen by an independent trial with probability p of green. In the previous hypothetical experiment, each new ball's color is determined by random choice among the colors which have occurred so far. To an observer of the balls deposited in the boxes, the two experiments are indistinguishable[3]

Before proceeding to biological examples. I want to emphasize what almost sure convergence to a nondegenerate limit random variable looks like to people participating in an experiment with this property. With increasing time, each man's experiment settles down to systematic, regular, and lawful behavior; his graphical plot of proportion green as a function of time wiggles at first but smooths out gradually to a steady flat line. However, if he repeats the experiment or gets a friend to do so under identical conditions, where the curve of the replicate experiment levels out seems to bear no relation to the original.

It is only after a change in the level of analysis–only after considering an ensemble of replications–that regularity and simplicity reappear. It is the law of the limit random variable that is simple.

In retrospect, Bernoulli himself made just such a change in the level of analysis. If each copy of Kerrich were to toss a coin just once, then Kerrich1 might get heads, Kerrich2 tails, Kerrich3 again heads, and so on without apparent pattern. As the size of the ensemble of copies of Kerrich increases, however, the proportion of copies

---

[3] Violet Cane (1973) has discovered an equally surprising, and closely connected, observational equivalence of models for negative binomially distributed counts, such as accident statistics.

whose single trial results in heads approaches the limit p near one-half. Already Bernoulli knew that life is lawful to the ensemble though chaos to the individual.

BACK ON THE FARM

Now suppose that a breeding stock on a pig farm is maintained by mating a boar and a sow each generation. One male and one female from the offspring are chosen to mate in the next generation. Suppose there is a single gene locus at which, in the initial generation, both parents are heterozygotes. For example, each has genotype Aa. Since each off-spring receives one allele chosen at random from each of its parents, there is positive probability that both offspring will have the genotype aa. If this happens, all future offspring will have the same genotype at that locus. There is an equal positive probability that both offspring will have the same genotype AA, with the same consequence.

Sooner or later both offspring must become homozygous for the same allele, and geneticists have calculated the rate of approach to homozygosity under the regime of inbreeding just described. The offspring of a particular pair of heterozygous parents will fixate on the genotype aa with probability one-half and on the genotype AA with probability one-half.

Aside from their good looks and intelligence, pigs are bred for characteristics of commercial interest such as quantity of edible meat. These quantitative characters are believed to be controlled by the additive effects of genes at several loci. Suppose, for the sake of illustration, that weight is controlled by five independently assorting loci with alleles A, a; B, b; and so on up to E, e. Let homozygosity for the capital letter at a locus correspond to an increase in one kilogram over the heterozygote and homozygosity for the small letter at a locus correspond to a decrease in one kilogram below the heterozygote.

If a breeding line is started with parents both of genotype AaBbCcDdEe, then eventually the descendants in that line are certain to drift to homozygosity, at each locus the same for both male and female. The weight of pairs in successive generations will cease fluctuating eventually, all else being equal, and will be the same for both members of the pair. Their weight at fixation will be 5 kilograms above those of their initial ancestors if all five loci fixate at capital letters, 3 kilograms above if four of the five loci fixate at capitals. 1 kilogram above if three of the five fixate at capitals, or symmetrically below the weights of their ancestors. The weight at fixation of another line of descent might differ. As the size of an ensemble of lines of descent increases, the proportions of lines at each weight approach the probabilities calculated from a binomial distribution with parameters 5 and 1/2 (roughly, a bell-shaped histogram with its highest values symmetrically placed on either side of the ancestral weight).

When a selective breeding program uses a finite stock of pigs (and infinite numbers of pigs have not yet been observed), this underlying drift, due to random sampling of genes, sets limits to what selection can accomplish. Moreover, drift to a nondegenerate limit random variable sets different limits in different replications of an identical breeding program. As Robertson (1960, p. 244) observed: "If $u(q)$ [the chance of fixation of an allele whose frequency at the beginning of a breeding program is q 1 is very different from unity for many genes, we will notice that replicate lines from the same initial population will be very different in the limit they reach." In our example, $q = 1/2$ and $u(q) = 1/2$ for all five loci, so Robertson's warning applies.

A failure to recognize the nondegeneracy of the limit random variable to which polygenic characters drift has practical consequences. Hill (1971, p. 294) points out that some authors estimate realized heritability in a single selection program "by fitting a linear regression to cumulative response and cumulative selection differential each generation. But with genetic sampling (drift) the variance of the population mean increases each generation, and these means become correlated. In standard regression analysis the observations are assumed to have equal variance and be uncorrelated, so that the estimates of variance of realised heritability obtained by. . . . using standard regression techniques are biassed downwards. In other words, the observed variance among heritability estimates from a replicated experiment would exceed the variance predicted from a single replicate." Hill gives an explicit quantitative analysis of what nondegenerate drift does and what to do about it in an important series of papers (most recently, 1974).

While geneticists have long known of genetic drift and have recently assayed its practical impact on breeding programs, other areas of population biology seem to have remained in bliss. Suppose two bacteria, say a wild type and a mutant, are distinguishable by some marker but are absolutely identical with respect to growth in a particular culture, which is sufficiently favorable to growth that no deaths occur. After a while one or the other of the bacteria will divide, giving three bacteria. Then, one of those three will divide, each one being equally likely, and so on. If we ignore the interval between divisions and advance an artificial clock by one unit at each division in the culture, we obtain exactly the Polya urn model. If we identify the wild type with green balls, and the mutant with blue, then after a long time, since the proportion of green balls converges to a limit, so will the proportion of wild type bacteria in the culture, and to the same limit. The chance that this limit is exactly one-half is zero. If the limit is p, the culture would behave as if each new bacterium added were wild type with probability p. Blackwell and Kendall (1964, p. 295) state succinctly: "This might lead the incautious observer to attribute a real difference to the ... clones in respect of their growth mechanism, although in fact they are in all ways identical." The same phenomenon might lead incautious observers to infer that a genetic change affecting growth had occurred if they attempted to replicate the experiment and found, as they must, a different limiting proportion of the wild type.

Similarly, suppose that individuals of a growing population fall into one of k age categories, where k may exceed 2. Under certain assumptions (Athreya and Ney 1972, p. 206), which may even be defensible in some real situations, the proportions in each class will approach proportions which depend only on the fertility and mortality, but not on the initial numbers of individuals, of each age class. Moreover, the population will (with positive probability) eventually grow exponentially at a rate which also depends only on the fertility and mortality of each age class. If total population size is plotted on a logarithmically scaled ordinate against time on the abscissa, the graph will eventually fall along a straight line. The point at which this straight line intersects the time axis is where a deterministically growing population with the same growth rate would have had to begin growing exponentially in order to fall into step alongside the stochastically modeled population. Call this point the lag time. It is a nondegenerate random variable. Though the laws of growth are the same, the lag times, or times to apparent exponential take-off of growth, of initially identical populations obeying this stochastic model are different.

No deterministic interpretation of such differences in limiting proportions or lag times could possibly be right, though the differences are real enough. The variation in an ensemble of replicates must become the object of study when the limit random variable of an individual replicate is nondegenerate. Luria and Delbrück (1943) practiced this precept in their classic experimental proof that phage-resistant mutants arise randomly.

But population biologists who study macroscopic populations seem less inclined to this view of nature. Here are a few heretical possibilities Is it possible that differences in successional changes and in so-called climax state in apparently similar habitats are not to be explained as due to any causal difference between the habitats, but should be interpreted as variation in an ensemble of such habitats? Is it possible that the differences in species composition of apparently similar islands result from the operation of identical forces which produce regularity only in an ensemble of islands? Could observations that one animal population cycles with a period of 4 years and another with a period of 3 or 13 or 17 years become intelligible if the ensemble of periods of cycling animal populations were examined? Could the differences in the sizes of prides of lions or in the social organization of troops of Japanese' macaques reflect the inherently but lawfully variable outcome of identical underlying stochastic forces, rather than deterministic ecological differences?

In behavior, for example, is it possible that some of the significant differences among mother-child interactions, which are obvious by the time a child reaches five years of age, are due neither to inherent differences among individuals nor to environmental differences, but to sequentially dependent random forces applying equally to all mother-infant pairs? (Which hand the mother first holds the infant with is random, perhaps, but that choice affects the skill with which she performs tasks with her remaining hand, which affects the infant's response, which affects....) Could nondegenerate stochastic limits provide useful models of what students of plant development (Evans 1972) call "ontogenetic drift"?

The possibilities I raise will leave cold or enrage people who believe they *know* that deterministic factors explain some of the differences I cite. They may well be right in part. All I suggest is that there may be variation which deterministic factors do not usefully explain and that the possibility of understanding phenomena is preserved by redirecting attention to a lawful-looking ensemble.

CHAOS AND COSMOLOGY
May (1974, p. 645) has emphasized the ecological interpretation of the mathematical fact that very simple deterministic difference equations can have astonishingly messy trajectories, including "cycles of any period, or even totally aperiodic but bounded ... fluctuations." The recent, still unpublished work of several people shows that

many (though not necessarily all) difference equations studied by ecologists can act so weirdly. Implicit is the suggestion that the apparent variability of population fluctuations may represent the working of a simple deterministic mechanism. The behavior predicted by this mechanism is so sensitive to the values of the parameters in at least some ranges that it will probably be necessary to compare observations with a probabilistic approximation. Thus, the apparent variability of population fluctuations, for example, is interpreted at two levels in the models May and others consider: first, in the complexity of the trajectories predicted with fixed parameters (including initial values); and second, in the impossibility of estimating exactly, and the likely actual fluctuation of, the parameter values. These models do not attempt to account for uncertainty or fluctuations in parameters but assume, at the kernel of phenomena, a simple determinism.

It seems impossible to reject with any data an affirmation of faith that a deterministic mechanism could supply sufficient apparent variability to describe a real population whose parameters were known and constant. The preceding biological examples, and others which could be cited, suggest an alternate view: At least some biological processes incorporate stochastic elements that can cause long-term behavior which appears lawful only in an ensemble of replicates. The empirical program suggested by this view is to examine such ensembles.

Worn exclusively, the deterministic glasses of Laplace and the stochastic glasses of Charles Sanders Peirce give equally roseate views of the world. In the interest of fair advertising, I have to admit that the strategy of moving from the individual to the ensemble to find order in variability will not always work. There are stochastic processes which approach a limit (any kind of limit, degenerate or not) only with a probability 0. Some misanthropes claim experience is like that, too: Some parts of nature simply change more slowly than others, they say, and those parts that change slowly compared to the time scale we are interested in serve us as points of reference, or limits built on sand. Such misanthropes may be right.

Having speculated thus far, let me raise and answer a metaphysical question prompted by Polya's urn which would, I hope, have amused Peirce as it amuses me on dark nights. If you and I had been- born in another universe which had started from exactly the same initial conditions as our present one and which had been subject to the same dynamics, would we necessarily infer the same Jaws of nature as we (in the collective sense of civilized thought) infer for this universe? I take the existence of genetic drift on pig farms as establishing a stochastic element in the dynamics of the universe, and therefore have no guarantee that the apparent lawfulness in this copy of the universe would take the same form in any other. The order of this universe may be an irreproducible result.[4]

ACKNOWLEDGMENTS
This paper is based in part on a talk given in the Provost's Lodge, King's College, Cambridge, 25 April 1975. For extraordinary hospitality during 1974-75, I thank

---

[4] Anne Whittaker points out that Ray Bradbury has dramatized possibility (see Bradbury 1962).

members of King's College Cambridge and the University of Cambridge Statistical Laboratory. I thank Robin Sibson for excellent teaching; and John Hajnal, David Kendall, and the referees for helpful comments. This work was supported by King's College Research Centre and the U.S. National Science Foundation.

REFERENCES CITED

Athreya, K. B., and P. E. Ney. 1972. *Branching Processes.* Springer, New York.

Blackwell, D. and D. G. Kendall. 1964. The Martin boundary for Polya's urn scheme and an application to stochastic population growth. *J. Appl. Probab.* 1(2): 284-296.

Bradbury, R. 1962. A sound of thunder. Reprinted in *R Is for Rocket.* Doubleday, Garden City, New York, 195 2.

Cane, V. R. 1973. The concept of accident proneness. *Académie Bulgare a$$ Sciences Bulletin de l'nstitut de Mathématiques* 15: 183-188.

Eggenberger, F., and G. Polya. 1923 Uber die Statistik verketteter Vorgänge. *Zeuangew. Math. Mech.* 3: 279-289.

Evans, G. C. 1972. *The Quantitative Analysis of Plant Growth.* Blackwell Scientific, Oxford.

Hill, W. G. 1971. Design and efficiency of selection experiments for estimating genetic parameters. *Biometrics* 27: 293-311.

———. 1974. Variability of response to selection in genetic experiments. *Biometrics* 30: 363-366.

Kerrich, J. E. 1946. *An Experimental Introduction to the Theory of Probability.* Einar Munskgaard, Copenhagen.

Loève, M. 196 3. *Probability Theory,* 3rd ed. D. Van Nostrand, Princeton.

Luria, S. E., and M. Delbrück. 1943. Mutations of bacteria from virus sensitivity to virus resistance. *Genetics* 28: 491-511.

May, R. M. 1974. Biological populations with nonoverlapping generations: stable points, stable cycles, and chaos. *Science* 186: 645-647.

Robertson, A. 1960. A theory of limits in artificial selection. *Proc. R. Soc. Land. B Bioi. Sci.* 153: 234-249.

www.ingramcontent.com/pod-product-compliance
Ingram Content Group UK Ltd.
Pitfield, Milton Keynes, MK11 3LW, UK
UKHW051654071225
9419UKWH00030B/713